First printed 1983 — Second printing 1984 — Third printing 1986 — Fourth printing 1988 — Fifth printing 1992.
Printed by Times Printing Company, Random Lake, WI.

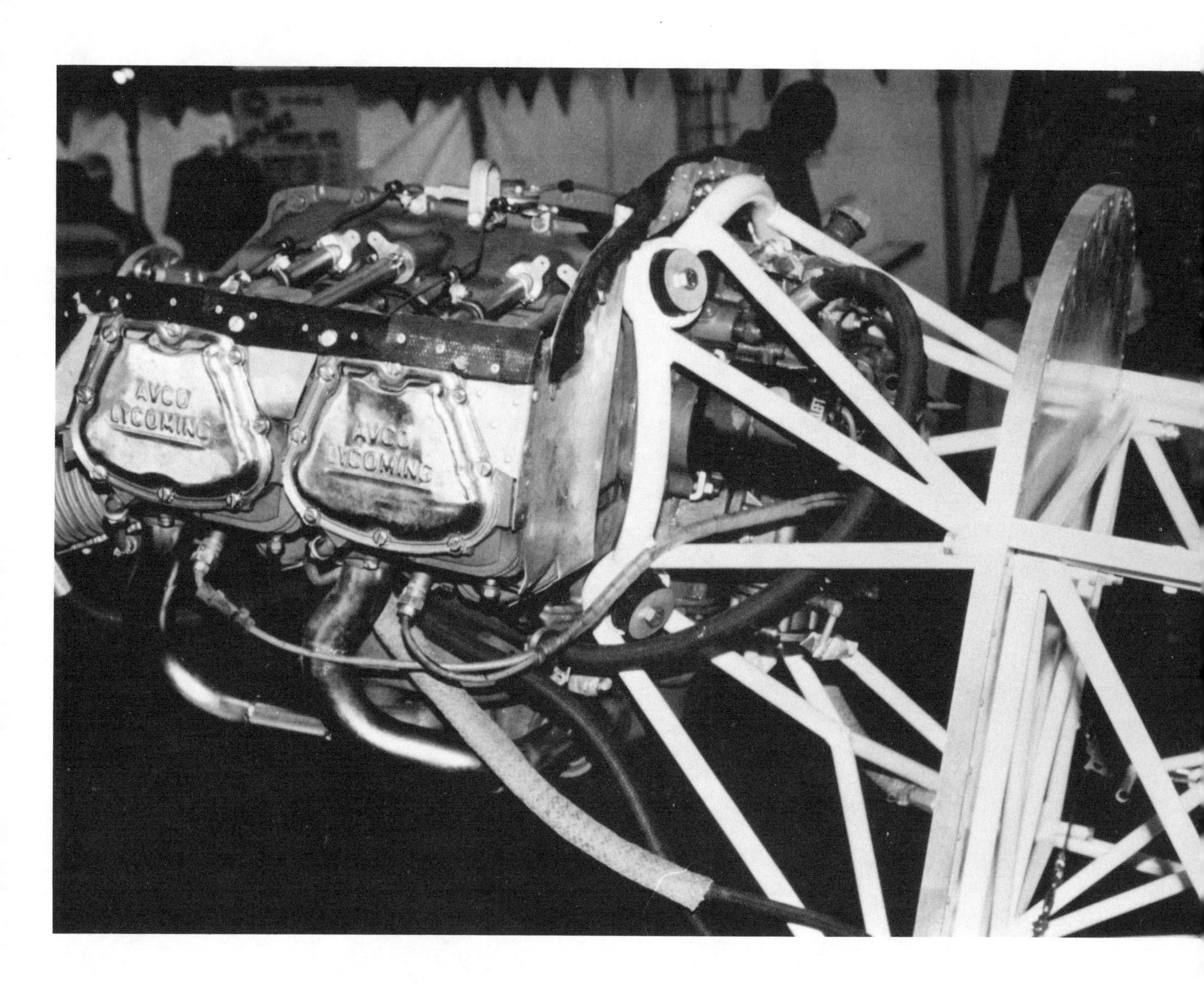
AVCO
LYCOMING
AVCO
LYCOMING

FIREWALL FORWARD

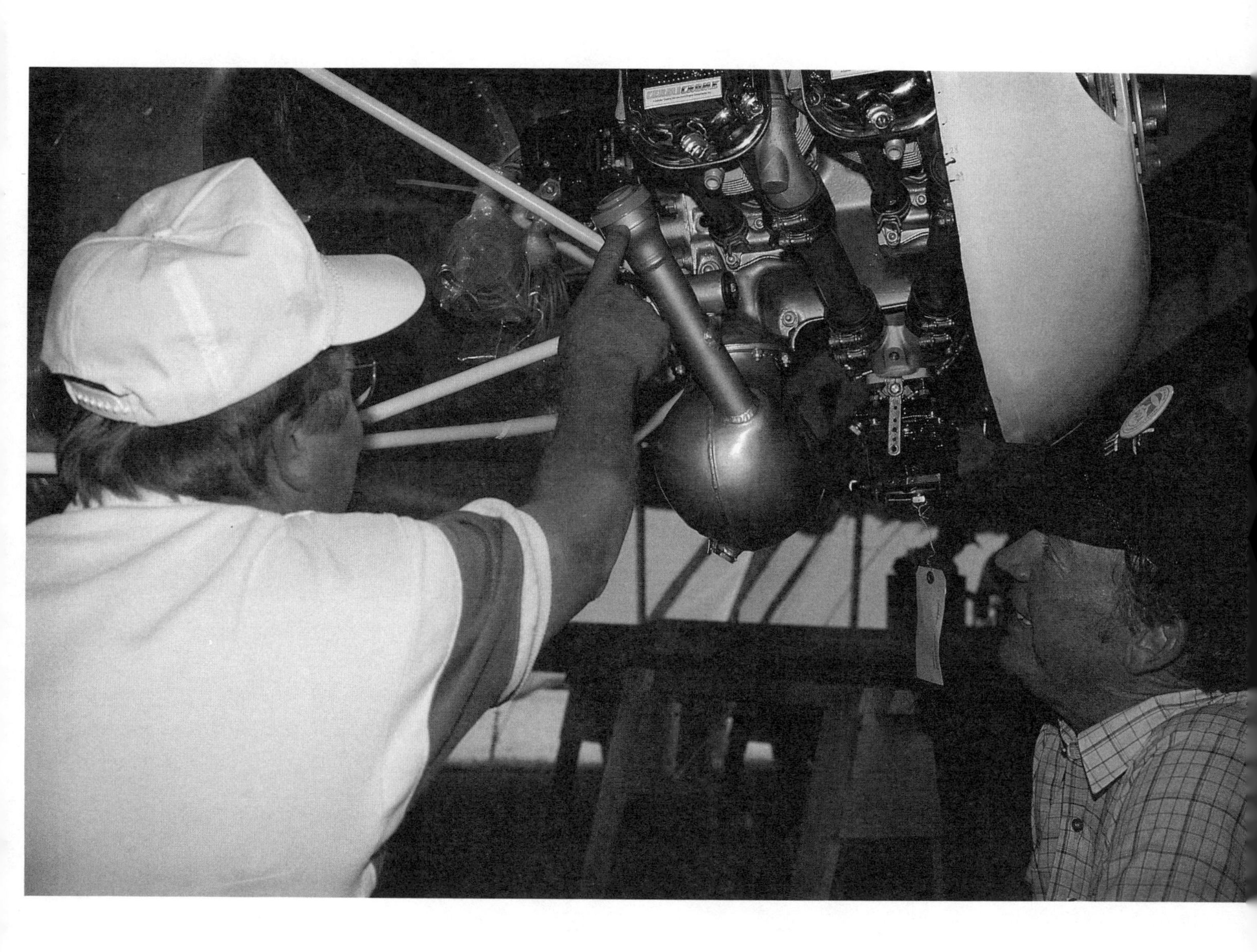

Firewall Forward

ENGINE INSTALLATION METHODS

tony bingelis

EDITOR
David A. Rivers

COVER DESIGN
Joan N. Rivers

Illustrated by the Author

EAA
Aviation
FOUNDATION
®

DEDICATED TO MY WIFE MORINE

Acknowledgements

For years I have relied considerably on designers, aeronautical engineers, mechanics, amateur aircraft builders and other knowledgeable individuals, who have helped me stockpile the information I now share with you in this book. How could one possibly thank each and everyone of these fine folks for their interest and for the time they have taken to answer my questions and patiently explain the salient features of their aircraft? I will, however, single out a few individuals who have made this book easier for me to write. Thank you Seth Hancock, Don Childs, Bert Brundage and Leonard Dobson, Austin, Texas; Walter Lane, Downey, California; Francois Lederlin, Grenoble, France; Joe Locasto, Aneheim, California; Ricky Rouse, Beaumont, Texas; Major Bill Gunn, USAF; Frank Luft, Central Point, Oregon; and K. W. Johnson, AVCO LYCOMING Corp. Thanks, also, to the FAA and its fine staff. The FAA, of course, has been the authorative source for much of the technical information, and all of the regulatory material, which has had such a profound influence over all builders and flyers. The Experimental Aircraft Association and its members also receive my gratitude. After all, where would sport aviation be today without it?

A very special thanks to friend David A. Rivers, who edited this book, and to Joan N. Rivers, who designed the striking cover. My compliments to you both.

THE SPORTPLANE BUILDER
Tony Bingelis
FIREWALL FORWARD
Tony Bingelis
SPORTPLANE CONSTRUCTION TECHNIQUES
Tony Bingelis
A Builder's Handbook

Forward

This manual discusses piston engine installations in amateur-built aircraft — hence the name Firewall Forward. It was written after some prodding and encouragement from loyal readers of my first book, The Sportplane Builder: Aircraft Construction Methods, and the technical articles which I write regularly for SPORT AVIATION magazine. The book's purpose is to provide information about engine installation practices that have proven effective and practical. It picks up where most aircraft plans leave off, faithfully detailing engine hook-up methods long followed by homebuilders, antique restorers and those engaged in rebuild projects. As a builder myself, I know how important even small details can be to the successful completion of a homebuilt airplane. With this in mind, I have attempted to guide the reader through the whole process of powerplant installation, from helpful tips on selecting an engine to mounting, wiring, and actual operation. In short, this manual offers a screwdriver and hammer approach to engine compartment work that I believe will be welcomed by homebuilders who have been unable to find this sort of data in any single source until now. The information, the photos, the extensive detailing and illustration of engine and propeller installations presented here were culled from my files and notes on the thousand and one homebuilts I have inspected and admired over the past two decades.

Let me add that the inclusion of any procedure, method or practice in this book does not constitute endorsement of the same. Moreover, nothing presented in this book is intended to substitute for specific instructions issued by the manufacturer of any engine, accessory, part or product. The manufacturer always knows his product best and his operating guidelines and instructions should always be followed.

Tony Bingelis

Contents

6. COWLS & COOLING

7. LINES & HOSES

8. FUEL SYSTEMS

9. IGNITION & ELECTRICAL

10. POWERPLANT INSTRUMENTS

11. ENGINE INSTALLATION

12. PROPELLERS & SPINNERS

13. CENTER OF GRAVITY

14. ENGINE OPERATIONS

engine selection

Deciding on an Engine

Gone are the days when you could buy an engine for $500, bolt it on and leap into the wild blue yonder in your newly built homebuilt. However, if you are a patient builder, a persistent builder, a lucky builder, you can still find a suitable engine at a "reasonable" price. Before you start your engine search in earnest, however, you must decide what kind of engine you would like to use — or rather, can use. Is the engine you want the engine you need?

What do you want in an engine? What does anyone want in an engine — low price, low weight, low maintenance cost, a good power-to-weight ratio, reasonable rpm at full power for propeller efficiency, and easy availability.

Although any builder would rather install a brand new engine in his homebuilt, for most of us, unfortunately, this is seldom possible. The price of a new engine often exceeds the entire cost of the aircraft structure. It is no wonder that we builders have to be satisfied with a more economical runout or used conventional aircraft engine. However, there is no harm in dreaming that someday, if the airplane proves out, it will be re-engined with a new and possibly more powerful powerplant.

The engine availability problem has always been a tough hurdle for an airplane builder and the problem seems to be getting worse. There are only two major aircraft engine manufacturers in this country and most of their production, limited as it is, is earmarked for large corporate aircraft engined with powerplants too large for homebuilt use. The few models of smaller engines more suitable to our use, are also in limited supply and are priced beyond what most of us are willing to pay.

Things are not much better in the used engine market. The growing demand for used aircraft engines is undoubtedly being caused, in part, by the large number of people who are undertaking a homebuilt project for the first time. As a result, the used aircraft engine market is drying up at a rapid rate for certain engine models.

Obviously, engine demand is also influenced by a sudden popularity for a particular aircraft design. For example, after the VariEze design hit the homebuilt market, the Continental engines, particularly the O(OH)-200 (100 horsepower models) became scarce and expensive.

Years ago, when the homebuilding era began, the same thing happened....although not on as large a scale as today. Plans for the Baby Ace, one of the most successful of the early homebuilts, were made available to an eager aviation community through a national magazine. As a result, builders and would-be builders eagerly snapped up used 65 hp and 75 hp aircraft engines by the hundreds. Suddenly, these engines were in short supply and prices rose to an eye blinking level.

Then there were the VariEzes, Quickies, the Dragonflys and similar new generation composite airplanes advertised as easy to build and with complete kits for the builders wanting them. These designs all contribute to the dwindling supply of used aircraft engines, and yes, the short supply of newly converted VW engines.

More recently, the Long-EZ, a composite airplane designed for a specific engine, the Lycoming 115 hp (O-235 model), made a dent in the inventory of these engines....especially the used ones.

Hombuilders' Choice

Ranking	*Engine*	*Horsepower*	*Number in Service*
1st	*Lycoming*	*160 hp*	*1284*
2nd	*Lycoming*	*140 hp*	*1199*
3rd	*Lycoming*	*180 hp*	*987*
**4th*	*Continental*	*75 hp*	*1074*
***5th*	*Volkswagen*	*36 hp*	*1075*
6th	*McCullough*	*72 hp*	*984*
7th	*Continental*	*85 hp*	*720*
8th	*Continental*	*100 hp*	*536*
9th	*Lycoming*	*115 hp*	*346*
10th	*Lycoming*	*150 hp*	*224*

NOTE: * *The FAA apparently lumped the 65 hp and 75 hp Continental models together. It could be that the 65 hp may be in third place.*

** *FAA data does not list out different horsepowers for the VW engines, recording them as a mere 36 hp. Collectively, fifth place is deemed appropriate.*

TABLE I

Fortunately, most aircraft are designed to accommodate a variety of engines and this does tend to ease the engine problem slightly. But, it also imposes on a builder a need to exercise self discipline. For example, if the designer indicates that the aircraft can use engines ranging from 85 hp to 150 hp, should a builder consider using a 75 hp or 160 hp engine, you ask? What's wrong with that if you find a good deal on an engine? Plenty! An aircraft design that is limited to engines of 85 hp to 150 hp might not be structurally safe if a more powerful and heavier engine were to be installed. On the other hand, if a smaller than recommended horsepower engine were installed, the aircraft would probably be unsafe operationally and be plagued with poor take-off and climb performance. Additionally, it is important to abide by the engine horsepower limitations established for the design for reasons other than those mentioned....for aerodynamic and aesthetic reasons, for example.

A Quickie or Long-EZ would look odd (funny) somehow with a radial engine....if you could find one light enough. It would probably be slower too, wouldn't it? So, why even consider such a thing?

Another reason for abiding by the designer's engine recommendations has to do with the thrust line location. An engine other than one recommended by the designer could have a substantially higher or lower thrust line. Using it could cause unpredictable effects on the airplane's power-on and power-off behavior and could lead to serious problems.

Finally, since your choice of a design was or will be, in large part influenced by something that appeals to your eye as well as to your craving for the type of performance it is reputed to have, why change it?

ENGINES FOR HOMEBUILTS

Externally, one Pitts looks very much like another. All VariEzes look alike, too. Emeraudes, Baby Aces and Acro Sports look like other Emeraudes, Baby Aces and Acro Sports. Under those cowlings, though, there is a world of difference. One Pitts can be pulled up into a vertical climb and held there until it becomes a tiny speck hardly visible from the ground. Another in the same antic will stall and fall after a faltering climb of only several hundred feet.

Of course, we realize that the difference in performance is primarily the result of the difference in horsepower of the engine used.

At some point in the construction of every homebuilt, the builder must determine what size and make of engine to use. He may even try to find out what kinds of engines other builders are using in similar airplanes. This is a serious matter because, when you get right down to it, there is often but a small difference in engine weight-to-power that makes one engine better suited to a particular design than any other.

An engine that is too heavy affects weight and balance, and performance. An engine that is too powerful could tear itself out of the aircraft, or, at the very least, be so thirsty that it would gulp down all the fuel you could carry in less time than it takes to get lost. Indeed, there is a very small choice among powerplants for one best suited to your project.

Speaking of power, you may be surprised to learn that homebuilders have, in the past (up to 1979), preferred the larger horsepower engines, particularly Avco Lycomings. The three engines ranking highest in popularity (largest number in use by homebuilders) are the 160 hp, the 140 hp and the 180 hp Lycomings, in that order. (Table 1) Amazing!

With all the interest in and display of automotive conversions at Oshkosh (the world's largest annual gathering of sport aircraft), many people mistakenly believe that a large number of homebuilts are flying around powered by economical, easily converted auto powerplants. Alas, it simply is not so. At least not according to the official data contained in FAA's latest (as of this writing — 1982) Census of U.S. Civil Aircraft for the Calendar Year 1979.

This report shows a grand total of 251,516 aircraft, including commercial air carriers as well as general aviation aircraft such as rotorcraft, gliders, blimps, balloons and amateur-built aircraft.

A separate section of the census provides a detailed listing of the 11,488 individual amateur-built aircraft making up part of the total air fleet of 251,516. However, after laboriously reviewing the report, I could only identify 9,870 of the aircraft as homebuilts. Why such a discrepancy?

Frankly, in my opinion, the report is spotted with omitted essentials and erroneous groupings. For example, would you believe that the hundreds of Volkswagen-powered aircraft built and flying are all powered by 36 hp engines! That's right. At least that's what the report states. You and I know that such documentation is ridiculous because only a few builders have ever installed the early model small 1340cc VW engine in their aircraft. Most existing VW-powered homebuilts have the larger basic 1600cc engine, or a more powerful conversion developing up to 80 hp (some even more). The difference between 36 hp and 80 hp is considerable and can only reflect indifferent statistical reporting. But, that's not all that is wrong with the data.

A large number of aircraft are listed as having an UNKNOWN powerplant. Even more unbelievable, the governmental statisticians apparently didn't know what engines were installed in hundreds of aircraft manufactured under an ATC (Approved Type Certificate) either. A similar perusal of the Amateur-Built tabulations revealed that as many as 55% of the homebuilts listed had an engine model and horsepower also reported as UNKNOWN. Can you imagine such a lack of data would exist when the FAA requires all manufacturers and homebuilders to provide that information before they will issue a Certificate and Operating Limitations?

There is no doubt that at least 10,000 of the aircraft are properly registered Amateur-built aircraft. But, with so many of the listed aircraft being improperly categorized, figuring out how many homebuilts were using a particular engine posed quite a challenge.

Not at all intimidated by such uncertainties, I boldly concluded that most of the more than 3,600 (55%) homebuilts reported with UNKNOWN engines of UNKNOWN power, were surely equipped with engines just like others reported for the same design. This assumption made it seem reasonable to convert the UNKNOWN engines to known engine types by working with percentages. The method may not be scientific, but I'm certain it's at least as accurate as the document from which the data was extracted. In Tables 1 & 2, I report my findings regarding the identity of the engines (and their power output) that made up the homebuilt fleet in 1979.

It is a surprise, I guess that the AVCO Lycoming engines are used in greater numbers, almost 2 to 1, over all other manufacturers combined. In second place are the Continental Motors powerplants. These two engine makers, of course, represent all of the major aircraft engine manufacturers we have to choose between. Franklin, now defunct, never did capture much of the aviation market in the United States and the homebuilt tabulations show that graphically.

Perhaps, as previously stated, the biggest surprise to me, was that Lycoming swept the top three rankings with three of the more powerful engines being used by the homebuilders (Table 1). I always thought that the most commonly used engine was the 65 hp

Continental. I hate to bring this up, but the FAA virtually ignored the separate identity of the 65 hoss Continentals in its report.

It really means little, but what do you think the average horsepower rating is for homebuilts? 85 hp? 180 hp? Less?

Notice that the large number of Volkswagen engines in use is a clue that they are coming on strong, considering that their entry into the U.S. aircraft community has been recent when compared to the old-line Lycomings and Continentals.

The McCullough 72 hp engines make such a good showing because of their widespread use in gyrocopters....about 70% of the gyrocopter fleet uses McCullough engines.

There is already evidence that, as the tidal wave of fuel efficient ultra-lights hit the governmental number crunchers, engine identification will become totally a hit-and-miss matter because even the engines on some of these ultra-lights are being designed and built by individual builders. I suspect that some readers will challenge the absence of several engine types. Engines like the Subaru, Citroen, Mazda, Oldmobile, etc. But, don't blame me. I don't maintain the official records.

Many interesting and exciting automotive engine conversions appear each year at the world's largest aviation event --- Oshkosh. Most of them are displayed on test stands and in truck beds, not in aircraft. How come?

Tabulation of Engines

Number of Aircraft	Engine Installed	Horsepower
4398	Lycoming	55-260
2802	Continental	40-435
1075	Volkswagen	36
984	McCullough	72
187	Franklin	49-240
61	Evinrude	85
47	Warner	125-175
36	Hirth	100
33	Corvair	145
31	LeRhone	80
29	Pratt & Whitney	220-2500
23	Ford	60
16	Fairchild	200
14	Wright	150-180
12	Outboard	35
11	Menasco	125-150
11	Siemans	113
11	Hispano	180
8	Walt Minor	105
8	Salmson	40-135
8	DeHaviland	140
5	Triumph	40
5	Anzani	35
5	Jacobs	245-275
5	Mercedes	180
5	Rolls-Royce	100-1180
5	Renault	230
3	Heath AVN	25
3	Aeronca	45
3	Limbach	68
3	Gnome	160
3	Porsche	75
2	Micro Jet	800
2	Clerget	130
2	Lambert	90
2	Leblond	70
2	Honda	75
2	Keickhafer	80
1	Ken Royce	90
1	Argus	250
1	Pollman	40
1	Nelson	48
1	Breda	45
1	Universal	65
1	Mercury	70
1	Cushman	18
9870 **Total**		

TABLE 2

Some Engine Options

Well, what options do you have in selecting an engine for your project....assuming that you will abide by the horsepower range acceptable for your particular design?

Very few small, liquid cooled engines are available or acceptable for use in light aircraft. Your choice of an airplane engine will probably be limited to one of the air-cooled engines produced especially for aircraft. However, even among this family of engines there are three very different types to choose from:

* Radial engines — uncommon, large and heavy, but in demand by builders of larger homebuilts, classics and replicas.
* In-line engines — uncommon, but popular with warplane replica builders.
* Opposed-type engines — most common and most readily available.

Many of the types mentioned are older engines going around for the second or third time. The exception is to be found in some of the over-100 hp class of 4-cylinder horizontally opposed type engines. These are currently being produced for a few airplane manufacturers.

The older Continental engines of 65 hp to 100 hp have, of course, always been popular in the smaller homebuilts while airplanes requiring larger engines of 115 hp to 180 hp seem to be equipped, for the most part, with Lycoming engines.

Among the smaller and most sought after in-line engines are the Rangers and Menascos used in old military trainers. Some might still be found by the determined replica builder.

There are very few other aircraft engines to choose from except some vintage Franklin engine models. What other options there are would undoubtedly have to be imported from Europe.

It is no wonder that builders dream of using an economical, fuel efficient Ford or Chevy auto engine. Dream is right because "THE" auto engine that would be ideal for aircraft use has not yet come along. The VW engine is perhaps the closest we will get to a dependable inexpensive auto engine converted to aircraft use in large numbers (at least for the foreseeable future).

ABOUT AUTO / LIQUID COOLED ENGINES?

Liquid cooled engines are comparatively rare in homebuilt circles simply because there have been virtually no small liquid cooled aircraft engines manufactured through the years. The lone exception that comes to my mind is the RotoWay 100 engine, similar to the powerplant originally designed by RotoWay for use in their Scorpion helicopter. The engine ultimately reached a limited production in late 1979 or early 1980. Most other liquid cooled engines are the larger horsepower types (up to 2,000 hp) which were used in World War II military aircraft.

The only other engines of this type worthy of note are the automotive engines that have been converted to aircraft use. Among these are the Javelin Ford engines (you will have to build your own), the Corvair and the Geschwender Ford V-8 conversions. Outside of these few models there are special conversions of auto engines attempted by individual builders. Even so, these are rarely seen at Fly-Ins and places where homebuilt aircraft may be displayed.

You might wonder why you don't see more auto engines converted for aircraft use. One of the biggest drawbacks to modifying any type of auto engine is weight. Another is the comparatively high rpm the auto engine operates at in order to obtain its horsepower. Propeller efficiency suffers when an engine turns up much over 3,000 rpm. And, since auto engines operate well beyond this speed, the problem in most conversions is achieving some sort of propeller rpm-to-crankshaft reduction ratio.

There have been successful auto engine conversions. The old Model T and Model A were among the first. Then there are the VW, Corvair, Ford and Oldsmobile engines to name a few. Undoubtedly, there will be others in time....but not in great numbers, I'm afraid.

Steve Wittman's Oldsmobile V-8 powered Tailwind installation (1974)

A Corvair installation by James E. Mails for his Pietenpol (1982)

Chevrolet Vega-powered Pietenpol by Forrest Lovley (1972)

The versatile VW engine offers a builder many choices. It ranges in power up to 80 hp.

Radial engines are large and look better on antiques and big biplanes.

MOTORCYCLE ENGINES

It is no wonder that the plentiful and versatile motorcycle engines interest many would-be builders. They are light and cheap when compared with aircraft engines, but to date, there is no sign that any one of these engines is developing an established record for the durability and reliability required of an aircraft engine.

Motorcycle engines also turn too fast for aircraft use — 4,000 to 5,000 rpm and higher. This is simply too fast for propeller efficiency. Of course, reduction gears or belts can and have been resorted to in an attempt to keep propeller rpm down to a reasonable level. Unfortunately, the methods used to achieve this have usually been less than long lived or reliable. The point is, aircraft engines, auto engines and motorcycle engines are each designed and built for a specific mission in life. Their manufacturer is generally not interested in hybriding engines for other functions, sometimes because of product liability, but also because the market simply isn't there.

Since the manufacturers don't want to do it, attempts to convert auto or motorcycle engines to aircraft use have been made by homebuilders. However, should such a thought occur to you, be advised that it is an undertaking loaded with unexpected problems. That is not to say you shouldn't do it if you are determined enough, just that it might take more time and money than you would care to spend for an unproven powerplant.

Conversion kits are on the market for a few kinds of engines but not many. Perhaps the best example of auto-to-aircraft conversion kits are those offered for the VW engine.

ULTRALIGHT AIRCRAFT ENGINES

Modern ultralight aircraft are being powered by a variety of industrial engines, snowmobile engines, chain saw engines, homemade engines and numerous other small 2-cycle powerplants. These engines are inexpensive and plentiful. Unfortunately, they are not suited to aircraft we typically call lightplanes. Like motorcycle engines, these powerplants obtain their horsepower at a rather high rpm (approximately 8,000). As a consequence, engine reliability does not approach that of the slower-turning aircraft engines. Since these engines are used only in ultralight aircraft capable of being landed easily and safely almost anywhere, reliability has to be traded off for light weight, economy and simplicity of construction.

The variety of engines available and being used in Ultralight aircraft seems almost endless. Those most commonly used appear to be the Cuyuna (30 hp, currently the most popular), the Chotia (25 hp), Chrysler (18 hp), Yamaha (15 hp), McCullough, Honda, Onan and the Zenoah small air cooled engines. There are, of course, others currently ranging from 10 to 30 horspower.

It is important to realize that these small 2-cycle engines require some sort of propeller reduction to obtain the most efficient propeller performance. These reduction units are ordinarily of the belt-driven type because they are light and effective for low-

Would you believe it? Dual ignition in a one lunger?

A Cuyuna two-cylinder engine with a three-belt reduction unit.

CGS Powerhawk installation with typical belt reduction unit.

powered engines. The propeller reduction is usually on the order of 2:1 or 3:1, getting that 8,000 crankshaft rpm down to a usable 2,500 or 2,000. As for weight, these engines are feather light. The "big 'un", the Cuyuna (30 hp) is said to weigh in at a mere 64 pounds, complete, ready to run, with tuned exhaust, muffler and a recoil starter to boot.

THE VOLKSWAGEN AIRCRAFT ENGINE

You can't help but notice it. The gutsy little Volkswagen engine is the top contender for the honor of being the most popular auto engine for use in light amateur-built aircraft. Not everybody, however, is sold on the idea of using a VW engine in an airplane. Many builders distrust them and prefer to stick with standard aircraft powerplants, which they consider more reliable. They say: "A VW engine is nice, but....an aircraft engine it is not!" They say: "How many VW engines have gone 1,200 hours between overhauls? Some Lycomings are supposed to be 2,000-hour engines." Nevertheless, the humble little VW is appearing in increasing numbers of homebuilts and is constantly being subjected to significant evolutionary changes at the hands of hundreds of builders and engine specialists.

The VW engine was introduced into the homebuilt aviation element in Europe, where a number of aircraft like the Druine Turbulent, Pou Du Ciel, Jodel D-9 'Bebe', Luton Minor and other light aircraft were flown quite successfully on the modest 36 hp engine. But, 36 hp isn't much, even for a small single seater. Therefore, it wasn't until the more powerful 1500cc and 1600cc engines came along in plentiful numbers that the potential of this powerplant became apparent.

Designers responded with increased interest, and, in quick succession, many designs appeared that were specifically tailored to the use of the lightweight VW engine. (I was able to think of 27 such designs but I'm sure there are more.)

The promise of low initial investment as well as low maintenance costs, thanks to inexpensive parts obtainable from the neighborhood auto parts dealer, made the VW engine a very attractive do-it-yourself alternative to the expensive low power aircraft engines no longer in production and in short supply.

The VW engines being converted to aircraft use today are capable of putting out between 40 and 75 horsepower. And, perhaps surprisingly, the average builder has been able to make them perform quite satisfactorily.

The majority of current conversions are in the 1600cc displacement class, and produce a nominal 50 hp at an acceptable propeller rpm. As this is hardly enough power for the average two-seater, the aircraft designed for the VW engines tend to be the small, single-seat variety.

Without a doubt the most successful builders and those who are happiest with their VW powerplants are the ones who use a simple, basic coversion built around a good 1600cc engine. This is not to say that there is a universally accepted, typically standard 1600cc VW engine conversion for aircraft use.

Ultralight engines come in all shapes. However, even when used in dual installations, the combined horsepower is still insufficient to power the typical sportplane.

This VW engine has a plug-in Vertex magneto.

This VW engine has battery ignition.

This VW engine has a turbo-charged system.

A STANDARD VW ENGINE?

If you looked at 30 VW engine-equipped aircraft you'd probably see 30 different VW conversions. There are direct drive installations, reduction drive units, pushers, tractors and gyroplane variations. As for ignition systems, you'll find all sorts of components, ranging from the regular automotive distributors to a mix of aircraft magnetos such as Bendix, Slick, Eiseman, etc. Sometimes a builder will take the easy way out and plug a Vertex magneto into the engine's distributor hole.

Much of the variety in the VW conversions results from the use of non-VW engine parts. In the continuing attempts to convert it to a good aero engine, builders modify the VW's induction and fuel metering system by trying all sorts of carburetors (auto, aircraft, tractor, boat, motorcycle, updrafts, downdrafts, sidedrafts, injectors, etc.). Don't forget, multiple carburetors and turbo superchargers.

ENGINE WEIGHTS

You cannot substitute a standard aircraft engine in an airplane designed to take a VW engine without encountering a serious weight and balance problem. A drastic difference in the location of the center of gravity (CG) results from the use of a heavier aircraft engine. The A-65 Continental weighs 170 to 177 pounds and, since it has no electrical system, there is really nothing you can strip off to lighten it.

The VW engine, on the other hand, stripped of its automotive paraphernalia, weighs approximately 135 pounds. Even the larger (Limbach) 1800cc engine complete with starter, alternator, fuel pump and all weighs but a modest 164 pounds. Some VW engine conversions weigh as little as 117 pounds but, more commonly, around 145 pounds without electrical capability and certainly less than 170 pounds with.

The Lycoming O-145 series (50 to 75 hp) weigh between 157 and 177 pounds depending, in part, on the power output and whether a single magneto or dual magneto system is used.

Consequently, any builder who is considering the substitution of a Continental or Lycoming aircraft engine in an aircraft designed to accommodate a VW engine should think again as he will have a substantial weight and balance problem to solve. Adding to the problem of using the heavier standard aircraft engine in a VW design is the fact that while the relatively lightweight VW engine may be bolted directly to the firewall on spool-like spacers, the standard aircraft engine requires the use of a traditional engine mount. This extended engine mount positions the engine much farther forward adding to the CG difficulty.

The VW engine, particularly the 1600cc version, is still relatively plentiful and may be purchased at reasonable cost, another strong point for the VW engine. In addition, no small aircraft engine approaches the VW engine's weight per horspower advantage....and to the homebuilder that is important.

ENGINES FOR PUSHERS

Most older aircraft engines manufactured in the U.S. over the years were originally intended for use in conventional type (tractor) aircraft....those with the propeller up front. A lot of these same aircraft engines are still around and constitute the major engine supply source for homebuilders, who install them in a wide variety of lightplane designs. In reviewing the aircraft that have been built, we find that a surprising number of them are pushers. Consider the Breezy, Coot, Osprey, Volmer, Woody Pusher, Variviggen and, recently joining the aeronautical scene, the fabulous new generation of pushers....the Taylor Mini-Imp and the Rutan VariEze and Long-EZ.

Each time a builder undertakes the construction of one of these or some other pusher design, a new demand is created for an engine. Not just any engine. No, in most instances, it must be an engine capable of absorbing the reverse thrust imposed by a propeller that pushes instead of pulls.

Naturally, one big question begs an answer: "What engines can be used in the pusher I am building?"

Of course, there are the Continental and the Lycoming engines. Undoubtedly, they will be the first to come to mind as both brands can generally be obtained more easily than any other type of powerplant. However, can you use one of these engines in a pusher without modification? Unfortunately, your engine manual will probably be mute regarding the engine's adaptability to pusher configuration.

As most pushers are designed to take the smaller powerplants, I have selected a few representative engines within the 85 hp to 150 hp power range for comment in regard to their adaptability.

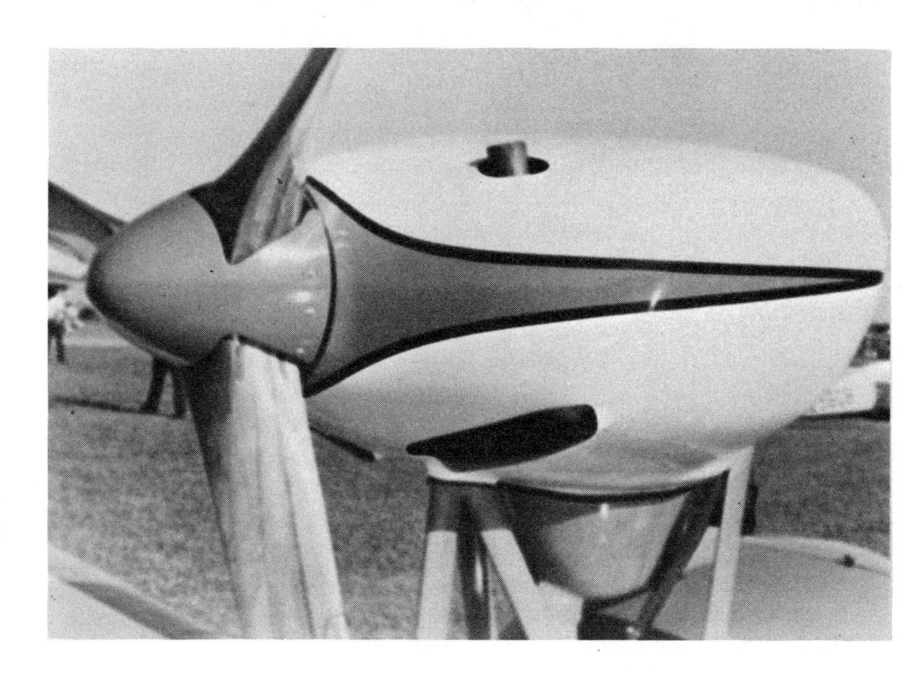

Pylon-mounted engines pose unusual installation problems, not the least of which are longer engine controls and cabin heat ducting.

If the following information does not cover your engine, there is one sure way to find out if that particular engine model is suitable for pusher use — write the engine manufacturer and ask him. Be brief, specific and be sure to give the engine type and model.

THE LYCOMING ENGINES

According to Lycoming, their O-290 engine, which incidentally is no longer in production, can be used in pusher installations without modification. This is welcome information for many builders as these engines will be around for years to come. The ever popular O-320 Lycoming is also suitable for use in pushers without any changes.

As a matter of interest, AVCO Lycoming now lumps all of their direct drive aircraft engines together in one overhaul manual with only a brief reference here and there to some feature unique to a particular engine model....beyond that, one must assume that all other details are uniformly identical.

THE FRANKLIN ENGINES

It is too bad that the Franklin engines are no longer produced in the U.S. However, some have made the homebuilt scene and apparently a few are being used in pusher amphibians. A few years back, there were plans to produce a reverse rotation version of Franklin's Sport 4 engine. It was called the Sport 4R. The reverse rotation capability permits the utilization of a standard propeller in pusher aircraft. This results in considerable savings when buying a propeller.

THE CONTINENTAL ENGINES

1. Continental O-200 Series Engines: It is common knowledge now that Teledyne, Continental Motors Aircraft Products Division no longer manufactures the O-200 engine. However, not as well known is the fact that it originally came in two varieties and that only the O-200B model was FAA and factory approved for pusher operation. In other words, if you are building something like a Volmer, or any other pusher that can use one of the 100 hp Continental O-200 engines, it should be an O-200B. This engine is identical in every respect to a normal O-200A engine internally and externally with the exception of the thrust bearing and crankshaft design. Thus, as far as the manufacturer is concerned, the standard O-200A engine is not intended to have the capability for pusher operation and should not be installed for that purpose in most designs. It is interesting to report, however, that the standard 0-200A has been utilized in the Vari-Eze pusher aircraft quite successfully and without modification of the bearing, according to the aircraft's designer Burt Rutan.

2. Continental C-90 Series Engines: Some C-90 engines may have been modified into the reverse thrust C-90 series and may be used in pushers. The C-90s were built a long time ago — in the late forties and earlier. In the years since, most of them have been overhauled many times. It would, therefore, be an impossible task to identify, in all instances, a C-90 that has been modified into the reverse thrust C-90 series without actually pulling it down.

However, these engines, if modified into

A Lycoming -- easily the most popular type engine with homebuilders. It is readily recognized by its large starter ring up front.

A Continental? No, by golly, its an 0-200 Rolls Royce! Only the data plate and rocker box covers differ.

pusher configurations can be identified by the letter "P" after the model number on the data plate. For example, C-90-8F-P. Here is the data and the parts requirement for the two pusher versions of the C-90 engines (electrical and non-electrical):

C-90-8F-P Spec. 27 (no starter, no generator)
Crankcase # 531434
Bearing # 40338
Crankshaft # 535804

C-90-12F-P Spec. 34 (with starter and generator)
Crankcase # 531432
Bearing # 40338
Crankshaft # 535804

Note that the crankshaft and the bearing are the same for either engine. The engine case differs as it provides the necessary pads for electrical accessories for one and not the other. The specification numbers provide further identification for this engine series.

3. Continental C-85 Series Engines: A lot of builders will be glad to learn that ALL of the C-85 engines are approved for pusher configurations. They require no modifications. But, unlike the C-90 series engines, don't expect to find a "P" identifier letter after the model number on the C-85 engines to show it is OK to use as a pusher engine....there won't be any.

Someone is sure to ask whether the tapered shaft or the flanged shaft C-85 is the one to use in a pusher. The Continental people do not discriminate in this regard. Either type crankshaft can be used.

Incidentally, of Continental's C series engines, only the C-75 and some of the C-85 engines were equipped with the tapered crankshafts. The C-90 and the 0-200 engines utilize flanged shafts exclusively. This in itself is of little consequence to the builder unless he intends to equip his engine with a propeller extension. In that event, the flanged shaft engine better lends itself to that purpose. There are some experienced engine men, however, who frown on the use of the prop extension on C-85 engines.

Just one more thing before leaving the C-85. In a pusher installation, the gas tank frequently must be located in the structure beneath the engine. A fuel pump installation, therefore, becomes mandatory. In this situation, either the Stromberg NA-S3A1 or the Marvel-Schebler MA-3SPA carburetors will give excellent service for your C-85. It might be prudent, however, to provide a backup system for any engine-driven pump system. Often selected by homebuilders for such functions are the Chevrolet and Buick submerged electrical fuel pumps.

WHAT ABOUT ENGINE ROTATION?

In order to utilize a standard metal propeller, some pusher builders dream of changing the direction of rotation of their standard aircraft engine, but this is quite a task, one that the ordinary builder should shun like the plague. In changing the normal direction of rotation, not only are internal engine changes such as valve timing necessary, but you must remember the magneto rotation will also probably have to be changed. If this is true, the magnetos will have to be retimed in their distributor section after the new "E" gap position is attained. Naturally, any magneto to be equipped with an impulse coupling will require a new opposite rotation coupling, and, of course, there would be other changes. No, this is not the answer for most of us.

Well, what does all this really mean? It means the manufacturers consider that there is a difference between tractor and pusher engine requirements. It also means that several popular aircraft engines can be used in pusher configurations without modification and with the manufacturers' blessings. And last, but not least, there are plenty of interesting designs in which you can install one of these trusty and proven powerplants.

COMPARATIVE ENGINE DATA

(CERTIFICATED SMALL AIRCRAFT POWERPLANTS)

AVCO LYCOMING DIVISION
AVCO CORP.
WILLIAMSPORT, PA 17701

MODEL	COMPRESSION RATIO	HP	RPM	FUEL	HEIGHT / WIDTH / LENGTH (DIMENSIONS)	DRY WEIGHT (LBS)
0-235-C	6.75:1	108/115	2600/2800	80/87	22.40 x 32.00 x 29.56	213
0-235-L	8.50:1	105/112/118	2400/2600/2800	100	22.40 x 32.00 x 29.05	218
0-290-D	7.00:1	135	2600	80/87	22.68 x 32.24 x 29.56	264
0-320-A, E	7.00:1	140/150	2450/2700	80/87	22.99 x 32.24 x 29.56	244
AEIO-320-E	7.00:1	150	2700	80/87	23.18 x 32.24 x 29.05	258
0-320-B, D	8.50:1	160	2700	100	22.99 x 32.24 x 29.56	255
0-320-H	9.00:1	160	2700	100	24.46 x 32.68 x 32.26	253
IO-320-B, C	8.50:1	160	2700	100	19.22 x 32.24 x 33.59	259
LIO-320-B, C	8.50:1	160	2700	100	19.22 x 32.24 x 33.59	259
0-360-A	8.50:1	180	2700	100	24.59 x 33.37 x 29.56	265
0-360-E	9.00:1	180	2700	100	21.08 x 33.81 x 32.26	269
IO-360-B	8.50:1	180	2700	100	24.84 x 33.37 x 29.81	270
LO-360-E	9.00:1	180	2700	100	21.08 x 33.81 x 32.26	269
AEIO-360-A	8.70:1	200	2700	100	19.35 x 34.25 x 29.81	299
AEIO-360-B	8.50:1	180	2700	100	24.84 x 33.37 x 29.81	275
IO-360-A, C	8.70:1	200	2700	100	19.35 x 34.25 x 29.81	293
LIO-360-C	8.70:1	200	2700	100	19.48 x 34.25 x 33.65	306
TO-360-C	7.30:1	210	2575	100	21.02 x 36.25 x 34.50	343

TELEDYNE CONTINENTAL MOTORS
AIRCRAFT PRODUCTS DIVISION
MOBILE, AL 36601

MODEL	COMPRESSION RATIO	HP	RPM	FUEL	HEIGHT / WIDTH / LENGTH (DIMENSIONS)	DRY WEIGHT (LBS)
A-40 (Series 2-5)	5.25:1	37/40	2550/2575	72	20.44 x 26.44 x 27.94	144/154
A-50	5.40:1	50	1900	73	27.81 x 31.50 x 31.00	170
A-65	6.30:1	65	2300	73	27.81 x 31.50 x 31.00	170
A-75	6.30:1	75	2600	73	27.81 x 31.50 x 31.00	177
A-80	7.55:1	80	2700	80/87	27.81 x 31.50 x 31.00	177
C75-8	6.30:1	75	2275	73	27.86 x 31.50 x 30.50	168
C75-12	6.30:1	75	2275	73	27.75 x 31.50 x 32.20	168
C85-12F	6.30:1	85	2575	73	28.75 x 31.50 x 31.26	169
C90-12F	7.00:1	95	2475	80/87	28.75 x 31.50 x 31.26	169
0-200	7.00:1	100	2750	80/87	23.18 x 31.56 x 28.50	188
C-125 (6 cyl)	6.30:1	125	2550	80/87	23.41 x 31.50 x 41.56	257
C-145 (6 cyl)	7.00:1	145	2700	80/87	27.41 x 31.50 x 35.46	268
0-300	7.00:1	145	2700	80/87	27.41 x 31.50 x 35.53	268

TABLE 3

OTHER INFORMATION

- MODEL DESIGNATORS -

  G = GEARED
  I = FUEL INJECTION
  L = LEFT HAND ROTATION
  O = OPPOSED CYLINDER ARRANGEMENT
  S = SUPERCHARGED
  T = TURBOCHARGED
  AE = AEROBATIC ENGINE

- NUMBERS FOLLOWING LETTERS INDICATE DISPLACEMENT IN CUBIC INCHES (LYCOMING MODELS)
- ENGINES WITH COMPRESSION RATIOS OF 7.30:1 AND HIGHER USE 100 OCTANE FUEL.
- LYCOMING PRODUCES OVER 200 VARIATIONS OF THE BASIC ENGINE MODELS.
- THIS DATA SUMMARY IS PROVIDED AS A GUIDE FOR ENGINE SELECTION. REFER TO APPROPIATE ENGINE MANUAL FOR DETAILED SPECIFICATIONS AND DATA.

Locating a Suitable Engine

First, ask around locally. You may be surprised to learn of engines in the area that you had not even heard about. Projects are being abandoned and sold all the time and the engine you need or can use may be sitting idly in a corner just waiting for you to come along. At any rate, let people know you are looking for an engine. The more people who know about your need, the better your chances are of finding the engine you want.

Don't overlook the obvious. Talk about your needs to EAA (Experimental Aircraft Association) members, fixed base operators, local mechanics and pilots. Be sure to notify the local EAA chapter newsletter editor. He might be willing to run an "Engine Wanted" notice. Most chapter newsletters are exchanged with other chapters as well, thereby expanding your search area.

Another useful action you can take is to run an advertisement in the local newspaper. It will only cost you a few dollars and it might pay off. Other advertising outlets include supermarket bulletin boards and the "Shoppers Want Ad", often a free service for consumers, especially in larger cities. It's a long shot but worth the effort.

A builder who lives in the vicinity of a number of airports or aircraft companies engaged in general aviation activities has a much better chance of locating a suitable engine. Make the rounds and get acquainted. Even if you can't locate that engine, the contacts you make may be able to help you later when you need other parts and supplies.

EXPANDING THE SEARCH

The next, or rather concurrent action would be to subscribe to or obtain current copies of that venerable aviation publication TRADE-A-PLANE. (Crossville, Tenn. 38555 - Tel. 615-484-5137)

TRADE-A-PLANE is the world's largest listing of aircraft, parts and services. It is published three times each month and is an individual's single best source for aviation related items.

TRADE-A-PLANE's classified advertisements are so numerous that the columns are broken down into large major category listings such as Engines, Radios, Propellers, Cessnas, etc. In addition, illustrated advertisements are sandwiched between its many pages of classified copy. A persistent search through this tabloid yields leads that generally produce results.

You could consider running your own advertisement in TRADE-A-PLANE, SPORT AVIATION magazine, or in one of the other aviation magazines partial to sport flying and amateur-built aircraft. One advantage of such an advertisement is that any response would be directly to you. You would be the first to hear about a certain deal. Unfortunately, homebuilders complain that those "Engine Wanted" ads in the national press often fail to produce results.

When you respond to someone's "Engine For Sale" ad you should realize that others have also seen it and you may find yourself in competition with them for the same engine.

Engines and engine parts .... a shopper's paradise for the serious builder. And where can you find this sort of open air "Fly Market" or "Country Store"? At one of the major Fly-Ins held around the country (Oshkosh, Sun 'n' Fun, Kerrville, etc.)

If the engine is still in the aircraft and operational, your chances of establishing its actual condition are much better than if it is sitting in some dark corner in the hangar.

You might just find a good used engine on the bench in some maintenance shop. Its owner may have wanted a larger, more powerful installation for his bird.

Obviously, the "good deals" will be snapped up quickly. With this in mind, the serious bargain hunter may be willing to pay for a first class rather than third class subscription, thereby getting his TRADE-A-PLANE a day or two earlier than much of the competition.

Don't confine your search to only the "Engines For Sale" section. Other listings — Rebuildable Aircraft, Homebuilts, etc. — may also have just what you are looking for. You will need many other parts for your project, for example, wheels, instruments, propeller, hardware and radios. You might, therefore, consider purchasing an entire aircraft to canabalize for its engine and parts....and get it for very little more than you would expect to pay for the engine alone. Obtaining an engine that is complete from the firewall forward will represent a tremendous savings in time and money, as the engine is already baffled and set up for you.

Yes, it is true, often you can buy a whole airplane for a few hundred to a thousand dollars more than if you bought that engine alone. Usually it would be a bargain only if you needed many of its parts. On the other hand, you could remove what you want and sell the rest of the airplane. Or, if the structure is in poor shape, you could canabalize it, sell the parts individually, and the rest as scrap metal. In the end you could make enough to clear your cost and realize a few bucks to boot. Be warned though, such an undertaking might be a time consuming diversion from your building project.

Remember this if you go the national ad route, however! Buying an engine locally is often better than buying one that is in some distant state.

In any case, don't "want to buy" an engine more than the gent "wants to sell" it. Reasoning processes seem to cease functioning efficiently when the buying fever hits.

Even if you have memorized what you want in an engine, it's a good idea is to write out what you really need and can safely use, and how much you are willing to pay. This helps clear your mind of "impulse buying fever" wonderfully.

....AS ADVERTISED

Buying an engine from an advertisement lead is sometimes difficult and risky. This is especially true if you live in Texas and the engine is located in Alaska or Maine or some other distant land.

Should you even consider such an engine? Think of the problems it could create for you. Should you try to go there and take a look at it? Or, maybe buy it on impulse, taking the seller's word for it? Some guys can be very persuasive over the phone.

If you do buy it, Will the seller crate and ship it for you? Obviously it would be cheaper to have the engine packed and shipped than to drive a great distance to pick it up yourself. But shipping costs are high too, and don't forget, all this adds to the cost of the engine.

Wouldn't it be better to get excited over an engine in your own or nearby state, even if it is slightly more expensive? At least you could look at it and inspect it carefully before you buy. As a general rule of thumb, buying an engine without looking at it is about as wise as buying interplanetary mining stock by mail order.

BEFORE YOU LOOK AT THAT ENGINE....

No single purchase you make for your project entails so many unknown elements of risk as the engine acquisition. You are literally putting your life on the line. Present day aircraft engines are reliable but only if they are in reasonably good mechanical condition. Your objective is to locate as good an engine as you can find. But how can you separate a good engine from a bad one? The advertised engine may be a good deal or you may be buying somebody's troubles.

It would be a wonderful thing to be able to do business without fear of being ripped off with a worthless assembly of scrap metal when you thought you were buying an operational engine. There are plenty of conscientious and honest people out there doing business but the odds against your having the pleasure of doing business with one are such you might not want to risk the gamble.

There is no guarantee that the engine you buy will be as good as the seller says it is. Even the seller may not know the real condition of the engine. It could be that he bought the engine for his own use years ago and never installed or even ran it. If so, what is its condition internally? Was it properly prepared for storage? This points up the importance of asking questions, a lot of questions.

Don't allow yourself to be rushed into making the purchase. If the buyer doesn't want to answer your questions or appears evasive forget the whole thing and look for another engine.

Most of us grew up in a mechanical environment. We were literally surrounded by cars, boats, motorcycles, lawn mowers, household gadgets and machinery of all sorts. Can there be anyone building an airplane who has never worked on a car? Although most of us could probably do a pretty good job of evaluating an aircraft engine without the aid of a good mechanic, wouldn't it be better to get an experienced friend to go along to look at the engine. Paying an A&P (Aircraft and Powerplant mechanic) to check out the engine for you isn't a bad idea either.

ABOUT ENGINE PRICES

When we examine the used engine marketplace, we find that certain engines are in greater demand and command higher prices than others. Thus, a higher price is not always a reflection of the excellence of a particular brand or its horsepower, but rather more a manifestation of the demand and supply status of that particular engine.

Another element that directly affects demand is time on the engine. You would expect to pay more for an engine that is new or for one that has just been through a major overhaul than you would for an engine that has, say, 1,900 hours in its Engine Log...and you generally do.

Engine prices usually vary in direct proportion to the amount of time remaining before a major overhaul is required. The number of hours on an engine is, therefore, a reasonable yardstick because a used engine is

Not damaged, simply neglected. The owner of this aircraft may be very eager to sell.

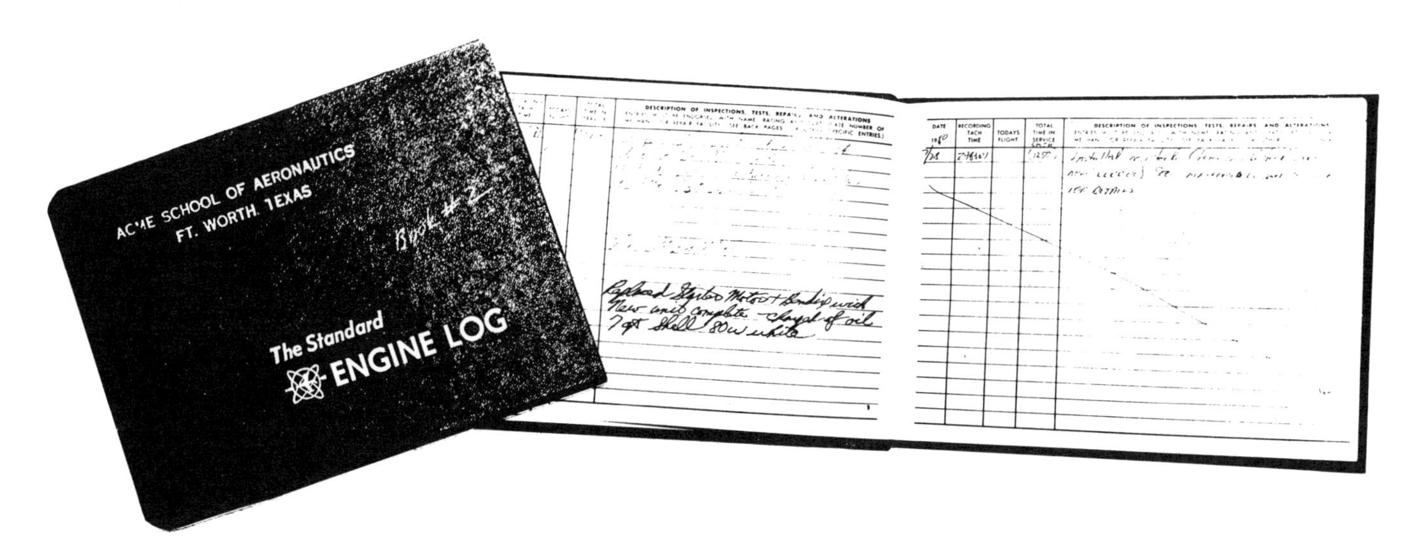

simply a used engine. It no longer has its original life expectancy and it comes with no factory guarantee that it will make it to its natural run-out time before an overhaul is mandated.

Manufacturers establish the estimated useful life for their engines by designating a certain Time Before Overhaul (TBO). Some small reciprocating aircraft engines are expected to have an operational life of 1,200 to 1,800 hours before a major overhaul is deemed advisable. Many small Lycoming engines have a 2,000-hour life before overhaul. But these estimates are no guarantee that any given engine will reach that magic number without some sort of a major malfunction or breakdown. Still, you can expect the advertisements to play up the low number of hours as a major selling paint.

There are no foolproof rules that guarantee you'll pick a good engine at a good price. Price is important, of course, but it is no assurance of quality. When evaluating an engine consider its advertised price only as a starting place....then, read on.

ENGINE LOGBOOK

Any engine installed in a type certificated aircraft must have an Engine Logbook. This log should document the maintenance and inspection history of that engine as performed by FAA licensed mechanics or authorized repair stations. When an engine is sold the logbook should accompany it. On the other hand, there is no Engine Logbook requirement for an engine installed in an amateur-built aircraft because it does not have to be maintained by FAA-licensed mechanics. An engine used in a homebuilt cannot, ordinarily, be returned to use in a certificated aircraft.

This leniency on the part of the FAA has opened up a new market for people who deal in engines of questionable mechanical condition. In this regard you may see advertisements similar to the following:

> CONT. A75 ENGINE O SMOH, includes prop hub and Bendix mags. No carburetor. No logs. Suitable for experimental. $1,600 or make offer.

Engine advertisements such as this should arouse your suspicion and distrust because, if you think about it, it seems to imply that the engine is not good for a certificated aircraft but is good enough for a homebuilt. The fact that the above referenced advertisement indicates that the engine has zero time since major overhaul (O SMOH) indicates only that it might have been "overhauled" by an unlicensed mechanic or, perhaps, the guy next door who may or may not have a talent for such things. This doesn't mean that such an engine is automatically not airworthy....just that it has no credentials, no history and, certainly, no guarantee.

Always begin your engine evaluation by examining the Engine Logbook, if there is one, and at the same time question the seller to find out what he knows about that particular powerplant. It is important to realize that, although the number of hours on an engine, as advertised, were actually logged as reflected in the Engine Logbook, they might be far fewer than the total hours the engine has been operated. In other words, all of the hours may not have been logged. People are not too good with the paperwork and unless the engine is still installed in an airplane and its has been currently logged with an hourmeter or a recording tachometer, its total time may remain an unknown element....a figment of the owner's imagination.

If a logbook is available, look for continuity in the entries. This is the best clue to conscientious maintenance. The lack of detailed entries should alert you to the possibility that the engine did not get much TLC (tender loving care).

WHAT THE ENGINE LOG CAN TELL YOU

In addition to containing a description of the engine (including its serial number) the engine log contains special data such as manufacturers' notes, test data, reconditioning information, etc. Regular entries for oil change, including the type of oil and quantity used should give you an idea of their frequency or infrequency. There should also be entries indicating that annual inspections or 100-hour inspections were made at the proper time.

Just as important are the occasional notes that a Service Bulletin has been complied with. It would be well to check with some maintenance people to determine how many Service Bulletins apply to the engine you intend to purchase. Buying an engine that has not had all the latest Service Bulletins taken care of might mean considerable expense later as many of these affect the airworthiness of the engine.

And last but not least, the logbook should detail the maintenance and repair work performed on that engine throughout its life. Replacement of parts and accessories can give you a clue as to the remaining life of those units.

Unfortunately, many logbooks have but a few entries and those few are often indefinite and lacking in useful information. Nevertheless, the absence of a logbook could further complicate your evaluation of the airworthiness of the engine.

Everything else considered, an engine with a good logbook is an excellent candidate for acquisition for your own use.

INSPECTING THE INSTALLED ENGINE

Begin with the Engine Logbook and try to determine the engine's history. Where did it come from? Was it used up in student training? Has it already been through one major overhaul? Will you have to overhaul it before you use it?

Most of us rarely fly a homebuilt more than 50 to 100 hours each year so a mid-time engine with 500 to 1,000 hours on it, if in good condition, could last another 5 years before it would have to be overhauled. Actually, an engine that has been trouble free for 1,000 or so hours may be a better engine than one that has been freshly overhauled by some unknown individual. The first few hours of operation after a major overhaul can be critical.

If the engine is still in the aircraft and operational, your chances for establishing its actual condition are much better than if if is

Visit all of the small airports in your area and you could chance upon a scene like this. Undoubtedly you could purchase this engine for a song. The catch is that a major rebuild and a new crankshaft will hike up the eventual total cost.

The markings on the rocker box covers are results of a recent differential compression check. You'll find it much easier to inspect an uninstalled engine if you can hoist it off the ground to a convenient location.

sitting on the hangar floor or in some dark corner.

One of the best tests you can perform on an operational engine is a differential compression check, preferably immediately after a flight. If this cannot be accomplished, at least turn the propeller through so that the presence of compression in each cylinder can be established. All cylinders should offer a similar amount of resistance when passing through the compression stroke.

Take a long look at the engine, carefully checking the crankcase, crankshaft and flange areas. Note whether any of the cylinders have been damaged. You cannot spare the loss of any of the cylinder fins as engine cooling could be seriously affected. Examine the oil filter and spark plugs. If the engine is being maintained in a certificated aircraft, check the log book for frequency of oil changes and maintenance accomplished.

The engine's external appearance may provide a good clue as to the amount of loving care that has been lavashed on the old mill. If the hoses, baffles and wiring are in excellent condition, that can also be taken as a good sign.

Logbook statements such as "Top Overhaul" mean nothing unless backed up by a detailed listing of the work accomplished. Verbal claims for a top overhaul or a recent major overhaul, likewise have no meaning unless the work can be verified. What really counts is exactly what has been done, why it was done, and when.

INSPECTING THE UNINSTALLED ENGINE

Ask the seller how long the engine has been inoperative and how it was prepared for storage. A freshly removed engine should show some signs of its recent use. There may be some grime and oil smudges here and there but you won't be able to discover serious oil leaks and drips.

A recently overhauled engine that has not been run is highly susceptible to rust — much more so than a well used engine whose cylinder walls have been oil glazed with a varnish-like sheen. An engine that has not been operated for months or even years may have serious internal corrosion problems.

How does the engine Look? Is it clean as a whistle? That's good, or is it? Could the seller have had it "sanitized" just to sell it? A clean high-time engine may be misleading, as any sign of chronic leaking will have been removed. Of course, if it has been freshly overhauled, it should be nice and clean.

Has the engine been subjected to sudden stoppage? Technically, sudden stoppage refers to an immediate, and possibly catastrophic cessation of engine rotation, because the propeller has hit a hostile and uncooperative object, like the ground, another airplane, car, building, etc.

An engine subjected to sudden stoppage has been overstressed and may suffer a bent crankshaft, flange or other external or internal damage. Such an engine must be regarded as suspect until otherwise proven damage free.

When buying an engine you may be influenced more by its external appearance than by its reported history and Engine Log entries. If so, you should at least check out the following items before making that final decision:

* Rotate the crankshaft and listen for grinding noises. Be alert to any abnormal sensation or vibration transmitted to your hands.

* Take a close look at the vulnerable crankshaft flange (or spline) for evidence of heavy scrape marks, warpage, cracks or other deformation.
* Attach a dial indicator to the engine and rotate the shaft slowly in 90-degree increments to check the shaft for run-out. Every manufacturer gives a run-out tolerance for each of its engines. For example, the maximum flange runout for the Continental O-200 engine is only .005-inch.
* Remove the oil sump drain plug and check for metal chips or other foreign material. Don't be surprised if you find that the person selling the engine has had the oil drained and the plug cleaned. However, your finger may still pick up enough residual film for you to determine its condition at the time it was drained.
* Visually inspect the entire exterior of the engine case. Look for signs of long term or fresh oil leakage and for cracks. Give close attention to the propeller thrust bearing area around the nose case section.
* Examine the engine mount lugs for cracks....one or more may even be broken off if the airplane the engine came out of was in an accident.
* Inspect the cylinders for cracks and oil leaks. Check the cooling fins for cracks or broken sections. Be sure to investi-

gate the cause for any indication of cracks, oil leaks or other visible damage.

So much for the exterior. What about the interior? As a rule you will probably not have an opportunity to remove the cylinders so that the counterweight blades can be visually inspected, nor would you normally disassemble the engine to make a magnetic particle inspection of the crankshaft and other steel parts, or a dye penetrant inspection of the non-ferrous crankcase areas — around its various holes and openings, for example.

That's right, you may not have the opportunity to do that and yet you may be looking at an engine that has been subjected to sudden stoppage and may have been damaged in the areas mentioned.

Some builders are content when the crankshaft does not have excessive runout and assume that everything is all right internally. Unfortunately, this may not always be true. If you do not check the crankshaft flange runout and later find that it is beyond limits, you may be faced with replacement of the crankshaft at a probable cost approaching one-third of that paid for the entire engine.

LOGICAL ASSUMPTIONS SUMMARIZED

* It is a fairly low-time engine. A low-time engine is better than a high-time or runout....but not always.
* It is currently operational and still in a certificated aircraft, or it has just been removed by the owner for the installation of a larger engine modification.
* It came out of a wind-damaged aircraft and not out of one that has experienced sudden engine stoppage due to an accident.
* It has not been stored for a long period without being run, particulaly after a major overhaul.
* It was not used for student training.

THE BOTTOM LINE(S)

* If you want an engine that can be easily bolted to a standard engine mount, the quickest, most reliable solution for you is to use a conventional aircraft engine ....even if it is a used one.
* If you are determined to use a conversion, the VW engines have proven, by far, to be the most successful for homebuilt use.
* If the engine you want is for a pusher, be advised that not all engines are capable of absorbing reverse thrust loads, plus you will need a "pusher" propeller.
* If you must build up your own engine, be prepared for unanticipated surprises, problems and costs, and no guarantee that the end result will be what you expected.
* 'Caveat Emptor'... Let the purchaser beware!

NOTES

handling & storage

2

Coping with that uninstalled Engine

So you think you're fortunate because you already have an engine even though your project is not yet complete. Maybe so, but look around at your crowded workshop. Need I remind you what an inconvenience it is that the engine is always in the way, always in the wrong place? Furthermore, you've probably noticed it's a major undertaking just to move it. Even if you get help, it's so ungainly that there seems to be no way to get a good hold on it, much less lift it.

Just think! That awkward assemblage of metal will be around for a long time....a year or two, maybe more. Fortunate as you are to have already acquired the engine you need, I'm sure you'll agree that you have also acquired some unusual handling and storage difficulties. Now that you have that powerplant, you will have to look after it and protect it from deterioration and accidental damage.

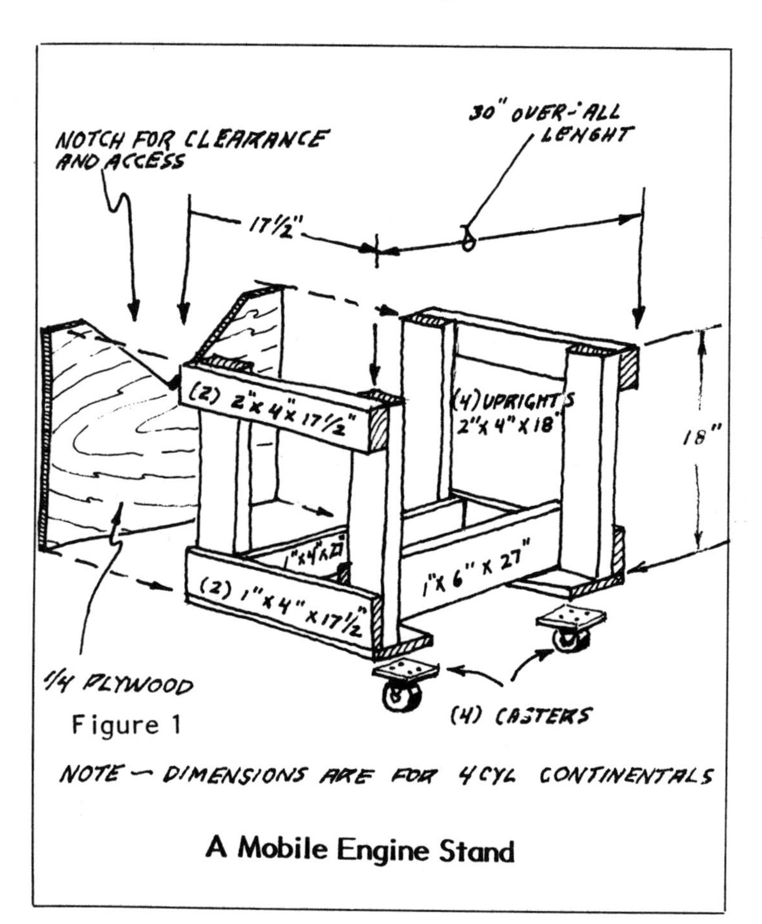

Figure 1

A Mobile Engine Stand

SHIPPING CRATES AND OLD TIRES

Once the initial joy of engine acquisition has subsided you will probably be receptive to almost any helpful suggestion for coping with the space problems it has introduced into your tidy workshop. If you visit a few builders who already have their engines, you will see first hand how some of these guys cope with an uninstalled engine. Mostly they just drape a plastic sheet over it and shove it into a corner. Sometimes you'll spot it in a corner draped only in dust.

Engines which were bought out of town and shipped in will probably be imbedded in some sort of a shipping crate, usually made of scrap lumber and about the size of an imported economy car. There is a great temptation to just let that engine be until it is needed. Certainly, you can't move a crate around all by yourself. Unfortunately, since the engine is difficult to examine, it might well languish there for a long time, unattended by all except the corrosive forces of nature.

Sometimes you will see an engine impaled, nose down, on a homemade engine stand similar to those used in engine build-up shops. The engine may be very accessible on such a stand, but it is also very difficult to move out of the way unless casters are a feature of the stand. Also, such a stand may be rather top heavy and easily tipped over in a careless or hectic moment.

Then there is the builder who has his engine resting upside down on an old auto or truck tire. Maybe he can drag it off to another location if it gets in his way, but with the engine upside down, he can't do much else with it or to it.

Perhaps the most important elements missing thus far in these illustrations has been a reasonable degree of mobility and accessibility for the stored engine.

BUILD A MOBILE ENGINE STAND

Your engine can be given mobility even in storage. The photos in this chapter illustrate a very compact, very useful, and very

In the foreground is an engine shipping crate after it has been stripped to the bare minimum base. Can you imagine having this monstrosity, with an engine on it, taking up residency in your crowded workshop? Compare with the engine stand in the right-hand corner of the photo.

Give your engine mobility, even in storage with this easy to build, useful and very compact engine stand. It takes up less space than the engine. Its greatest asset is the accessibility it affords to the engine.

efficient engine stand. It takes up less space than the dimensions of the engine and it provides the mobility that is desparately needed in any workshop full of airplane components.

The drawing provides the dimensions suitable for the small 4-cylinder Continental engines. The Lycomings and others can be similarly accommodated. The stand requires only scrap material and a set of swivel dolly wheels. Almost any kind of swivel wheels or casters will do as no single wheel will ever have to support more than 75 pounds. Even roller skate wheels would be adequate for that matter, but castered wheels provide greater maneuverability in crowded areas (Figure 1).

Notice how accessible everything is. The basic design of the stand could even be altered somewhat if a great deal of work had to be done on the engine while it was stored. However, even as it is, the stand provides convenient access to the engine for fitting the baffles and attending to the many essential pre-installation chores still remaining.

Note that the stand permits the engine to be stored in its upright position, making it an easy matter to affix a hoist to the lifting eye for transferring the engine from the stand to the aircraft.

My own stand was constructed of scrap 2-by-4s, a piece of leftover 1/4-inch plywood, and a couple of old cedar fence boards. Both nails and bolts were used indiscriminately in its assembly. No use making a big production out of it.

I expect that a few qualified engine folks might question the wisdom of supporting the engine in the manner depicted in the photos. However, I have studied the effects on the engine and weighed the various arguments pro and con and have decided that, for me, the advantages gained and the problems solved far outweigh such concern. You can be sure I want the best possible environment for my engine. So far, the rig has been invaluable to me.

HOISTING THE ENGINE IN YOUR SHOP

Ever try to connect a hoist to an engine that is upside down on the workshop floor.... with nobody around to help? Worse yet is the anguish suffered when trying to rotate that engine to its upright position while, at the same time, trying to ratchet it off the floor.

You Have an Engine.... Now What

It is almost as bad to get an engine too soon as it is not to have one when your project is ready for it. With one or two exceptions, an engine acquired during the early stages of a project is likely to sit around the shop for a long time. It does an engine no good at all to be sitting in a corner of your workshop, particularly if it has not been properly preserved.

Unfortunately, most engines acquired by homebuilders have not been prepared for long-term storage, although they should have been. At any rate, by the time a builder acquires it, the engine may already have been sitting idly for several years, and that is BAD! But, even after a builder takes possession of the engine, too often, he will not give it the proper attention and care it needs. You know how it is, when a builder is busy with his project, every hour counts. He hates to spend even a few minutes in "non-productive" pursuits. Time goes by and before he realizes it, as much as a year may have whizzed past without him ever looking after the engine he shoved over to a corner of his workshop. Wouldn't it be a shame if he allowed it to corrode after the big investment he had to make in obtaining it?

Is it too late for him to do something about it now? It would be a tremendous blow to learn that a complete disassembly and overhaul was necessary on what had been a serviceable engine when it was first acquired.

The proper time, the most effective time to preserve an engine that will not be in use for a long time is immediately after it has been shut down for the last time before storage and the still-warm oil can be drained out of the sump.

However, even if this opportunity was lost it would be better to take whatever preservation measures that are possible. NOW! DON'T WAIT!

A well used aircraft engine will survive storage or inactivity better than a newly majored engine. This is because a freshly overhauled powerplant, with newly honed cylinder walls, can show signs of rust in just a few days if precautions aren't taken.

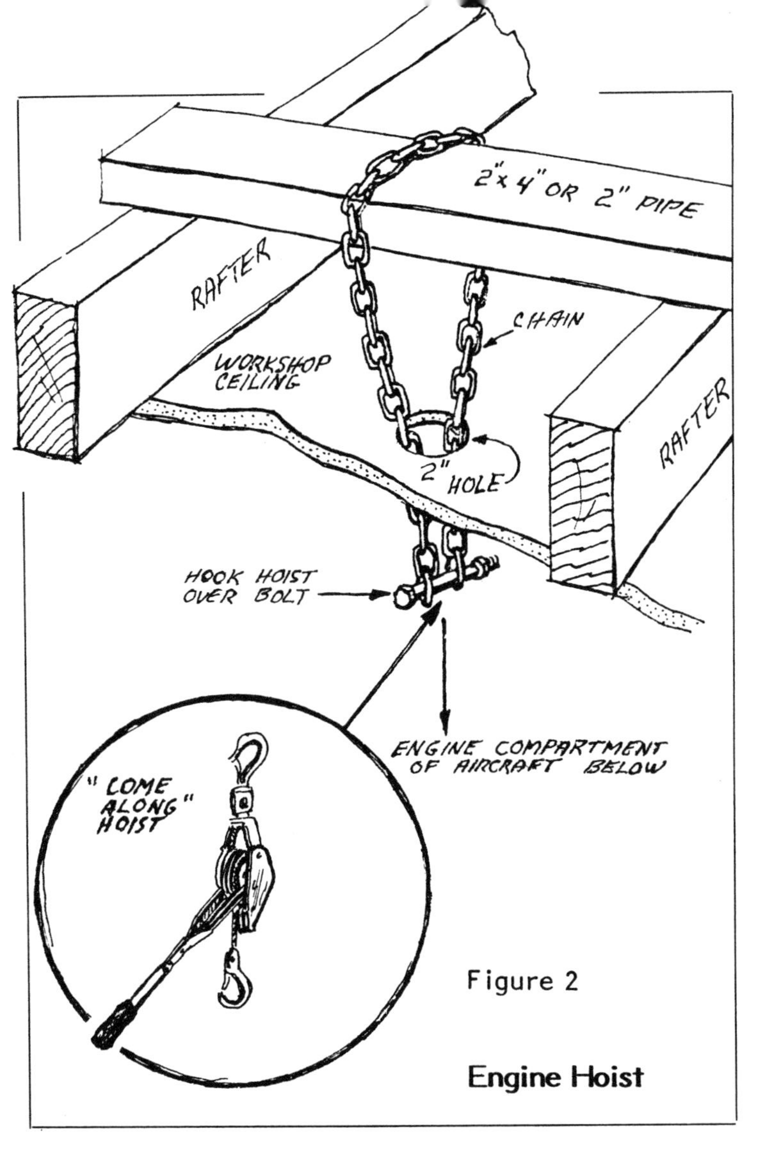

Figure 2

Engine Hoist

The rational builder would not try to do this alone. With the engine resting in its custom-built mobile stand, however, the entire job could be accomplished single-handedly. Of course, you would have to have some sort of mechanical hoist to furnish the lifting brawn for the effort.

You could rent a low profile mobile engine hoist from a tool rental shop. Perhaps one of your friends has a portable "A"-frame that could be used to lift the engine to the aircraft's engine mount. However, most home workshops and garages do not have sufficient ceiling height in the shop to accommodate such a rig. Naturally, you could perform this function out of doors. It would be easy enough to wheel out a castered engine stand, wouldn't it? However, one problem to consider when working outside (besides the weather) is the possibility that you might not complete the engine installation that day. You certainly wouldn't want to leave the fruit of your labors outside and unprotected overnight.

As an alternative, consider the use of one of those economical portable cable hoists sold at virtually all discount stores. Different models of these devices have mechanical advantages ranging from 10:1 to 40:1 with capacities of 500 to 1,500 pounds. This is more than enough to make handling a 250-pound engine a snap (oops). Because this kind of hoist is compact — approximately 18 inches between hooks — it is practical to hang the hoist from the workshop ceiling and still have sufficient working height to hoist the engine into place in the aircraft.

To ensure that you do have sufficient lift range, the hoist should be mounted as close to the ceiling as you can get it. If your garage shop walls and ceilings are finished, you don't want to screw big hoisting eyes all over the place. Even so, sometimes it is difficult to find adequate structure in the ceiling where you need it. So, some other method of hoist attachment might be better. (Figure 2)

A simple solution to the problem is to cut a 2-inch hole with a hole saw into the ceiling through the plaster board over the exact location you intend to suspend the engine. A short piece of large diameter pipe, or a 2-by-4, can then be placed across a couple of rafters in the attic and a short chain draped over it in the manner depicted. Hook up the hoist and ratchet away.

After the project is completed, you can replace the plaster plug in the ceiling and refinish it to as good as new. Oh yes, one more suggestion. Cut the hole between rafters rather than into one of them.

Engine hoists (A-frames, etc.) take many forms. This homemade hoist uses a small, inexpensive hydraulic jack to provide the muscle.

Most aircraft engines are hoisted by using a lifting eye attached to the upper crankcase flange. If yours is missing you can make your own.

Clever idea but how are you going to hoist it? And, what if you need to use your creeper?

An unbelievable amount of material has been written about engine storage, much of it complex and conflicting in nature. Except for the military, large aviation organizations and shippers of engines, the subject and processes described here are largely ignored, particularly by homebuilders.

The large aircraft engine manufacturers, Teledyne Continental Motors and AVCO Lycoming, have long provided approved engine preservation guidance for owners of powerplants they build. These detailed instructions should be followed for the best results. However, homebuilders are using all kinds of engines, many of which are not supplied with the kind of engine preservation manuals and information they need. With this in mind, it seems to be worthwhile to devote some space to preservation treatments suitable for infrequently used or stored engines.

Aircraft that are infrequently or irregularly flown tend to develop engine cylinder wall corrosion, whereas aircraft that are flown often do not have this problem. As builders, we are naturally more concerned over the long term preservation of an engine removed from the aircraft and stored separately than we are with an installed engine in an aircraft. Nevertheless, a few suggestions regarding the preservation and protection of any infrequently flown engine may also be useful information to have for later use....after your airplane is completed and flying.

Some people shrug off the need for safeguarding an engine from corrosion and never have problems. Others are not as fortunate and their engines are virtually ruined by internal rust. Builders should be aware of the industrywide categories given to the different types of preservation efforts.

MATERIALS FOR ENGINE PRESERVATION

The basic corrosion preventative compound used by major aircraft engine manufacturers and most maintenance facilities is the same as what just about everyone else uses. It is identified as MIL-C-6529, Type I or Type II and is obtainable commercially.

Type I is a concentrate that has to be mixed with oil, preferably a grade 1100 mineral aircraft engine oil, in a ratio of three parts corrosion preventative compound to one part oil. Always follow the manufacturer's instructions for use. It is not a good idea to add this corrosion preventative compound directly to the engine oil. Always mix this concentrate externally with oil before using.

Type II is premixed and is ready for use directly from the container.

DESICCANTS (DRYING AGENTS)

Two types of desiccants are used in preparing an engine for storage. One is the dehydrator plug and the other is the silica gel bag or envelope.

DEHYDRATOR PLUGS

Dehydrator plugs are similar in appearance to spark plugs except that they are a clear plastic plug and are filled with a cobalt-chloride-silica gel substance. This is a desiccant or drying agent which serves as an indicator of the moisture level inside the engine or area in which it is used.

The plug stays a bright blue in low moisture conditions but begins to turn pink when exposed to an increase in relative humidity. If the humidity is very high it will eventually turn to its natural white color. As long as the deyhdrator plugs (installed in each spark plug opening) are blue, the engine's interior should be safe from moisture-induced internal rusting. FAA sources point out that corrosion is not likely to develop when the humidity is below 30%.

The same type of dehydrator plugs used in AVCO Lycoming engines can be used in Teledyne Continental engines as both types of engines (at least the ones we would be most likely to use) can accept the same 18mm size dehydrator plugs. Some aircraft parts departments do not stock dehydrator plugs so you may have to check the larger maintenance facilities.

Ask around locally for a source. At least two of the larger homebuilt suppliers carry them in stock: The Aircraft Spruce & Specialty Company, Box 424 Fullerton, CA 92632 and the Wil Neubert Aircraft Supply, P.O. Drawer 500, Arroyo Grande, CA 93420.

Dehydrator plugs are reusable. When they have turned pink remove them and pop them into an oven at low heat, about 200 degrees F to 250 degrees F overnight to dry out to their original blue color.

SILICA GEL (BAGS, ENVELOPES)

The substance used in silica gel bags and envelopes is the same as that used in dehydrator plugs and can also be re-used after drying in an oven.

OTHER MATERIALS TO USE

Other materials and equipment needed to prepare an engine for storage include waterproof paper or tape, wood blocks and plugs to seal engine openings left by the removal of accessories such as carburetor.

An airless spray gun with a long slender nozzle will also be needed for atomizing the sprayed preservative oil going into each cylinder. The nozzle or probe must be small enough to fit into the spark plug openings. If a regular spray gun is used, be sure it has a moisture trap installed. You don't want to introduce moisture droplets in the spraying process.

TYPES OF ENGINE STORAGE

* FLYABLE STORAGE: The airplane is flown only occasionally and reasonable protection is required for up to a month of non-use.

* TEMPORARY STORAGE: The airplane might not be flown for as long as 90 days. The preservative treatment used must be effective for at least that length of time.

* INDEFINITE STORAGE: The type of treatment in this case is for the long term....for the aircraft or engine that is to be out of service for an extended period.

Even so, the engine may have to be treated from time to time unless it is "mothballed" and sealed in a metal container.

In addition to these 'calendar' type preservation categories, other conditions will also influence the type of corrosion preventative treatment that will be required.

Humid environments — locations along either sea coast, the Great Lakes, or the Gulf Coast are very detrimental to engines due to their high levels of moisture and salt air. On the other hand, the dryer climates of more arid areas of the country are kinder to engines, even though effective preservation efforts will still be required.

FLYABLE STORAGE

An aircraft that has been or is going to be idle for longer than a week should be afforded some degree of corrosion prevention or protection and regularly inspected to guard against the development of rust inside the cylinders.

It is a traditional belief and practice to turn the engine through by hand about five propeller blades at least once a week as a means of preventing internal corrosion. Turning the engine over is supposed to re-coat the cylinder walls with engine oil and wipe away moisture droplets which may have formed on the cylinder walls because of condensation. This practice has merit. If you intend to rely on this method, you should repeat the process at least once each week to 10 days, more frequently in humid coastal areas and perhaps less frequently in more favorable climates.

After an aircraft (engine) has been idle for as long as a month, it should be flown, or at least be given a good ground run-up. The run-up must last long enough to get the oil temperature up to operational range (in the green) to ensure the evaporation of internal moisture. Otherwise, you may be doing more harm than good to the engine.

Some operators located in areas of the country that enjoy a low humidity (arid conditions) suggest that it might be safe to go as long as 60 days before having to make a run-up. Even so, any aircraft remaining inactive after a month — two at most — rates an inspection of its internal cylinder walls with a light to determine their condition.

TEMPORARY STORAGE

An airplane that is going to be inactive for more than a month deserves better protection than that afforded those in Flyable Storage. A deliberate planned corrosion treatment is your best assurance that the engine will be serviceable when you need it months from now. It requires a little extra effort, but it is really a simple process.

1. Remove the top spark plugs and spray in a mixture of corrosion preventative through each top plug opening while its piston is at its bottom dead center position. After each cylinder has been sprayed, turn the propeller slightly so that the crankshaft position is such that no piston is at top dead center.

2. Re-spray each cylinder internally without turning the propeller from this moment on.

3. Re-install the spark plugs.

4. Spray a couple of ounces of preservative inside the oil tank filler neck.

5. Cover or seal all engine openings. Use waterproof tape, plugs or whatever is needed. Your objective is to keep out air... moisture. Add red warning streamers to each location as a reminder for later removal.

6. Attach a sign to your propeller stating: DO NOT TURN....ENGINE PRESERVED, or some similar warning to yourself and others. You might also put the date on the sign.

(NOTE: Some people swear by the merits of WD 40 corrosion preventative spray being squirted into their non-flying or shop-bound engine periodically, and the crankshaft rotated by hand a few times. I don't have sufficient evidence that this is either helpful or harmful.)

RETURNING AIRCRAFT TO SERVICE:

1. Remove all tapes, streamers and plugs.

2. Remove the bottom spark plugs and turn the propeller through several revolutions to purge the cylinders of excess preservative. Replace the spark plugs.

3. Start and run-up the engine.

4. After shutdown, clean and inspect the engine. It's ready for flight again.

INDEFINITE STORAGE

The extended storage preparations normally used are identical for both an engine that is still installed in an aircraft and an uninstalled engine stored separately.

Prior to being placed in extended storage, an uninstalled engine should have been operated and brought up to normal operating temperatures before it was prepared for storage. However, unless this was done before you obtained it and while the engine was still in an aircraft, you have no choice but to treat it yourself now as best you can.

Before you start your preservation treatment, first clean and wash the engine exterior and dry it thoroughly. Then, inspect all external areas for signs of corrosion and rust. Inspect the condition of the internal cylinder walls through the spark plug holes using an inspection light and mirror. (It might also be a good idea to remove the rocker box covers and inspect the rocker assembly inside.)

CAUTION: The corrosion treatment you give

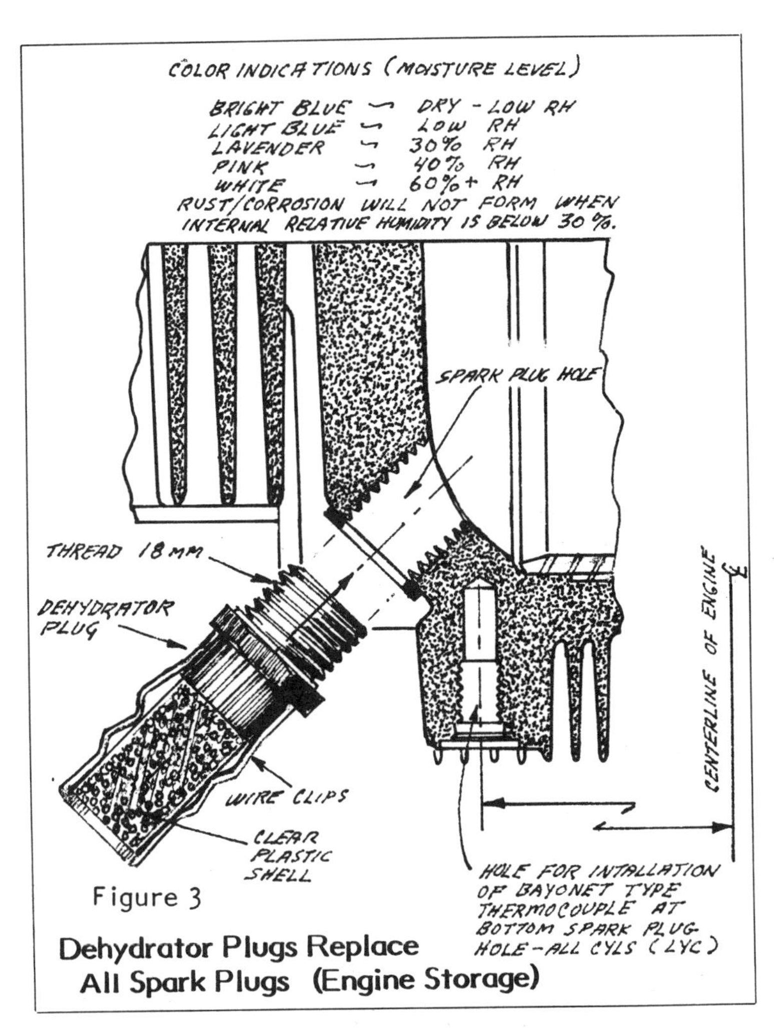

Figure 3

Dehydrator Plugs Replace All Spark Plugs (Engine Storage)

your engine will not stop corrosion that is already present any more than a coat of paint will stop rust that is underneath it. Oh, it might slow down the rate of deterioration, but it will not eliminate its cause. Contaminated areas will continue to fester and grow. Corrosion that hasn't progressed to the point where the cylinder walls are pitted (damaged) with rust can and should be removed first. To remove corrosion from inside a cylinder, a complete disassembly of the jug is necessary.

If your engine passes this inspection and you have corrected any existing minor corrosion problem, you are ready to prepare your engine for extended storage. This is the general procedure to follow:

1. Drain the engine oil and replace it with a corrosion preventative mixture (Mil-C-6529, either Type I or Type II).

2. Remove all of the spark plugs.

3. Spray the inside of each cylinder, while its piston is at the bottom center position.

4. Next, rotate the crankshaft so that no piston is at the top position and re-spray the inside of each cylinder. From this point on, the crankshaft must not be turned lest you wipe most of the preservative mixture off the cylinder walls. Place a sign on the engine as a reminder and warning....DO NOT TURN CRANKSHAFT....ENGINE PRESERVED, DATED ______.

5. Spray in and around the exhaust ports and exhaust valves (exhaust gas residue is extremely corrosive).

6. Seal the exhaust port openings or plug the exhaust pipes and tape them over to keep moisture out. (You could slip a silica gel bag into each pipe before closing it.)

7. Plug or tape over the crankcase breather and all other openings in the engine. Remember, air contains moisture in varying degrees and you want to keep that out of your engine.

8. Install a dehydrator plug in each spark plug opening if possible. Otherwise, replace the original spark plugs after coating their threads with some of the preventative mixture. Actually, dehydrator plugs are better moisture indicators than they are moisture traps —still they are the preferred storage plugs. (Figure 3)

9. A carburetor equipped engine should have its air intake sealed over. If the carburetor is not installed, make a wood pad to fit over the mounting pad and secure it in place with nuts after slipping in a silica gel container.

That's about it. If you want to go all the way, you should place a few silica gel bags or envelopes around the engine and then encase it in a sealed plastic container. If the plastic is clear you should be able to monitor the appearance of the silica gel containers and replace them when they show signs of pink.

WHERE TO KEEP THE STORED ENGINE

Store your engine inside a building if at all possible. A heated or insulated building is best, but even a T-hangar is better than nothing. Of course, if you can keep it in your workshop it would probably enjoy a more uniform temperature and get more of your attention than elsewhere.

Your greatest enemy during the engine's hibernation period is, of course, humidity. It is therefore recommended that you take an occasional look-see through a spark plug hole with an inspection light whenever any of the dehydrator plugs begins to show a pink hue.

If you live in a humid region of the country be prepared to retreat the engine interior with corrosion preventative mixture as frequently as every six months. Using a heated preservative (about 200 degrees F) mixture may be more effective.

If you can't store the engine inside a building, cover it with a fitted waterproof material. However, as you well know, outdoor storage introduces the additional problem of weather extremes. These extremes, unfortunately, cause condensation to form on cold metal parts both internally and externally. Since this happens more frequently than it would in indoor storage, rust and corrosion susceptibility is increased. Additionally, the synthetic and rubber parts also deteriorate at a faster rate out of doors.

INSPECTIONS WHILE IN STORAGE

Your stored engine should be inspected at least once a month, more often if stored outside. If the engine is mothballed, that is, encased in a sealed container and some silica gel bags are visible to you, you needn't get into either the container or engine if there is no sign of excess moisture present. The presence of a high relative humidity around the engine would be indicated to you by the color condition of the silica gel containers. Remember, blue is the safe color, and pink is a warning hue....an indicator that moisture is present in excessive amounts. Pink indicators could also mean that it is time for you to retreat the engine.

NOTES

firewall preparation

3

Firewalls

The firewall is an important component in your aircraft. Essentially, it is a fire-resistant bulkhead that separates the engine compartment from the cockpit area. This special bulkhead must be constructed so that no hazardous quantity of liquid, gas or flame can pass through it. In theory, at least, should a fire occur in the engine compartment, the firewall would protect the aircraft's occupants from the flames long enough for an emergency landing to be effected.

In addition to providing protection, the firewall has other useful functions. It provides a convenient surface on which to mount accessories and other essential units that are normally located in the engine compartment. And, since sooner or later it may be necessary to remove the engine or to replace other parts, it makes a handy junction for the disconnection and removal of engine control linkages, fuel lines, and various electrical and ignition wires.

MUST I HAVE A FIREWALL?

Generally speaking, yes. From the foregoing introduction the reasons are obvious why certificated aircraft are required to have firewalls....so, why should homebuilt aircraft be excepted?

Although experimental aircraft need not comply with the same stringent federal regulations expressly established for production-line aircraft, most FAA inspectors will insist that a suitable firewall be installed in the aircraft they inspect. And, why not? Compliance with the portion of the Federal Aviation Regulations pertaining to firewalls is relatively simple and highly recommended.

Should you want to read the exact language, refer to paragraph 23.1191, Firewalls (FAR PART 23, Airworthiness Standards: Normal, Utility and Acrobatic). For your convenience, however, I have paraphrased the regulation.

1. Any engine operated in flight must be isolated from the rest of the airplane by a firewall or equivalent means.

2. The firewall must be constructed so that no hazardous quantity of liquid, gas, or flame can pass from the engine compartment to other parts of the airplane.

3. Each opening in the firewall must be sealed with close-fitting, fireproof grommets, bushings, or firewall fittings.

4. Fire resistant seals may be used on firewalls for small engines (1,000 cubic inch displacement or less.)

5. A firewall must be fireproof and it must be protected against corrosion.

ACCEPTABLE FIREWALL MATERIALS

The following materials may be used in firewalls WITHOUT BEING TESTED:

a. Stainless steel sheet—.015-inch thick. (.284 lb./cu. in.)
b. Mild steel sheet (coated with aluminum or otherwise corrosion protected)—.018-inch thick.
b. Terne plate—.018-inch thick.
c. Steel or copper base alloy fittings.

Composite aircraft have firewalls too. This is the early stage of firewall development in a Quickie design.

Other materials may be used if they can pass the fireproof test.

THE FIREPROOF TEST

If you want to use some other material for your firewall, you may have to prove its ability to meet certain criteria for compliance as a fireproof material. Simply stated, the material must be given a flame test, in which a piece of it, approximately 10 inches in diameter, is subjected to a 5-inch diameter flame. Firewall materials and fittings must resist flame penetration for at least 15 minutes.

That's all there is to that. Compliance is relatively easy and no homebuilder wanting to try a new firewall treatment should slight the requirement.

CHOOSING YOUR FIREWALL MATERIAL

Galvanized sheet is the most common, most economical of the acceptable materials used in small general aviation aircraft. It gets its corrosion resistance from a zinc or hot-dip galvanizing coating obtained by immersing the sheet steel in a mixture of molten zinc at a temperature of approximately 865 degrees F until the base metal temperature of the immersed steel sheet reaches that of the hot-dip bath.

Local sources for galvanized sheet include metal shops and air conditioning duct fabricators.

Don't be surprized if the folks operating these metal shops refer to the thickness of their galvanized sheets in terms of gauges rather than inch sizes such as .018-inch or .015-inch. If that be the case, 26 gauge is pretty close to what you want (about .018-inch thick). You could assure yourself that you are obtaining the correct thickness if you bring your own micrometer to check the various sheets. There are a number of so-called "standard gauges" and some variations in the actual thickness for the same gauge number may crop up.

No useful purpose would be served by using a heavier firewall sheet than that required to meet acceptable minimums. Never lose sight of the fact that the addition of unnecessary weight, no matter how slight, degrades the aircraft's performance. The firewall metal for a two-seat aircraft will probably weigh as much as 4 1/2 pounds anyway, so why make it seven?

In order of preference for firewall material, stainless steel gets top billing in both quality and appearance. It also costs two to four times more than galvanized sheet. However, as the 'Cadillac' of firewall materials, it is the first choice among builders who want a "show airplane." Builders with economy in mind will ordinarily use galvanized sheet for their firewall and may console themselves by claiming that galvanized sheet is lighter than stainless. I doubt that the weight difference

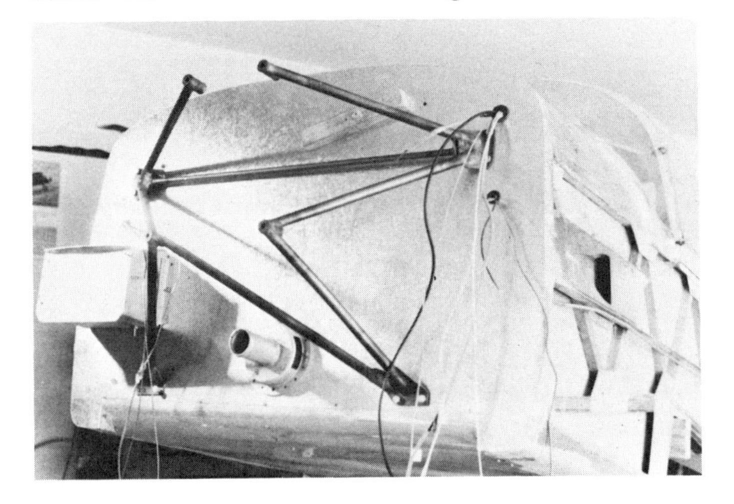

Equipping of the firewall on this fuselage is well underway. Note the location of the cabin heat valve unit, the battery box and the holes for some of the wiring (grommet protected). Firewall is galvanized sheet metal.

The firewall on this Falco is in the early stages of being equipped. Already installed are the engine mount brackets, nose gear brackets and the cowling attachment brackets. The stainless steel firewall over a sheet of asbestos weighs five pounds.

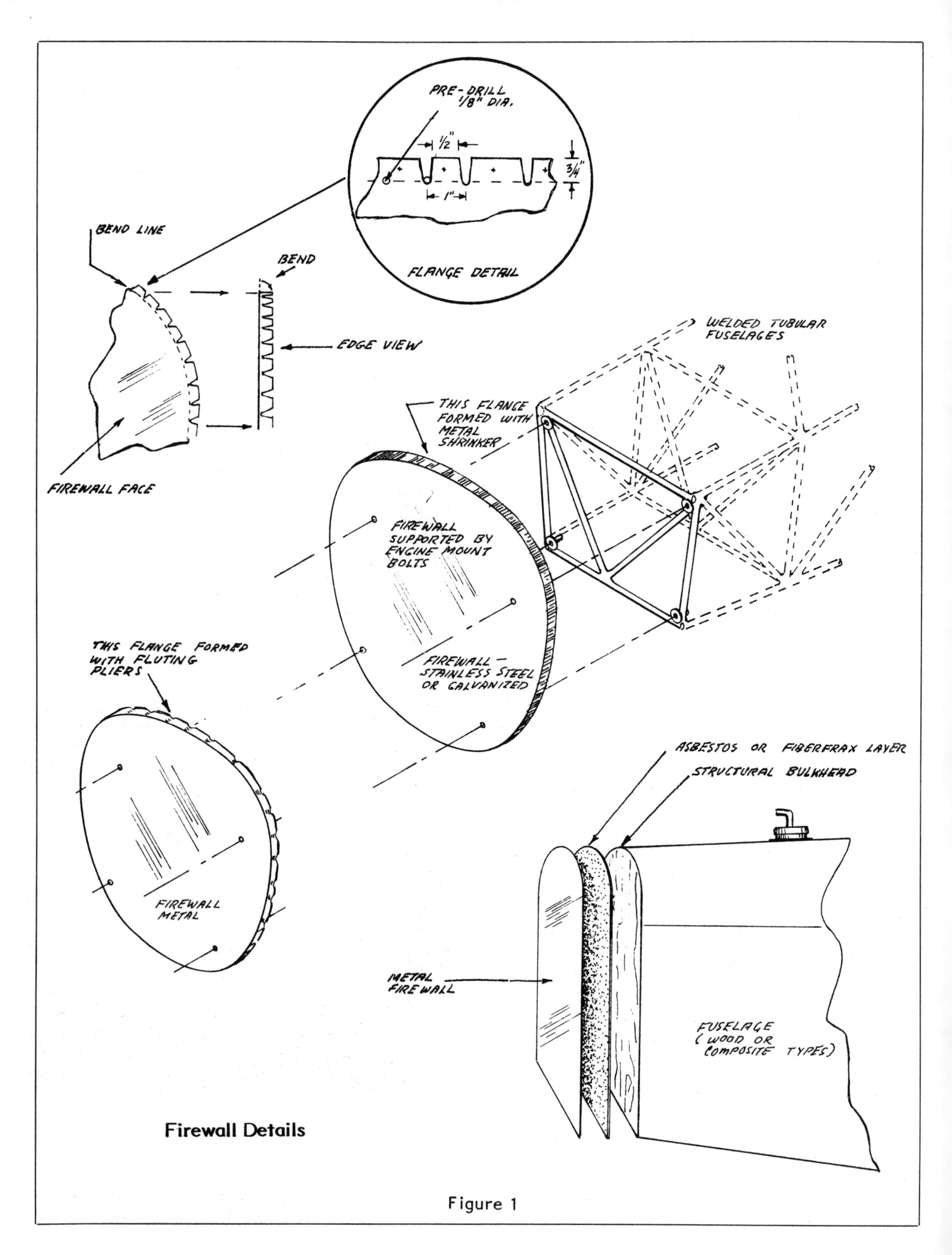

Firewall Details

Figure 1

Here we have a VW engine installed on a short, aircraft style engine mount. The structure is welded steel tubing and for the present, the firewall is held in place by the four engine mount bolts. Note the fluted edges. This treatment, in effect, shrinks the outer edges and allows severe forming of the flange.

A welded tube fuselage does not give much support to a firewall and "tin canning" vibration of the metal can occur. Note upright bracket to provide support for a battery installation. Firewall is initially supported by the four engine mount bolts.

could be measured on the crude scales most of us have around our shops, but there is that difference.

Another fireproof material, often used in the past and sometimes still found on older aircraft, is terneplate. It is not much used today because it isn't as commonly available as galvanized.

Monel and Inconel, although acceptable as firewall materials, are difficult to obtain. As far as homebuilders are concerned, these may be regarded as rare metals.

Concern over weight leads some builders to consider the use of aluminum. However, using aluminum sheet by itself for firewall material is not acceptable to most FAA inspectors. The official reason is, aluminum will not pass the flame test because its melting point is just slightly above 1,200 degrees F. In some applications....for very slow, light aircraft (ultralights) with small engines, it might be considered sufficient. However, before you decide to use it in your own aircraft, check it out with your local FAA inspector. Although the standards established by regulation for production-line aircraft do not necessarily apply to experimental aircraft, deviation from accepted practices may result in your having to get into a proof testing demonstration. Is it worth the trouble?

Recently, a new space age ceramic paper, called FIBERFRAX, has been developed and is finding its way into homebuilder circles. It can withstand temperatures of approximately 2,300 degrees F and may be considered as effective a fire barrier as stainless steel.

It has been used primarily in VariEze, Long-EZ and other light composite aircraft because it saves two to three pounds over the stainless or galvanized sheet normally installed.

Although FIBERFRAX is flexible and considerably lighter than steel, it is a ceramic material and is thus fragile. This may introduce in-service problems unless the material is somehow protected from abrasion and abuse.

The technique for using it seems to have been standardized as follows: The layer of asbestos cloth ordinarily used under the stainless steel firewall sheet in wood or composite aircraft may be omitted. Instead, a layer of FIBERFRAX is attached to the firewall and overlaid with a .016-inch sheet of 2024 T3 aluminum to protect it. Although aluminum is deficient when used alone as a firewall material, its use with FIBERFRAX is acceptable because it is really the FIBERFRAX that provides the fireproof barrier. (Figure 1)

FIREWALLS IN WELDED STEEL FRAMES

Before installing a firewall on a welded fuselage, you should seal the open ends of the longerons. This helps keep out rust-inducing moisture which can be very destructive over the years. Weld small metal washers or discs of .032-inch steel across each of the open ends.

Your firewall will undoubtedly consist of a single sheet of metal with nothing behind it for support except perhaps a diagonal tube welded across the front of the tubular frame of the fuselage. A large flat expanse of thin metal like the firewall is always likely to vibrate (called 'the oil-canning effect') noisily in flight. Firewall metal in this type of installation is, therefore, often made more rigid by embossing random beads or ridges across the sheet. This inhibits vibration by stiffening the metal and it virtually eliminates oil-canning noises.

At aircraft factories, technicians form the beads in hydraulic presses, but we amateur builders must usually hammer ours in by hand using a wood dowel and a back-up block with a recessed groove in it.

The firewall can be held in place initially by the four bolts securing the engine mount to the fuselage. Sometimes small tabs are welded to the firewall bulkhead and sheet metal screws, rivets or bolts are added to secure the firewall. Do not go overboard in this regard, however, because the firewall will take on greater security and rigidity as construction progresses. In addition, mounting the fuselage side skins will completely stabilize and immobilize the firewall.

Since there is no back-up structure behind the firewall metal in this type of installation, it will be necessary for you to reinforce parts of it before mounting heavy objects such as a battery, header tank, oil cooler or oil tank. If you don't do this you can expect to eventually find localized cracks that result from vibration.

....IN ALL-METAL FUSELAGES

The firewall sheet in an all-metal aluminum fuselage is ordinarily riveted to what comprises the engine compartment bulkhead. In effect, it becomes an integral part of the fuselage.

In this type of structure, as in the welded-tube type, the firewall metal has the same kind of oil-canning tendency. Heavy objects require localized reinforcement.

An unprotected plywood firewall is a NO, NO! Unless inadvertently overlooked, no FAA inspector would ever approve such an installation...not even in a homebuilt.

Production aircraft with large firewalls often have beads (ridges) pressed into them to stiffen the firewall and to eliminate oil canning tendencies. Few homebuilders bead their firewalls in all-metal aircraft although it is not difficult to do with a dowel, hammer and a grooved surface on which the metal can be hammered.

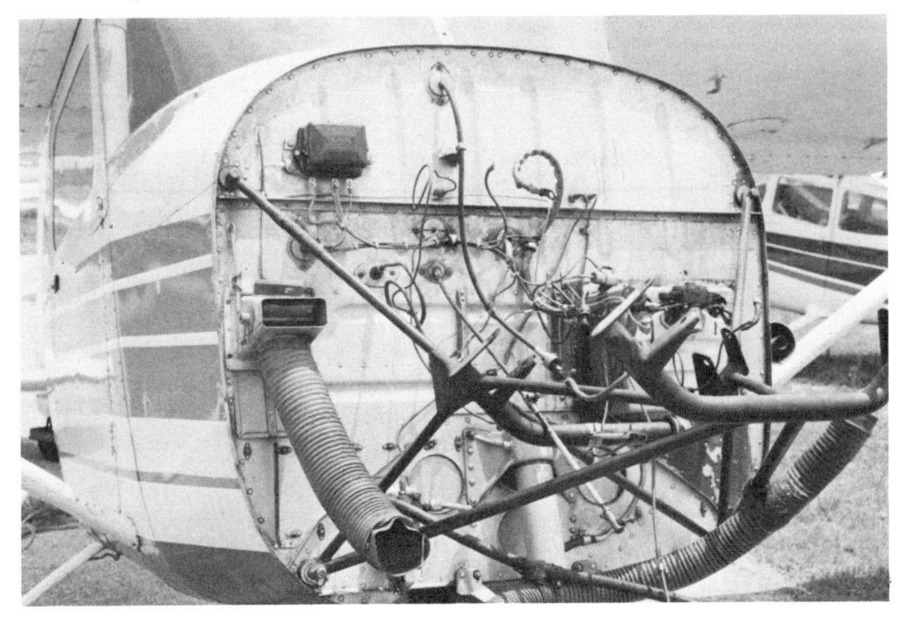

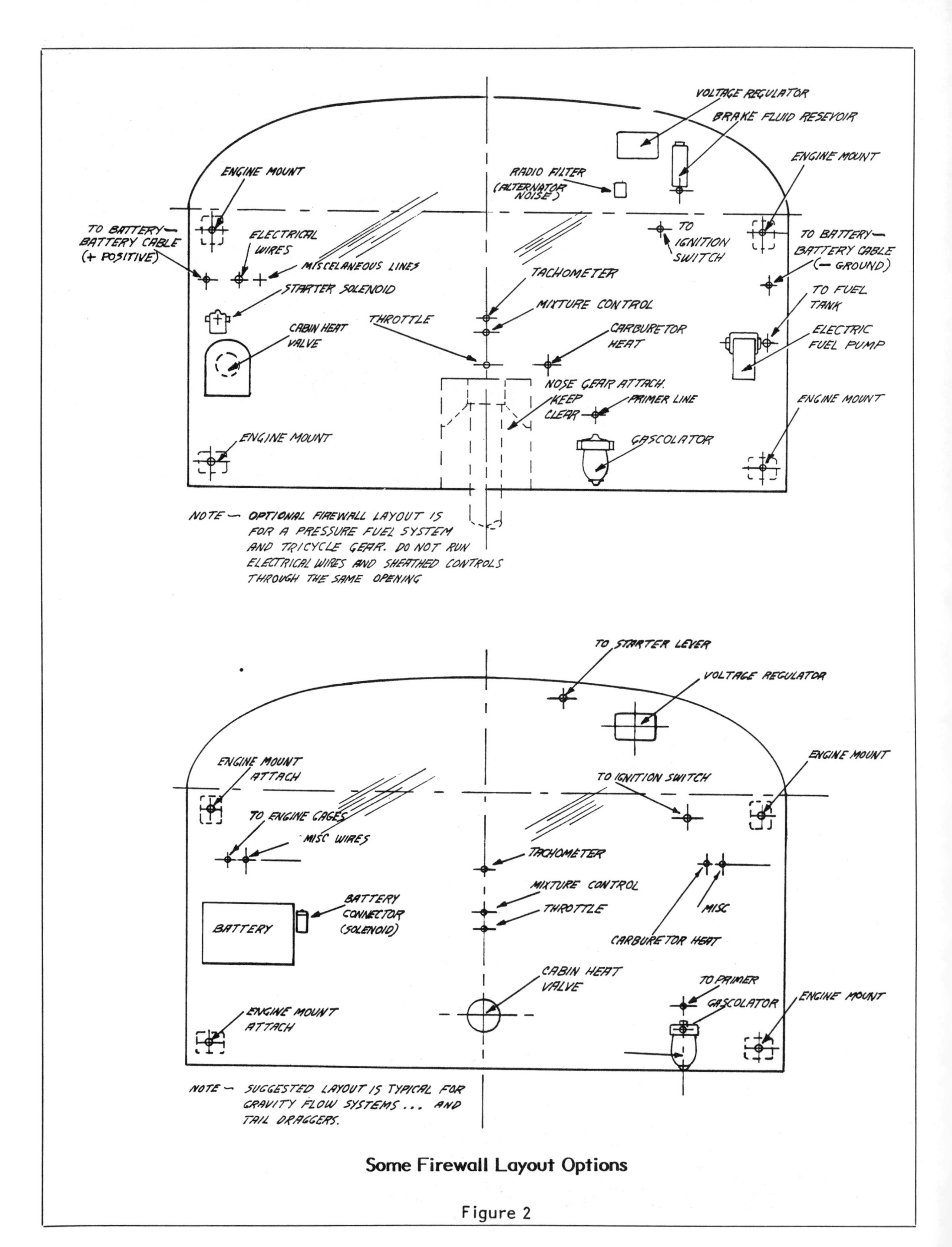

Some Firewall Layout Options

Figure 2

....IN WOOD AND COMPOSITE AIRCRAFT

An all-wood or composite aircraft is ordinarily constructed with a built-in engine compartment bulkhead to which protective firewall metal must be attached. The bulkhead, although structural in nature, is not fireproof because it is commonly made of wood overlaid with plywood.

Before a firewall is installed in a wood or composite aircraft, I recommend that the firewall bulkhead be overlaid with a protective heat insulating layer of asbestos, ZETEX™, or FIBERFRAX paper.

If asbestos is used, it may be stuck to the plywood with a coating of polyurethane varnish that has been allowed to get tacky.

Caution: Asbestos has developed a bad reputation and, according to federally funded studies, may constitute a health hazard. Because of its bad press, asbestos sheet is very difficult to obtain. However, it may still be found in some plumbing shops and hardware stores. As a precaution, avoid creating asbestos dust when cutting and handling the stuff. And, just to be sure, wash your hands afterward.

A safer material for insulating a wood firewall bulkhead is a fabric called ZETEX™ 800, whose manufacturers claim it provides "all the advantages but none of the health hazards of asbestos. ZETEX™ fabric can withstand temperatures of up to 2,100 degrees F and is unaffected by most acids, alkalies and solvents. It bonds well with resins and elastomers and is six times stronger than a comparable asbestos fabric. The woven material has a softer feel and is more abrasion resistant than asbestos. ZETEX™ 800 weighs about 24 ounces per square yard and is about .060-inch thick. Unfortunately, it is not commonly available and could be expensive in the small amount needed for firewall insulation. In time its availability should improve and its cost decrease if more homebuilders start using it. This material may be obtained from Newtex Industries, Inc., Railroad Ave., P.O. Box 25, Victor, N.Y. 14564.

If you choose to use FIBERFRAX in your wood or composite aircraft, don't waste time by attempting to bed down the material with epoxy as you would fiberglass. At best, wetting it all out would only add unnecessary weight. Instead, affix the FIBERFRAX sheet with a bead of silicone rubber laid around the periphery of the bulkhead. That should be sufficient to hold it in position until the firewall metal is installed with a few sheet metal screws. Use only enough screws to temporarily hold the firewall in place. Each firewall accessory and unit added will contribute to its security. You can always add fasteners later wherever needed to complete the job.

Being supported by a solid bulkhead, the firewall metal will not vibrate and oil-can as it would in an all-metal or welded steel tube installation so no particular pre-installation forming of reinforcing ridges or beads need be made.

Resist the temptation to close the firewall area (if construction method permits) for as long as possible to permit easy access and easy installation of rudder controls, wiring and other items.

All wiring, lines, controls and cables passing through a firewall must be protected against chafing by some suitable method.

Preparation

It is rather difficult to modify the firewall or drill additional holes in it after the engine is hung. Therefore, if at all possible, the various essential openings should be located and completed while access to the firewall is unhindered by the engine or engine mount.

Don't be too quick to drill holes through that firewall. Before drilling any holes or making any cut-outs in the metal be sure that you are putting them where they will be needed.

Too many builders, after making a hole for a control unit, learn to their chagrin that it is in the wrong place. Sometimes, after the engine is mounted, the builder finds that the hole is inaccessible or too small or large for its intended purpose. Proper planning here, is essential.

Obviously, any unused hole or opening presents a problem as it might, in an emergency situation, allow liquid, fumes or flames to penetrate the cockpit area.

Seal all cut-outs, gaps and joints with fire resistant grommets, bushings or fittings. It's a good idea to insulate the back side of the firewall in a welded steel tube fuselage as well as in an aluminum fuselage with a sheet of fire resistant material. This would serve the dual purpose of acting as a heat barrier and as a sound barrier.

FIREWALL GROMMETS

Fuel lines, primer lines and just about everything in the way of wires, cables and control linkages passing through the firewall need to be protected against the sharp edges of the firewall metal. Grommets, bulkhead fittings and similar devices are used for this purpose. However, the best protection is provided by a grommet made up of two parts — a metal retainer that is fireproof and a rubber grommet that chafe-proofs it.

Flexible synthetic rubber grommets (AN931 or MS 34589) are resistant to hot oil and coolant and may be installed around any tubing or wires already in place by cutting across one side of the grommet and working it into position. The cut is not required for grommets already installed in the firewall before cables or wires are passed through them.

Grommet selection is based on the hole diameter required to accommodate the size of the wire, cable or line that it is intended to protect. The support groove in the grommet varies in width but generally is rather narrow (about 1/16-inch) and is best suited for mounting in a sheet metal hole.

Unfortunately, plywood firewall bulkheads, for the most part, are rather thick and a regular grommet cannot be inserted. It is unlikely that you will be able to find grommets with grooves wide enough to accept a plywood or built-up firewall in a wood or composite aircraft. Some other protective and sealing means must be used for these firewalls. A number of methods used for sealing a variety of firewall openings is illustrated in Figure 3.

As you install each component on the firewall remind yourself that you may one day have to remove it. Will you be able to do so, alone, without having somebody in the cockpit holding a wrench for you? In fact, would the retaining bolts even be accessible? The cockpit side of the firewall can also get pretty crowded because the gas tank, instruments, radios and engine controls will have already been jammed in there somehow. You can prevent the problem of inaccessibility by installing everything that must be attached to the firewall with nutplates instead of regular nuts.

Incidentally, it is important to retain as much access to both sides of the firewall and the inside of the cockpit for as long as you can during construction. Avoid closing the fuselage sides at the firewall, if possible, until after you have installed and hooked up the engine. It will make your work much easier.

If you are working with a stainless steel firewall, you may find drilling it a very difficult task indeed. More information on stainless is provided in Section 5 of this book.

SOME VW FIREWALL MODIFICATIONS

Builders who want to mount a VW engine

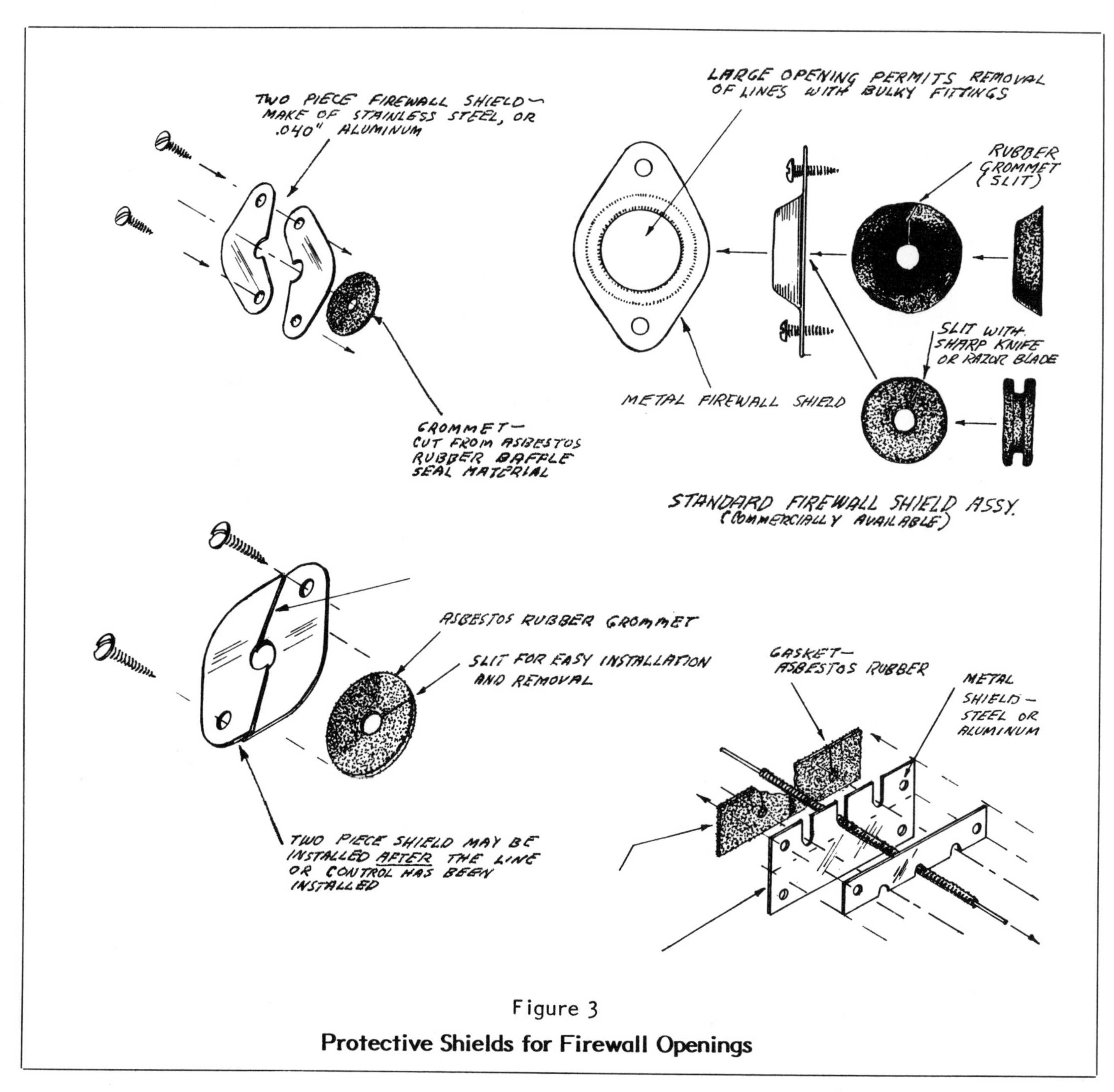

Figure 3

Protective Shields for Firewall Openings

directly to the firewall without using an engine mount, may find it necessary to make a large opening in the firewall to accommodate a rear-mounted magneto. Of course, if a Vertex magneto is plugged in up front in the distributor hole instead, this access hole can be avoided.

A fireproof cover must be made to go over and enclose the rear-mounted magneto and the hole in the firewall. A stainless steel pot of the correct size might be found in your kitchen (you married gents might be wiser to obtain one from the local discount store) adapted to do the job. Do not use an aluminum pot, it will not provide the protection necessary in the event of an engine compartment fire.

HOW/WHERE TO MOUNT ACCESSORIES ON THE FIREWALL

Don't lose sight of the fact that someday you may have to remove, relocate or add some accessory to the firewall. Will it take two people to do it? Will you have to have someone in the cockpit with a wrench while you manipulate some tool outside the aircraft at the firewall? It could even be worse —maybe the gent inside the cockpit can't reach the nut or bolt head on his side because the gas tank or other parts are in the way. How would you like to have to remove the instrument panel or console and gas tank simply to remove a defective voltage regulator from your firewall?

This should never become a problem in the future if you take the time now to install nut plates behind every accessory mounted on the firewall. As an alternative method, you could use some means to immobilize the individual bolt heads. This would permit you to install the mounting bolts from inside the aircraft. One advantage to this method of installation is that the ends of the bolts and the nuts are visible for inspection, and easy removal of the unit is assured.

Accessories that you might consider for this type of nut plate or secured bolt head treatment include items like the battery box, voltage regulator, cabin heat box, brake reservoir, solenoids, oil cooler mounting brackets and similar units.

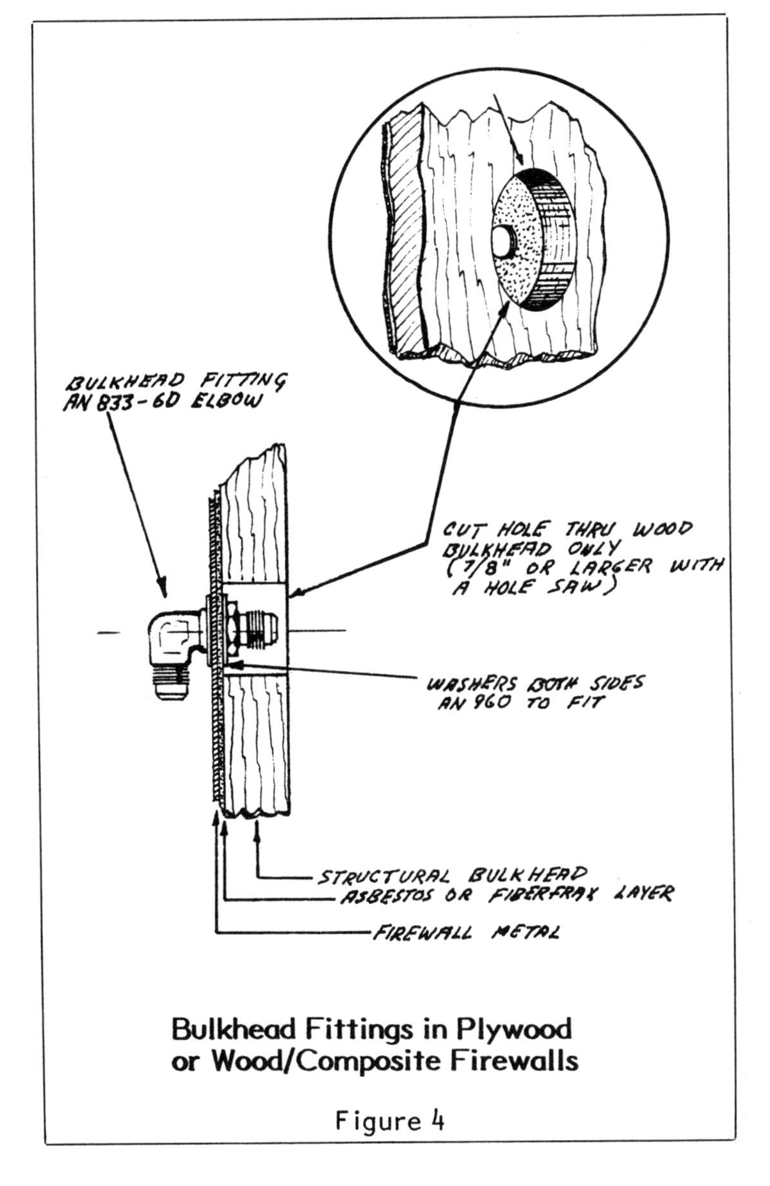

Bulkhead Fittings in Plywood or Wood/Composite Firewalls

Figure 4

BULKHEAD FITTINGS

It is preferable that lines and hoses which must pass through the firewall be connected to standard AN type bulkhead fittings rather than being forced to pass through unprotected holes in the firewall. This serves two purposes: it eliminates the likelihood that the line will chafe and fail at the firewall; and it provides you with a very convenient disconnect point for the system without disturbing the installation on the other side.

Wood firewalls are usually quite thick and bulkhead fittings designed for thin metal firewalls may not have the necessary reach to go through a wood bulkhead. This, however, should be no problem if a large hole, about 1 inch in diameter (use a hole saw), is first cut through the wood portion only. A regular (smaller) hole just large enough for a bulkhead fitting is then drilled through the metal part of the firewall and the bulkhead fitting is installed in the conventional manner. This treatment allows a wood firewall to retain its integrity as a fire barrier. (Figure 4)

GASCOLATOR INSTALLATION

All aircraft fuel systems should have a main line strainer or gascolator located at the lowest point in the system. Any one of several gascolator types may be installed in your airplane provided its ports are large enough for the engine you are installing.

Most aircraft engines of up to 200 hp require a gascolator that will accommodate 3/8-inch fuel lines. The size refers to the outside diameter measurement of the metal tubing used for fuel lines. Flexible (rubberized) fuel lines, on the other hand, are bulkier on the outside than equivalent metal tubing because of the nature of the construction used in the hoses. The inside diameter of hoses approximates in size the outer diameter of tubing. Both tubing and hoses are identified by the manufacturer (and parts departments) with dash numbers.

Just remember that metal tubing sizes are based on the tube's outside diameter while pipe (hose) sizes are based on inside diameter.

Gascolators come with and without a tapped port in the top cover for the installation of a primer line connection. All ports are tapped for fittings having pipe threads. If the gascolator you have has a tapped opening on top for the installation of a primer line, you may cap it with a threaded plug if you do not intend to hook up a primer installation.

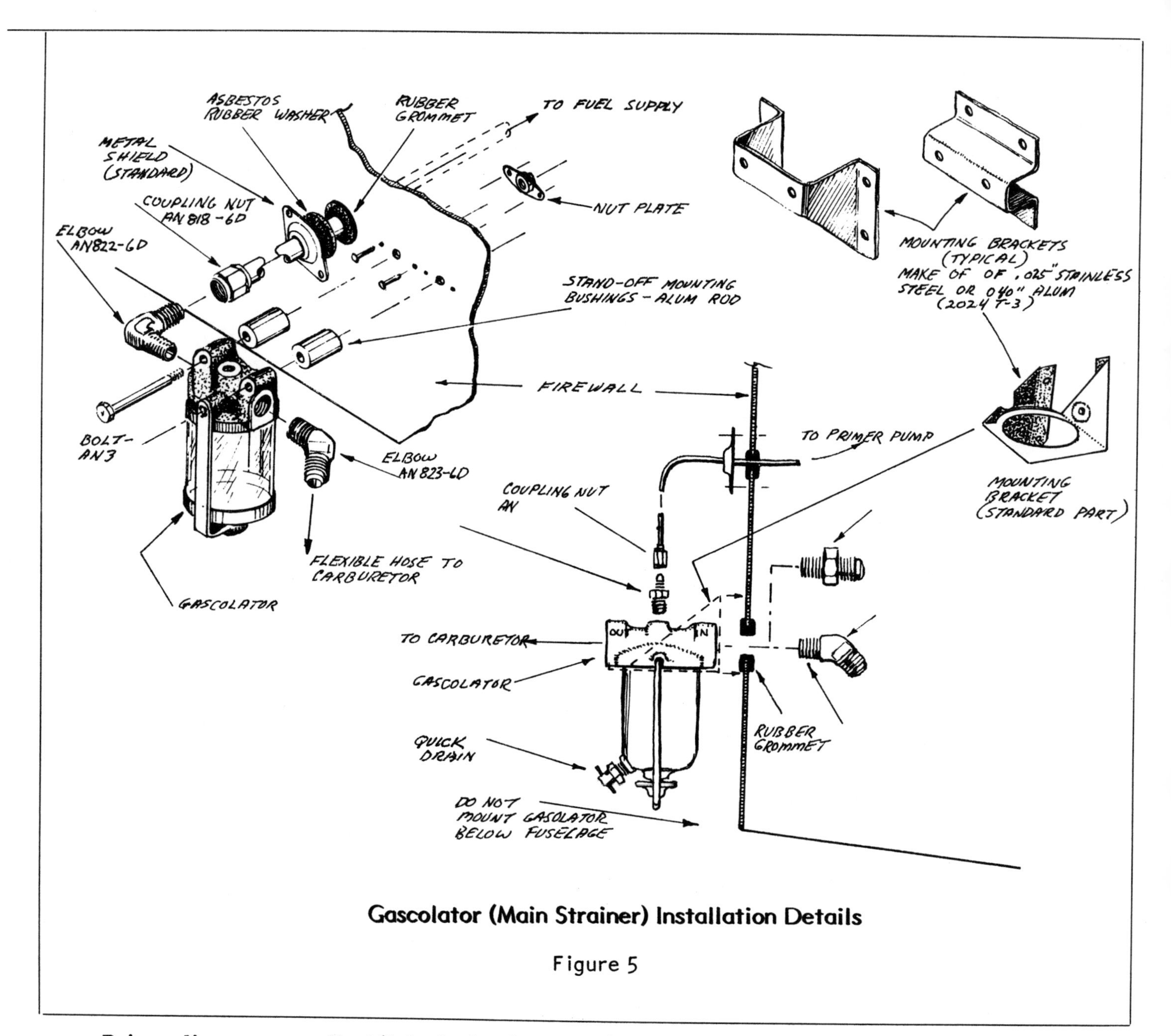

Gascolator (Main Strainer) Installation Details

Figure 5

Primer lines are usually 1/8-inch aluminum with flared tube ends which take the regular flare type AN fittings. However, some builders use 1/8-inch copper lines because they are readily available locally through auto parts stores. A few builders also use plastic tubing.

When using plastic tubing, the fittings are also generally plastic and especially designed for that tubing. Copper lines fabricated with automotive type fittings may make it more difficult to transition from automotive type fittings to aircraft fittings if that is your intention. At any rate, the preferred installation is that using 1/8-inch aluminum tubing and flare type fittings.

The tubing must be flared with an aircraft type flaring tool which has a 37-degree angle and not an automotive flaring tool which makes a more severe 45-degree flare. Flaring the ends of the tubing with an automotive tool will result in a slight misfit of the flared end of the tubing to the fitting. Forcing the tubing into conformance by torquing the fitting tightly will distort the tubing end and induce stresses in the connection that might cause it to fail later, even though it may not leak on initial installation.

The fuel line leading to the gascolator, which is rigidly mounted to the firewall, is ordinarily a 3/8-inch aluminum tube with a flare fitting nut. The fuel line from the gascolator to the engine-driven fuel pump (if installed) or, on the other hand, to the carburetor, must be flexible. When properly installed, it will run uphill or downhill without reversals enroute. (Figure 5)

SHOPPING LIST FOR GASCOLATOR ITEMS

You may not need all of the following, but this list should give you a good idea of the number and variety of small items that may be required in hooking up a particular unit.

1. Gascolator.
2. Mounting bracket for gascolator (or you can make one out of .050-inch scrap aluminum).
3. AN3 bolts (lengths as needed for your firewall thickness) with nut plates for mounting the gascolator bracket.
4. Fuel inlet fittings to gascolator.
   a. 1 AN816-6D nipple, flared tube and pipe thread.
   b. 1 AN822 (MS20822) -6-6D Elbow, flared tube and pipe thread.
   c. 1 AN816-2D nipple, flared tube and pipe thread for primer line connection.
5. Primer line tubing, 1/8-inch aluminum or copper tubing, total length as needed for your installation.
   a. 4 each, nuts, primer hook-up, AN8182D.
   b. 4 each, sleeves, coupling AN819 (MS20819) (primer hook-up) clamps, plastic (Radio Shack) to secure lines.
6. Quick drain valve (Curtis drain valve). Your gascolator will probably take one with a 1/8-inch pipe thread (better check though).
7. Primer — inlet and outlet may have 1/8-inch internal threads requiring an adapter nipple, 2 each, AN816-2D.

Make your own shopping list for the installation of the remainder of the firewall units.

VOLTAGE REGULATOR MOUNTING

Most voltage regulators are mounted directly to the firewall in some upper area of the firewall that will allow easy access to the unit. Here again, it is important to install the regulator so that it can be removed without having to have access from behind. This means you should use nut plates for the initial installation or devise a bracket that permits installation and removal from the front side.

The voltage regulator used must be matched to and compatible with the alternator or generator you have installed on your engine.

Most alternator regulators nowadays are of the solid state variety while the generator installations still take, for the most part, the familiar automotive type regulator.

Mounting the voltage regulator to the firewall metal ensures automatic grounding of the unit. In some shock mounted units, however, a short bonding strap may be required for this purpose.

A typical voltage regulator installed in the upper firewall area. When you attach yours, use nutplates so if you have to remove it, you won't have to also take out half the equipment in the cockpit.

BATTERY MOUNTING

A battery case mounted on the firewall must be located so that the battery can be installed and removed without interference from the engine mount or any accessories. Its location should be as far away from heat producing sources (exhausts) as possible. The brackets for attachment of the battery box are often riveted to the firewall. The assumption is that they will never need replacement. This is not a practical means of installation in a wood firewall and bolts may have to be used instead. AN3 bolts with nutplates behind will suffice. (Figure 6)

The area around the battery case will, in time, be subjected to corrosive effects of battery fluid and fumes. It would be worthwhile to protect the adjacent metal parts by painting them with an acid-proof paint (epoxy or a rubberized paint) upon installation.

CABIN HEAT BOX

Your aircraft should have a firewall-mounted cabin heat box to provide for a controlled entry of heat into the cockpit area.

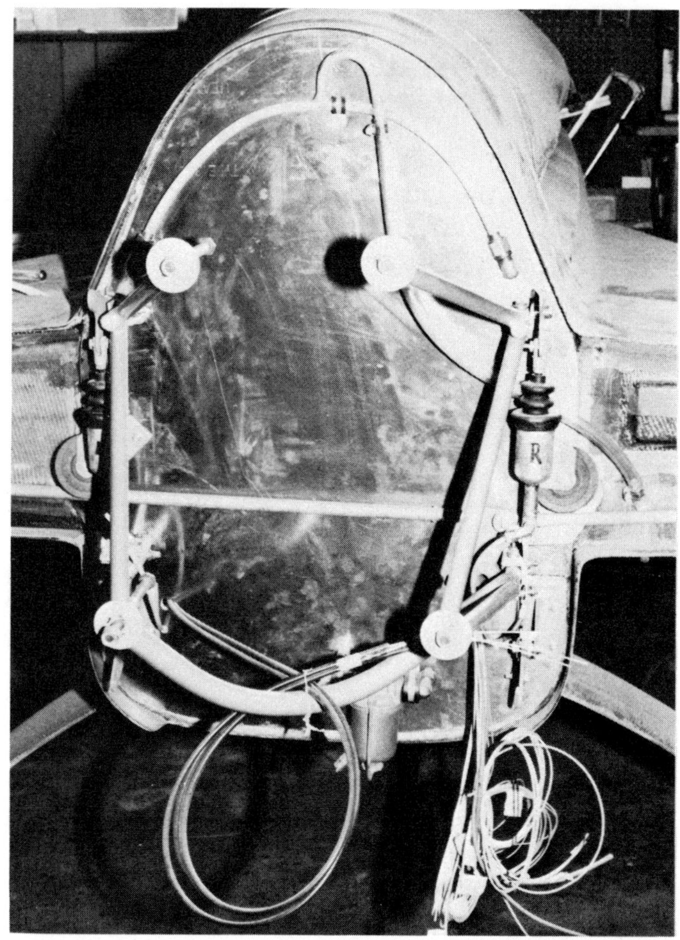

With the engine mount in place, this partially equipped firewall is just about ready for the engine installation.

That little black box with the elephant snoot attached is the cabin heat valve. Originally intended for a Piper J-3, it now graces the firewall of a Skybolt biplane.

The opening through the firewall for this device is ordinarily located along the approximate centerline of the aircraft at about foot level.

In two-seaters as well as in single-seaters, the inlet inside the cockpit should have a deflector to diffuse the entering hot air. This is usually better than allowing it to blow in a concentrated blast between your legs.

There is a variety of cabin heat boxes available. Whichever you choose, be certain it has an outlet for when the heat is shut off from the cabin. This "dump" is necessary to permit the hot air to exit overboard and not back up and become trapped around the exhaust pipe. Otherwise, the exhaust pipe may become excessively hot and suffer local damage. (Figure 7)

As with firewall-mounted accessories, use AN3 bolts or machine screws to install the cabin heat box. Don't forget the nut plates.

BRAKE RESERVOIR

Mount the brake reservoir, if one is required, in any convenient firewall location. The usual placement is around the upper left hand side of the firewall. This should be well above the level of the master brake cylinders inside the cockpit so that gravity works for you. How should you attach it? You guessed it, use a couple of AN3 bolts with nut plates behind. Incidentally, be sure that a small vent hole is drilled in the filler plug of the brake reservoir to equalize pressure.

The brake reservoir may be connected to the master brake cylinder by a rubber hose, plastic hose or aluminum tubing. Tubing size is usually 1/4-inch but some builders use 1/8-inch or 3/16-inch sizes to save weight. A short section of flexible hose is necessary at the rudder pedal/brake cylinder to allow pivot movement flexing to take place. However, from that point up to the brake reservoir the line could as well be a rigid (metal) one. Many builders prefer to run a flexible rubber aircraft hose all the way from the reservoir to the master brake cylinders when the distance is not too great.

MISCELLANEOUS UNITS AND ITEMS

Any other permanent attachments to the firewall can be accomplished using the same methods described for the larger accessories. Use screws or machine screws or bolts to attached small items such as solenoids, overvoltage regulators, noise suppressors and wire

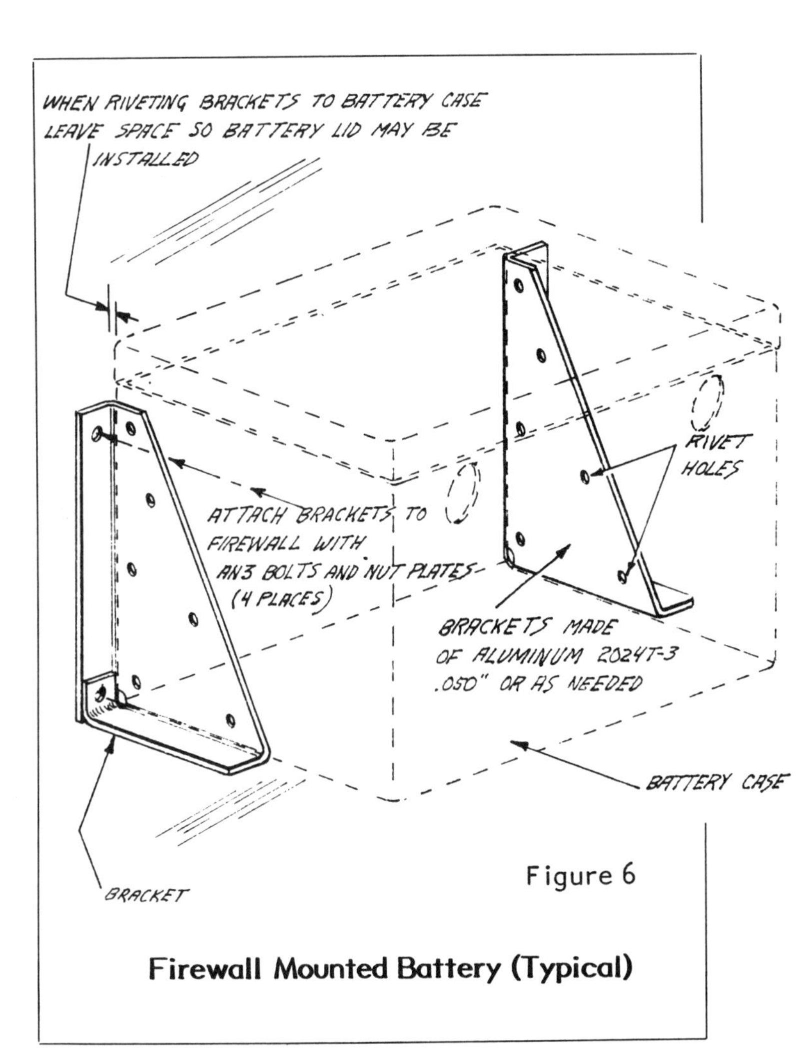

Figure 6

Firewall Mounted Battery (Typical)

bundles. Mount electrical units close to the accessory being served to minimize the length of wiring required.

ENGINE MOUNT ATTACHMENT

The engine mount attachment points must be very precisely located if you are building an airplane for which you intend to buy a ready-made engine mount. Otherwise, your dimensional control needn't be critical because you can build your mount to fit your fuselage attachment points as already located and/or installed.

Since the engine mount-to-fuselage attachment points are not shock mounted, it is all right to use self-locking nuts to bolt the engine mount to the firewall. Some builders may still opt to use castellated nuts with cotter keys.

FIREWALL OPENINGS

Virtually every hole in the firewall has some sort of wire, tube or line passing through it. These openings must be no larger than

Firewall preparations for a VariEze. It is already overlaid with Fiberfrax and now awaits its metal overlay.

This galvanized firewall has a flange riveted all around it. The flange was made from an aluminum angle (2024-T3) and was shrink-formed with a metal shrinker to get that angle to curve and fit firewall contours.

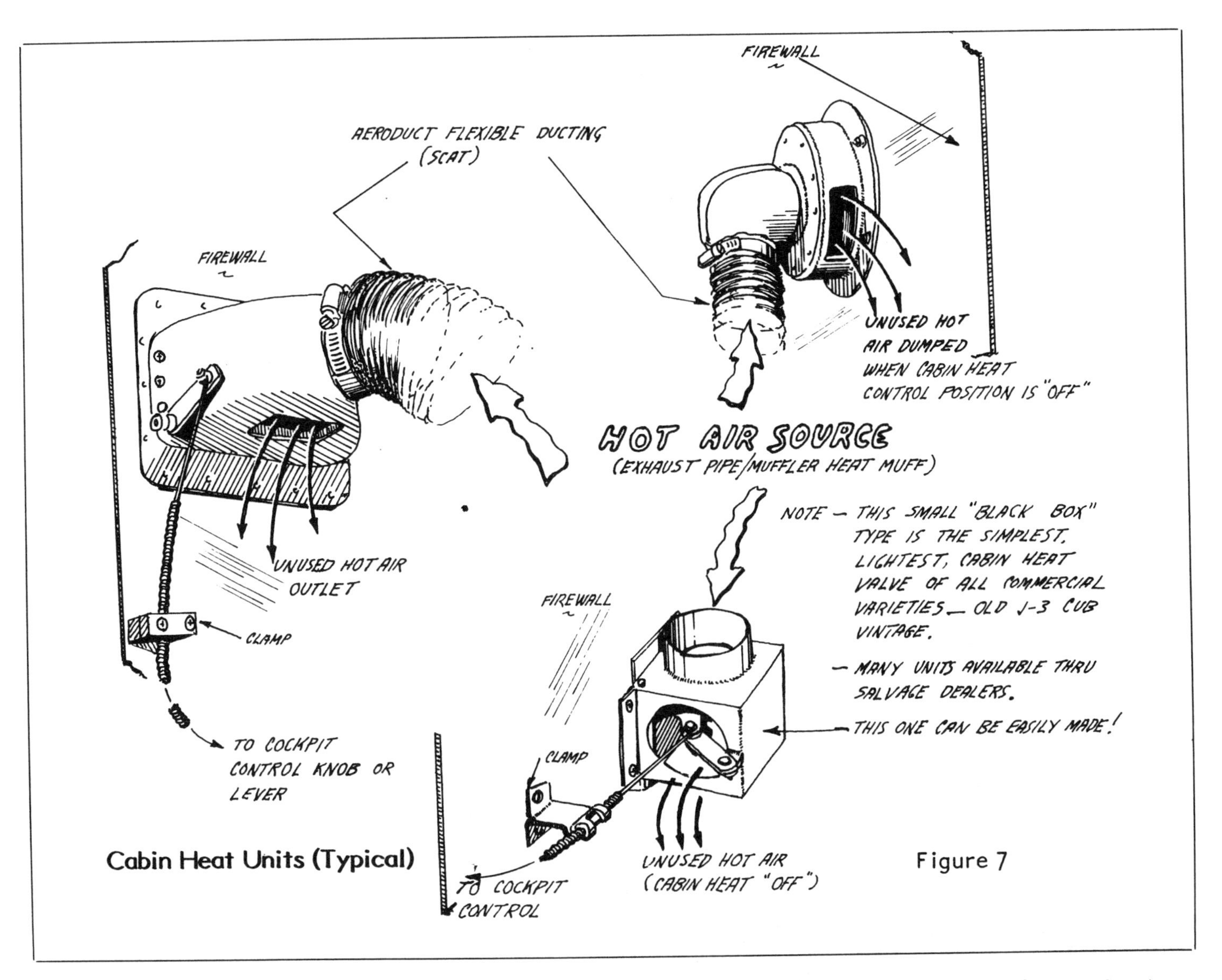

Cabin Heat Units (Typical) Figure 7

necessary and yet be of sufficient size to afford protection against chafing. Firewall grommets are made for this purpose and should be used wherever needed. Whenever practical use a bulkhead fitting for lines that must pass through the firewall.

COWLING ATTACHMENT LUGS

Many firewall installations require small lugs, flanges or angles for the attachment of the cowling. These lugs are conveniently spaced (about 6- to 8-inches apart) and are permanently riveted or screwed to the firewall. The side of the lug against which the cowling will be fastened should be fitted with a nut plate or a receptical for a Camloc or Dzus fastener. Remember, the cowling ought to be easy to remove so don't get carried away by installing too many fasteners.

The lugs or attachment brackets may be made from aluminum angle sections or bent from sheet stock approximately .050-inch thick. Either 2024T-3 or 6061-T6 aluminum angles are suitable.

FIREWALL SEALANTS

Standard light plane practice has been to seal all of the various small openings in the firewall with a putty-like material known around airports as 'dum-dum'. Be careful of what you use as some of the nonhardening caulking (dum-dum type) material you get, even from aircraft supply sources, may support combustion. If this characteristic is verified with a match, DO NOT USE IT.

Instead, you should place greater reliance in special metal plates and covers that you can make to close specific firewall openings. Some standard fireproof grommets may be obtained at homebuilder supply sources.

engine mounts

4

Types of Engine Mounts

The engine mount does more than just connect the engine to the aircraft's structure, it also serves to distribute the weight of the engine while diffusing the effects of thrust, torque and vibration that may be transmitted from the engine to the airframe. Only a well designed, properly constructed and properly installed engine mount can be expected to do all that effectively.

Tubular mounts are used almost exclusively in light aircraft. These mounts are built up of chrome-molybdenum (4130) steel tubing and are welded into a single lightweight unit. The welding methods most often used are the oxy-acetylene and heli-arc processes.

Some engine mounts are designed to impart a downward or upward thrust (relative to the top longeron or waterline of the airframe) while others offset the engine slightly to the right or left. The determination of how much the thrust line should be offset and in what direction is a design problem best left to the designer, who originally calculated the aircraft's aerodynamic requirements.

Only a few homebuilt designs call for engine mounts that impart a downward thrust to the engine. A few more require an engine mount that offsets the engine several degrees to the right to compensate for torque. (This arrangement is similar in effect to the offset fin seen on some aircraft. And, except for a few pylon-mounted engines, it is difficult to think of a single conventional homebuilt that imparts an upward angle to the engine's thrust.

To look at the typical spindly-legged aircraft engine mount you'd think it totally inadequate to support that big heavy engine bolted to it. But looks, in this case, are deceiving. The triangulation built into a well designed mount is the strongest structural arrangement possible for the weight.

STANDARDIZATION

Although the dimensions for the engine mount attachment points are standardized for a particular engine, there is no such thing as a standard engine mount, even for the ubiquitous Continental and Lycoming engines most often used by homebuilders. The reason, of course, is that the aircraft structure and fuselage attachment points differ greatly from one aircraft design to another.

In one aircraft the firewall might be 25 inches wide and in another it may be 30 inches or 36 inches wide. In one aircraft the engine may have to be mounted high on the firewall while, on another, the same engine may have a much lower profile.

There may be substantial center of gravity differences which dictate that one design have a very long mount while another places the engine right up against the firewall. These extremes make it very difficult to use an unmodified stock mount from a parts supplier or from some junked-out production aircraft.

If the designer of the aircraft you are building did consider the economy of using a stock mount typical of a popular production aircraft and designed his structure and firewall attachments points to the same dimensions, you are fortunate indeed. Of course, you might be able to purchase a ready-made mount for the design you are building. Ordinarily this will not be possible unless you are building a popular type of aircraft for which components and parts are readily available through homebuilder supply channels. Otherwise, you must build your own.

WHAT KIND OF MOUNT DO YOU NEED?

Most engine mounts for the average, flat 4-cylinder, horizontally opposed air-cooled engines look the same. There are some differences in dimensions naturally, and in the shock mounting systems used. Years ago builders mounted their engines directly to the engine mount and firewall without the benefit of any shock absorbing materials between the two. Without a doubt, this is the simplest engine mount to build and has been used of late on the basic Volksplane VP-1 design.

It consists, essentially, of four bolt holes in the firewall and four spools against which the engine is to be mounted. No welding or jigging is required outside of the accurate placement of the mounting holes. However, the resulting vibration in the airplane is annoying at best and sometimes anxiety producing or otherwise physically damaging. Not only does vibration induce greater pilot fatigue, particularly on cross-country flights, but it also increases the potential for structural failure of airframe components. (Figure 1)

From that simple mount type you go to the more complex built-up tubular engine mounts typical of most light aircraft. Now, instead of merely four bolts, you need eight, and you will have to devise a jig on which to fit and weld the mount. What makes the tubular mount construction so sporting is the challenge of having all the holes line up after the welding is completed. This can be a problem if you are not prepared to cope with it.

Unless you are a fair welder (not necessarily a whiz with the torch) you should not attempt to finish-weld your engine mount clusters. You can, however, build the mount and tack weld the different pieces together and then have a good welder finish the job. You will still have done 95% of the work of building the mount and end up with a good serviceable mount provided, of course, the guy you get to do the welding is good. If you are pretty good welder don't hesitate to do it all yourself.

You may be encouraged to learn that most engine mounts welded by amateur builders turn out to be pretty durable. Sometimes, even mounts constructed by aircraft manufacturers under controlled conditions do, on occasion, develop failures and other problems in use. Typically, the majority of the problems that develop are in the nature of a bent, cracked or completely broken engine mount tube adjacent to a welded cluster, and, most frequently, this problem is associated with nose gear installations. It seems that some pilots, when landing, tend to wheelbarrow the airplane in on its nose wheel (no tail-dragger pilot would do that). Such abuse imposes abnormal stresses on the mount, apparently a major factor in the development of this type of failure.

In addition to supporting the engine in correct alignment and reducing vibration and noise, shock mounts restrain the engine's movements in flight to reasonable limits. If you haven't noticed it before, the engine in a typical installation can be rocked on its shock mounts. You should, therefore, provide sufficient clearance between the engine and the cowling, the spinner and the cowling and allow at least one inch between the baffles and the cowling.

The vibration of a shock mounted engine has been known to cause severe damage to cowling openings and to the exhaust pipes protruding through them simply because the clearances were insufficient.

TWO BASIC SHOCK MOUNTS

The two basic engine shock mounts being installed in light aircraft these days are the conical and the dynafocal types.

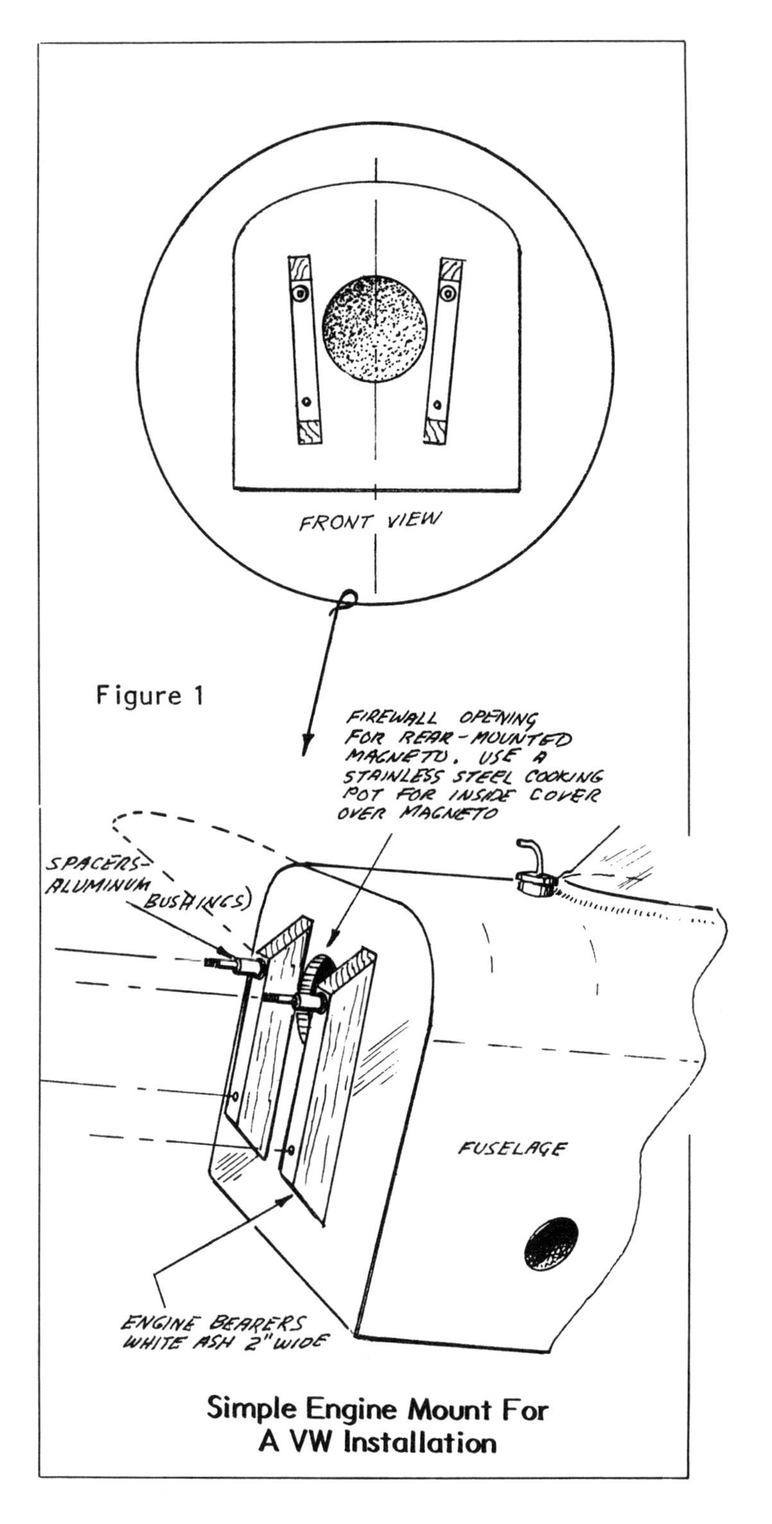

Figure 1

Simple Engine Mount For A VW Installation

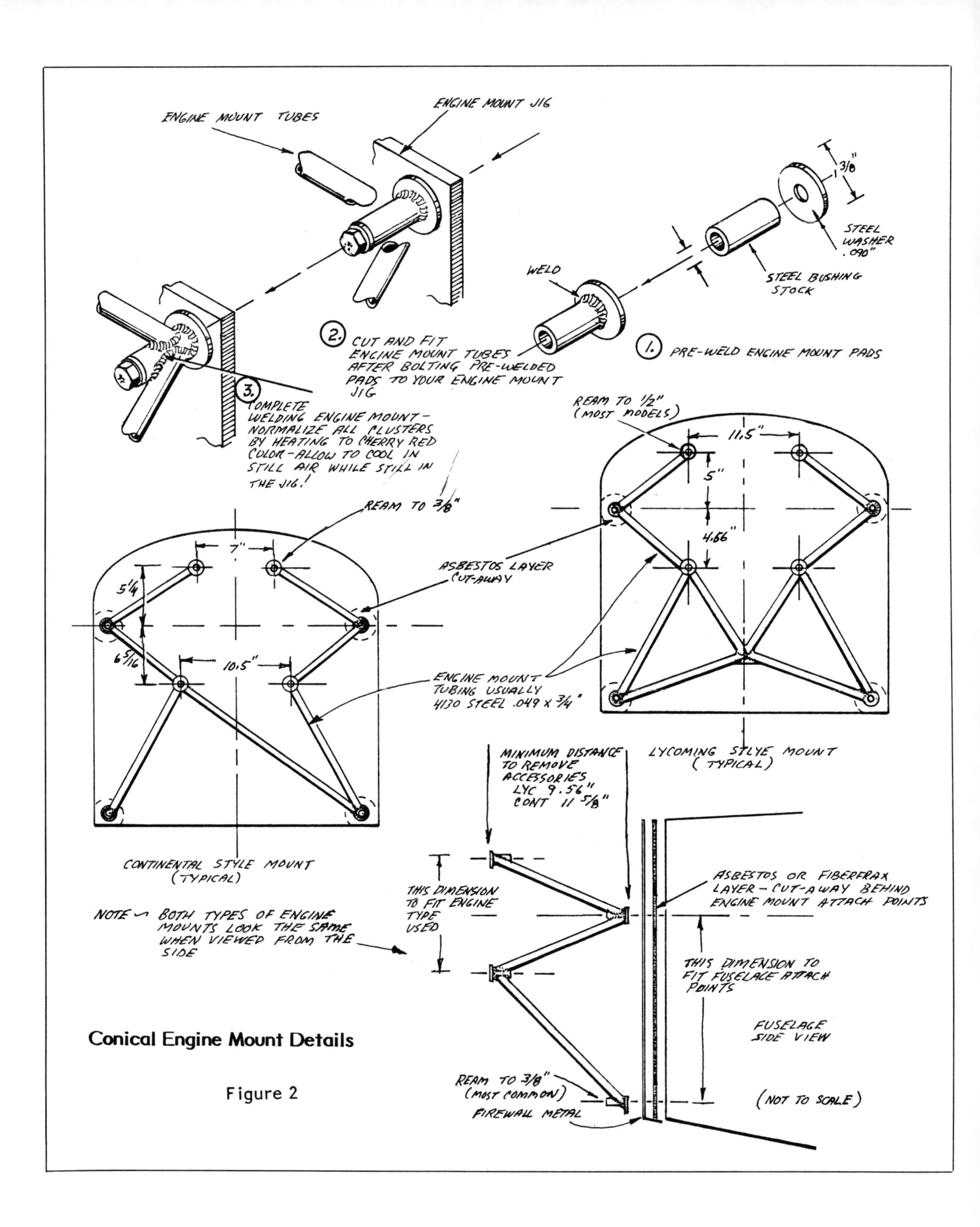

Conical Engine Mount Details

Figure 2

CONICAL ENGINE MOUNTS — The conical engine mount probably gets its name from the type of shock mount or bushing used between it and the engine. The bushing is cone shaped and made of rubber. Eight of them are required in the installation of the engine. (Figures 2 & 3)

This type of engine mount is, by far, the easiest of the two to build. Not only that, but its shocks or bushings, for both the small Continental and Lycoming engines, cost but a fraction of those for the larger C90-14F engines. However, even the greater cost of the shock mounting parts for these big Continentals looks fairly reasonable when compared to the cost of a modern day dynafocal installation.

DYNAFOCAL ENGINE MOUNTS — Yes, dynafocal mounts are the ultimate shock mounts for aircraft engines. They do an excellent job of cushioning engine vibration and reducing noise in the cockpit. The drawback, as you may have surmised, is that they are much more expensive and difficult to build than simple conical mounts. (Figures 5 & 6)

These mounts fit into dynafocal rings welded to the engine attachment side of the tubular mount. The rings or cups, must be cocked and jigged at the correct angle prior to welding. (The center lines of these rings typically intersect the center of gravity of the engine/propeller package). To complicate the matter somewhat, there are two different angles for dynafocal ring alignment: Type 1 and Type 2.

Type 1 is the most common and is used in the majority of AVCO Lycoming engines up through 180 hp.

The Type 2 dynafocal mount is used with the IO-320 and the IO-360 AVCO Lycomings ranging in horsepower from 150 to 200.

Dynafocal engine mount rings may be purchased through homebuilder suppliers. Each is pre-welded into a single unit with the correct angles locked in.

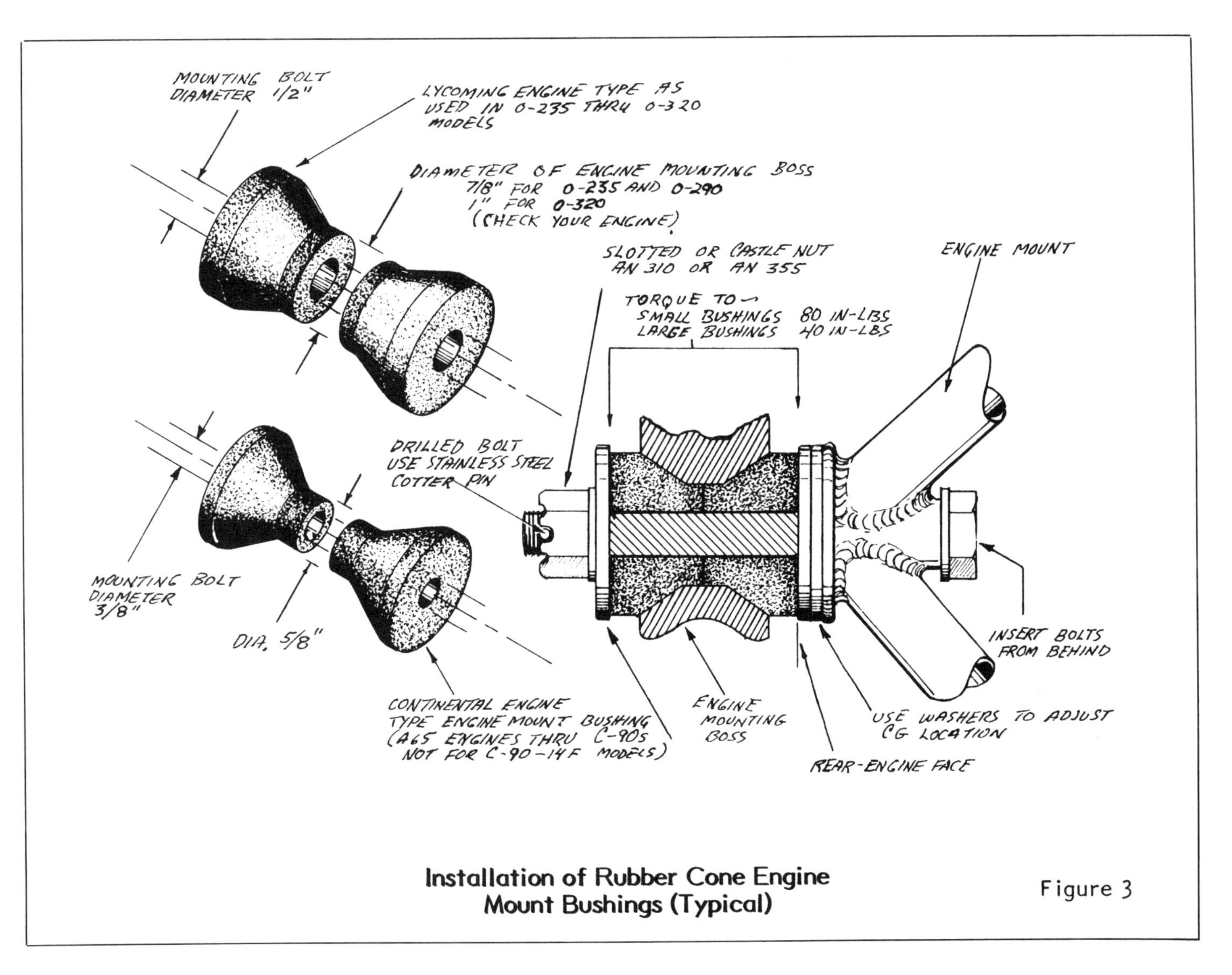

Installation of Rubber Cone Engine Mount Bushings (Typical)

Figure 3

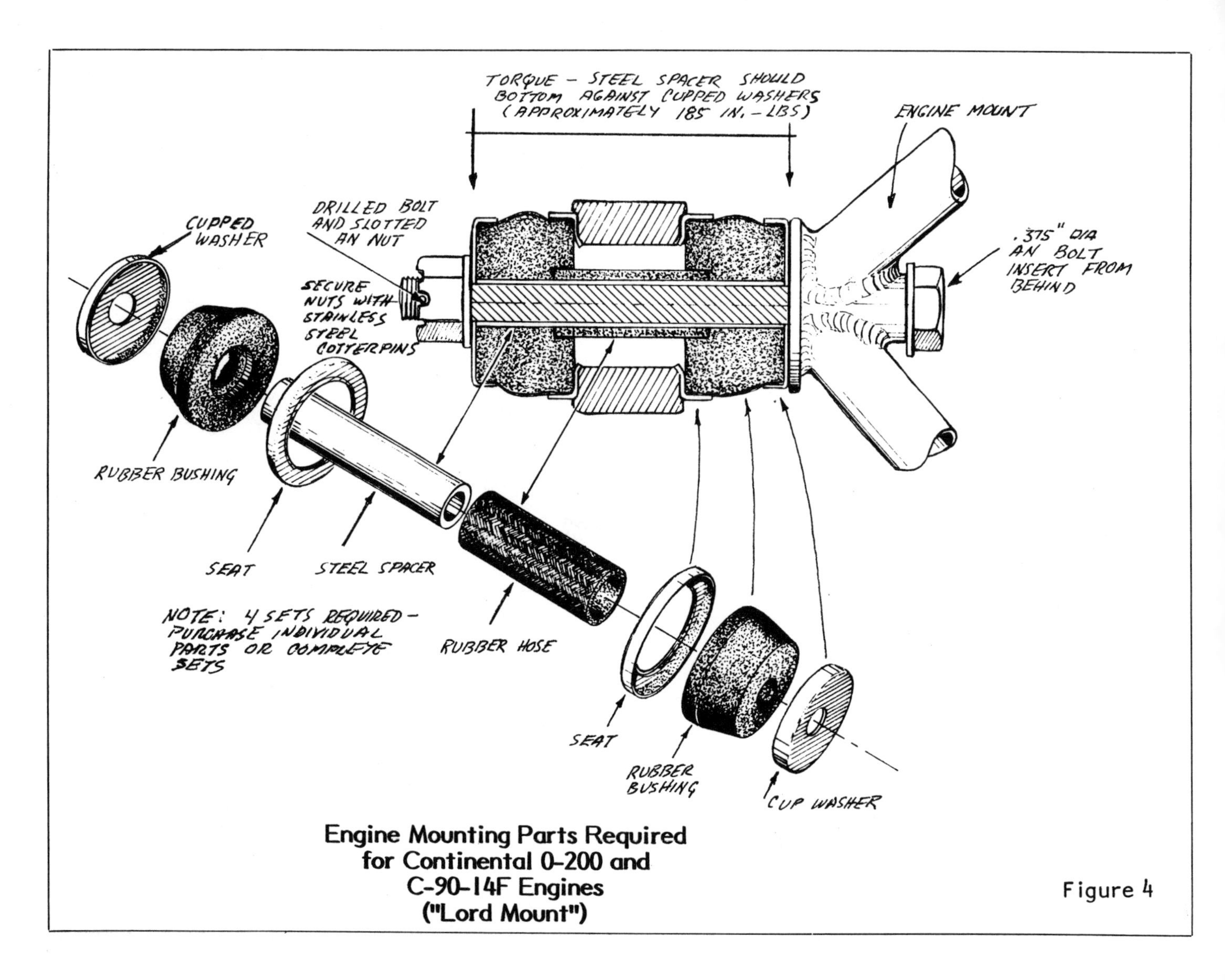

Engine Mounting Parts Required for Continental 0-200 and C-90-14F Engines ("Lord Mount")

Figure 4

When received, the dynafocal ring can be jigged into place and the tubing cut to fit around it. Welding the clusters will complete the assembly in the same manner as building up an engine mount from scratch.

The dynafocal ring type of engine mount is more likely than the conical mount to interfere with some engine accessories, particularly the magnetos. Often a minor modification can be made to provide the necessary clearance. Keep in mind, however, that the engine, being shock mounted as it is, will move around quite a bit. It is, therefore, essential that clearances be as generous as possible.

CONSTRUCTION....WHAT TO CONSIDER

Construction methods are the same for both dynafocal mounts and conical type mounts. Ideally, your mount should be long enough to space the engine away from the firewall a sufficient distance to permit you to install and remove various engine accessories. However, you may be frustrated in this respect because of weight and balance limitations. If your weight and balance problem is critical you may have to build a mount that is as short as possible. This could ruin any idea you may have had for the easy removal of accessories without first disconnecting and moving the engine away from the firewall. Should you be faced with such a dilemma, you might consider building a swing-out engine mount.

SWING-OUT ENGINE MOUNTS

Swing-out mounts may be used with any kind of engine mount or firewall construction. But, swing-out mounts have a couple of unique requirements. For one thing, they are more difficult to fabricate as they require twice as many pieces for forming the hinges and their mating attachment points. (Figure 7)

The design of a swing-out mount must take into consideration the need for the upper

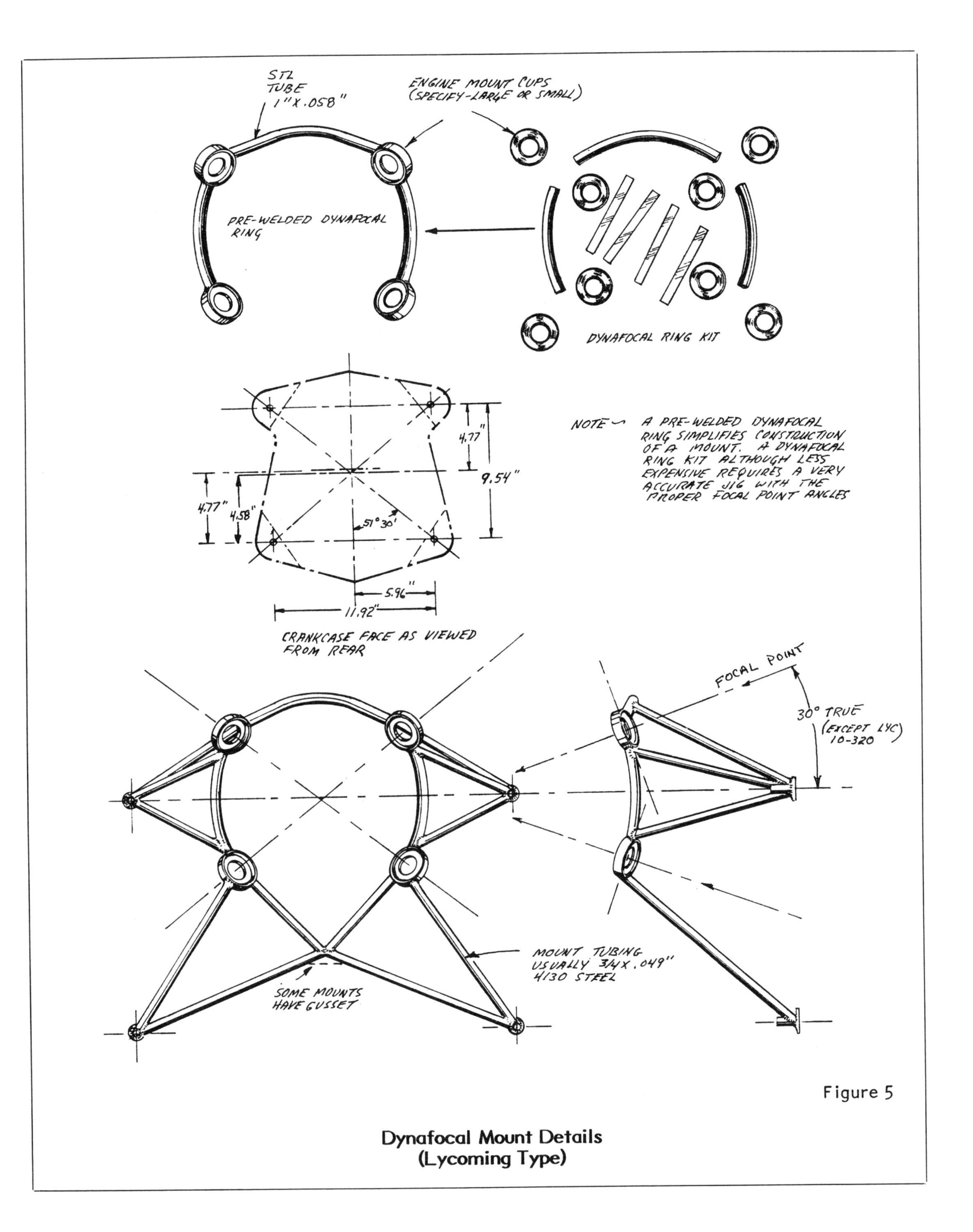

Figure 5

Dynafocal Mount Details (Lycoming Type)

Except for the prefabricated dynafocal ring, the rest of this complex structure is a jig for constructing a Falco engine mount. All this for an engine mount.

Here is the same jig after all of the engine mount tubing has been cut, fitted and welded in place. It is extremely important to do as much welding in the jig as possible if you want the mount attach points to fit. Distoration from welding is often extreme.

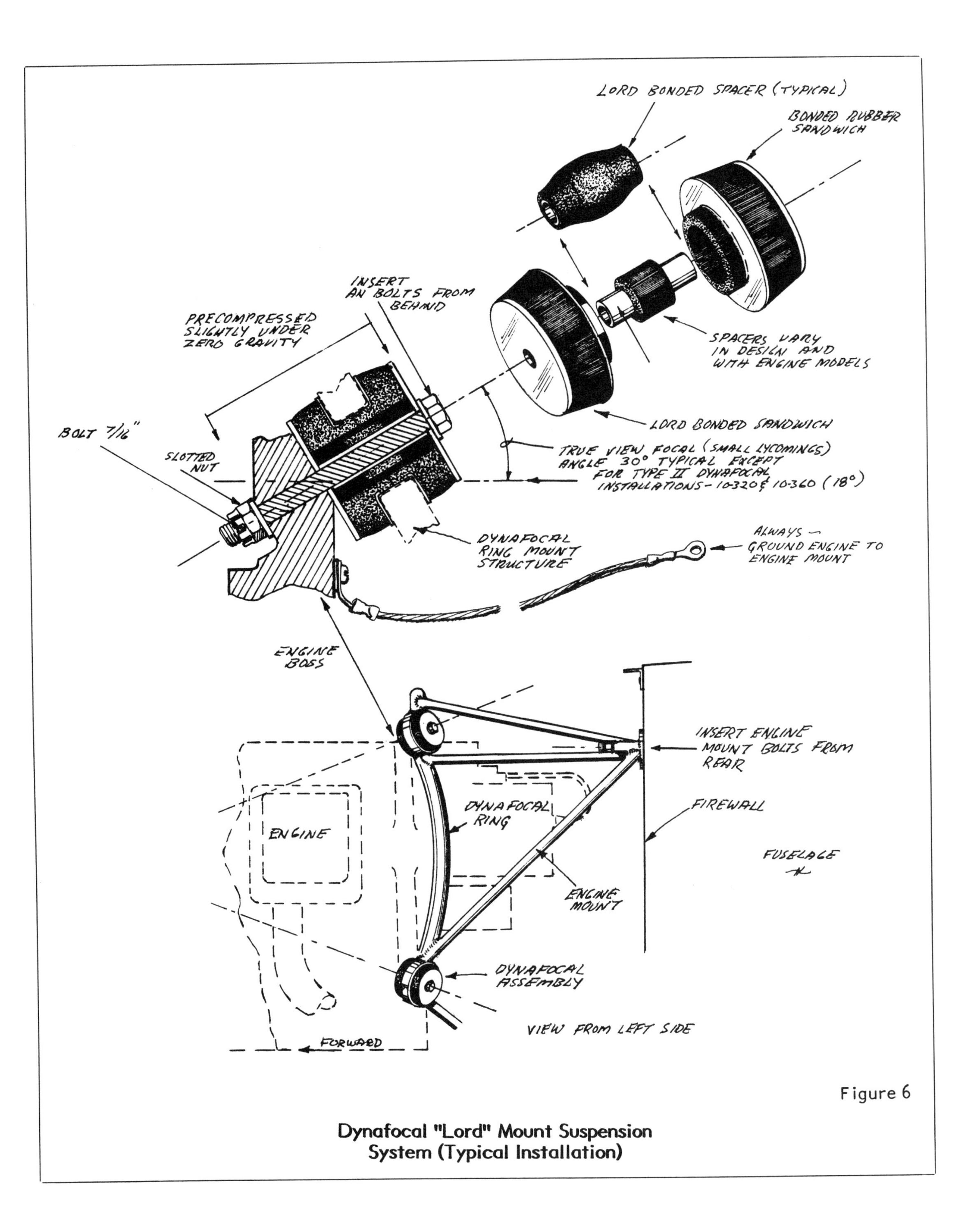

Figure 6

Dynafocal "Lord" Mount Suspension System (Typical Installation)

and lower hinge axes to be in alignment or the swing-out feature won't work. During welding, the additional precaution of maintaining the hinge alignment is introduced and added to the other alignment requirements.

A swing-out mount is much better in theory than in practice although there is no disputing the fact that it is much easier to swing an engine away from the firewall slightly than it is to pull one completely....just to remove a magneto, for example.

You should realize, however, that it will be necessary to relocate some standard firewall connections and to allow more slack in your wiring, lines and controls in order to enable the swing-out feature to work effectively. Even so, you will still find it necessary to disconnect some brackets and possibly a control or two to make the swing-out mount work.

Except for a critical center of gravity situation calling for the engine to be mounted as close to the firewall as possible, there is generally little justification for constructing a swing-out engine mount.

WELDING JIGS

The plywood jigs some builders use for fitting and welding their engine mounts are a poor choice and can result in the builder welding up a botched-up mount. Plywood jigs are usually inadequate because the heat of welding makes the tubing expand and contract considerably, causing the whole assembly to 'walk around'. Even though some pretty good engine mounts have been built in plywood and wood jigs, the best mounts are usually those fabricated in rigid metal jigs.

A wood jig invariably catches fire and the bolt holes in it burn out and become enlarged. Protecting a wood jig with asbestos is a noble thought and effort, but it still will not completely eliminate the charred hole problem. When the jig bolt holes become enlarged, they allow the tubing to creep, ruining the alignment.

In spite of all the precautions you take to protect your wood jig, you may find, when your mount is completed, that the bolt holes are off by as much as 3/8-inch. With a little luck, however, you might be able to reheat the clusters and gradually work out the distortion, but don't count on it!

If you are building a wood or composite airplane, do not drill the mounting holes for your firewall attachment fittings until after the welding is completed. Then use your completed mount as a drilling jig and locate the

All the elements of a dynafocal engine mount are very much in evidence. Notice how each engine mount bolt points inward toward some imaginary focal point. This is a hard mount to jig and weld because of the alignment problem.

firewall holes with the assurance that they will be perfect. But, what about the engine attachment end of the mount? Will those holes match the engine?

For these reasons, perhaps the best jig to make (and the cheapest in the long run) is one of heavy steel angles or 'C' channels. The heavy steel is difficult to weld with a gas torch so it should be assembled first, preferably with an arc welder. (Figure 8)

The exact design for a steel jig can vary considerably depending on the kind of scrap stock you can find in the junk yard. Essentially, you will need a base (firewall side) which can be assembled from two or more channels, a short center post (or four angle iron legs making up a center post), and a steel plate or angles to represent the engine side of the jig.

Sometimes a builder will use the firewall of his aircraft as a base for jigging the engine mount. He does this in conjunction with a stripped engine crankcase which has been supported the correct distance from the firewall and is properly aligned. The engine mount tubing is cut and fitted between the two and tack welded in place. This is a risky procedure unless the builder is pretty deft with the welding torch. I certainly don't recommend making the completed welds on the aircraft, or while the mount is attached to a perfectly good crankcase.

A safer method is to make a mock-up of the firewall and attach it to a wall or workbench. This makes it easy for you to shore up and align the crankcase to establish the engine attachment points. Then, with a sense of increased security you could proceed with the fitting and tack welding of the mount. However, you would still have the problem of completing the welds in a jig that cannot completely immobilize the parts. The net result could be a mount with some of the attachment bolt holes way off. DON'T SAY YOU HAVEN'T BEEN WARNED! Misalignment of mounting holes is the number one problem among builders who make their own mounts.

In a good steel jig, you can maintain alignment and an accurate bolt hole pattern during the entire welding process. Furthermore, you can normalize (stress relieve the mount without removing it from the fixture.

In making your jig, don't forget to get the engine alignment required, be it down thrust or a few degrees right thrust angle. This is as important as establishing the engine at the right firewall height and the correct distance from the firewall. Make a double check of the lateral alignment too. Although the airplane will fly just as well with the engine slightly tilted to one side, it will be hard to explain that strange twisted look to your friends, who are sure to notice it.

Always work from a center line in setting up the jig. It is the most accurate way to check on your jig and your work.

ENGINE MOUNT FABRICATION PRACTICES

A few suggestions for the preparation and welding of the mount tubing and fittings might be useful.

The typical attachment point for each bolt is formed of a stub tube or bushing to which a heavy washer is welded. One such stub unit is temporarily bolted to each bolt hole location in the jig. The engine mount tubes are then fitted to them and welded. If these stubs are poorly designed, excessively long bolts may be required in bolting the mount

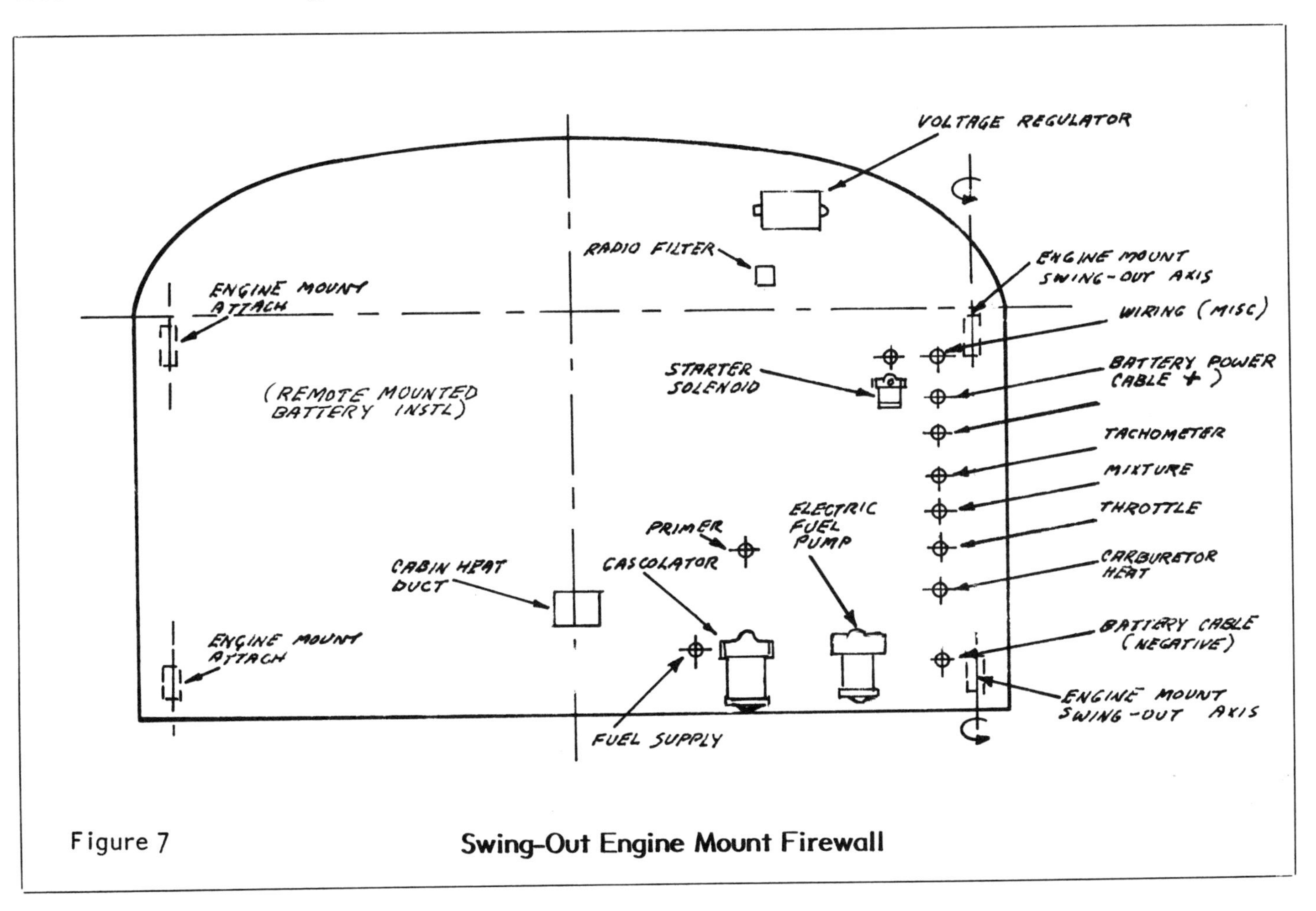

Figure 7 **Swing-Out Engine Mount Firewall**

Fitting the individual engine mount tubes. Jig must be designed to afford as much access to each cluster as possible. Note that this mount is designed to offset the engine three degrees to the right.

If you intend to mount your battery on the firewall you should check beforehand that you will be able to install and remove the battery without interference from the engine or engine mount. Note that the master solenoid is mounted on the battery box. Keep it close to the battery.

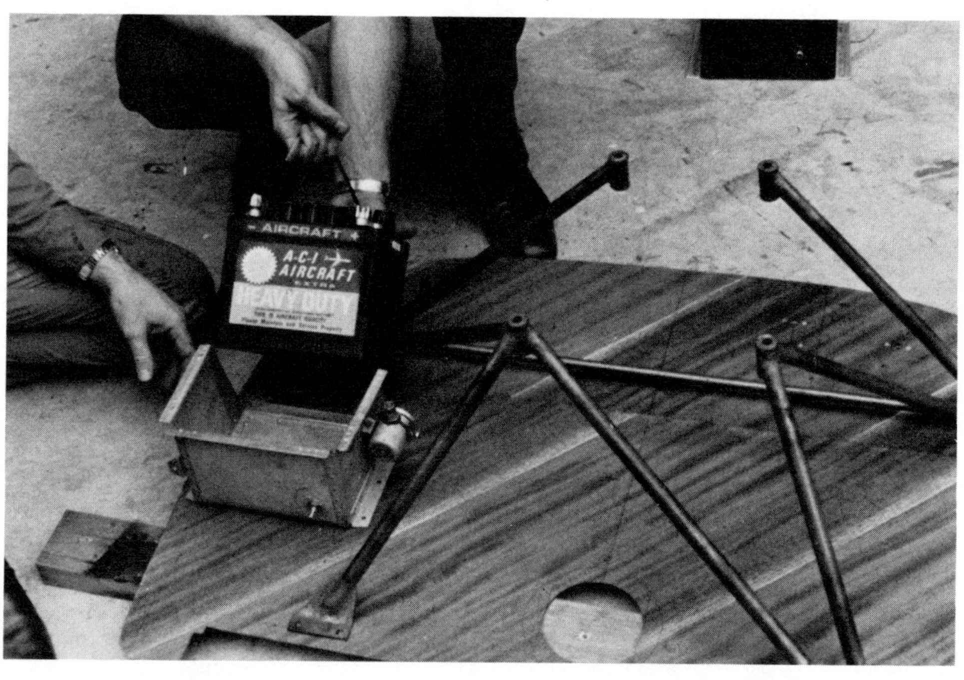

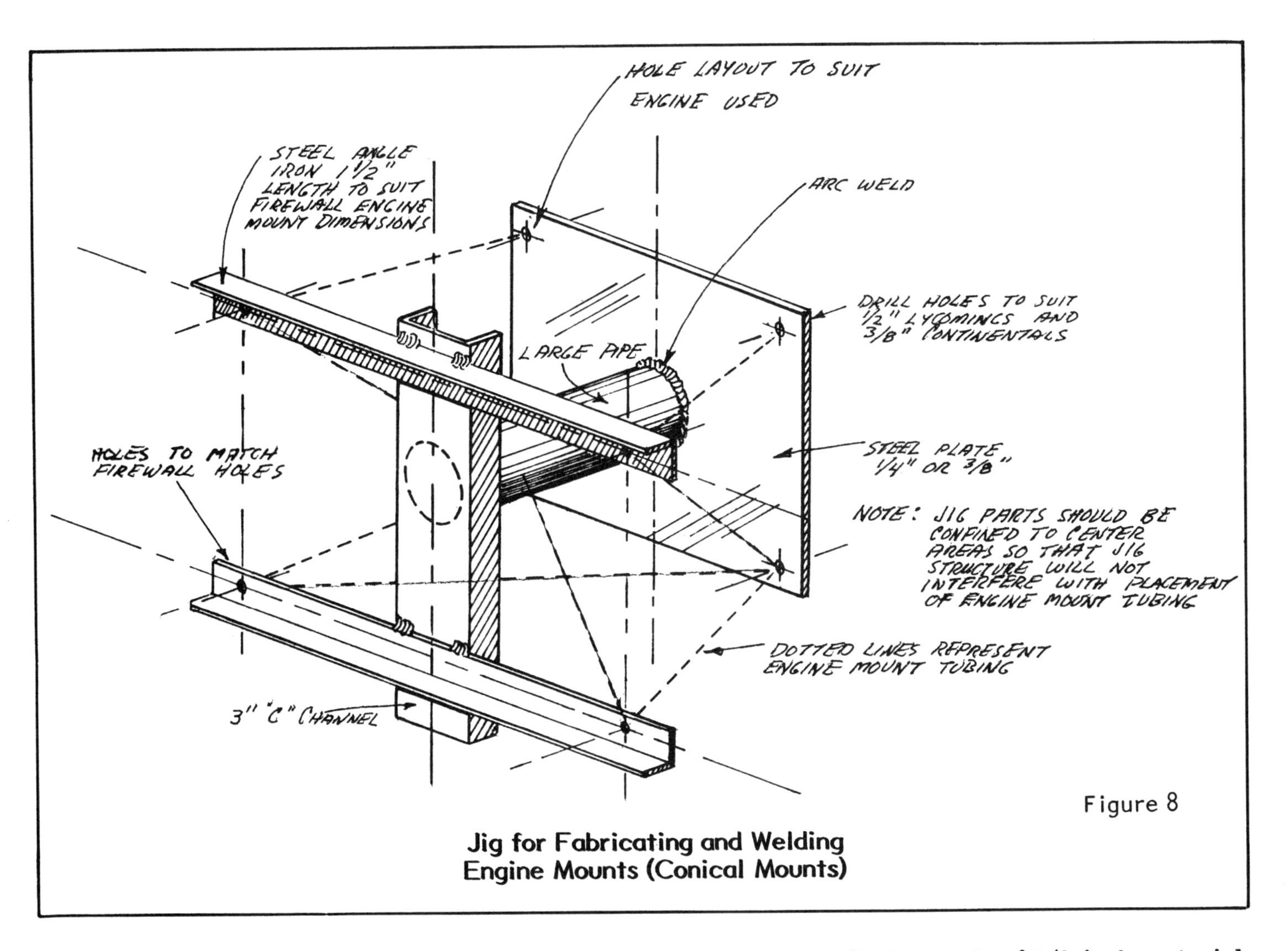

Figure 8

Jig for Fabricating and Welding Engine Mounts (Conical Mounts)

to the firewall and to the engine.

As some heavy welding of the tubing clusters is done around these attachment bushing stubs, the bolts may be difficult to remove from the jig once the mount is completed. You can ease this problem by grinding away a flat area on each side of each bolt to make it easier to remove. Naturally, you wouldn't be using your good aircraft bolts in the welding jig....would you?

Try for a good fit all around each tube. Don't force a piece in tightly. A uniform 1/16-inch space fit is just about right.

Tubing may be cut with a fine tooth hacksaw and the tube ends shaped on a grinding wheel. Some hand filing of the bevels may be done with a large 1/2-inch rattail file. Thin wall tubing, .035-inch, may be trimmed with aviation tin snips. Don't worry about getting smooth finishes on the trimmed ends....they will be melted away anyway.

Always remove all the rust and dirt from the areas of the tubes that will be welded. Get them nice and shiny.

If you do have to weld a thin wall tube to a heavy steel fitting made of 1/8-inch material, be prepared for trouble. You will do all right until you have to weld a cluster of a couple of tubes to that same place. A heavy fitting takes so much heat that by the time it is up to welding temperature you may have already started to melt away the thin tubing. The problem becomes even more acute when the backside has to be welded. (A heli-arc welder wouldn't even notice such a problem.) If you can't find a way around this difficulty, you might consider switching to a heavier wall tubing (yes, amigo, you will be adding weight) for the mount. Tubing with a wall thickness of .049-inch or thicker is much easier to weld under the conditions described.

If you intend to have your mount heli-arc welded, keep the joints extra clean and your tack weld spots, if any, small, as the heli-arc can get temperamental too.

Don't forget to stress relieve the entire engine mount after all of the welding has been completed. You do this by heating the welded clusters and adjacent areas of the mount to the required temperature (cherry red) and then

ENGINE THRUST LINE

A builder should never arbitrarily change the engine alignment in his aircraft unless he is willing to accept the consequences. Without an understanding of the aerodynamic considerations of the designer, any change for cosmetic reasons would be inadvisable.

Quite often, engines are offset to compensate for torque. An engine installation with its thrust line offset 3-degrees to the right, for example, presents an additional problem to the builder when he must fabricate his cowling. The resulting assymetrical shape is difficult for the builder to fair into the fuselage since one side of the cowling must be shorter than the other and the whole thing looks cockeyed.

Nevertheless, the builder will be a lot safer following the plans and aligning the engine as presented in the plans. Different types of aircraft have their peculiarities and it must be assumed that the designer has adequately coped with the engine thrust line problem.

allowing it to cool to room temperature in an area free of drafts. You should do this even if it was heli-arc welded.

The tubing interiors of the completed mount will have been sealed from the atmosphere. As a result, most builders do not bother to treat the mount internally with a rust inhibitor as they do their steel tube fuselages.

Don't forget to ream the bolt holes in the mount to size as there will be carbon build-up in them. This reaming may be hard on the reamer because the built-up stuff is very hard and brittle.

The next step is to perform a careful inspection of the mount and then sandblast it. Immediately afterward, you should prime it. The mount should be painted with a high temperature paint if any of it will be adjacent to hot exhaust pipes. Use a light color as it is very difficult to see cracks on anything painted black.

Wrapping asbestos around an engine mount tube to protect it from a hot exhaust is not a very good solution because rust can build up under the wrapping and not be noticed until the tube fails. An asbestos wrapping generally becomes a moisture trap that encourages the formation of rust. If you feel that an asbestos wrap is the lesser of two evils, be sure to make frequent inspections of the tubing areas under it just in case.

You can use self-locking nuts on the bolts used to secure the engine mount to the firewall. Most installations are made with the bolt head inside the aircraft and the nut visible from the firewall side. The engine, on the other hand, is secured to the mount with drilled bolts and castle nuts safetied with cotter pins.

Heavy objects, which might impose a bending stress on the engine mount tubes, should not be hung from them. If you do find it necessary to attach something like an oil cooler, make some provision for neutralizing the bending loads it will impose upon the tube or tubes.

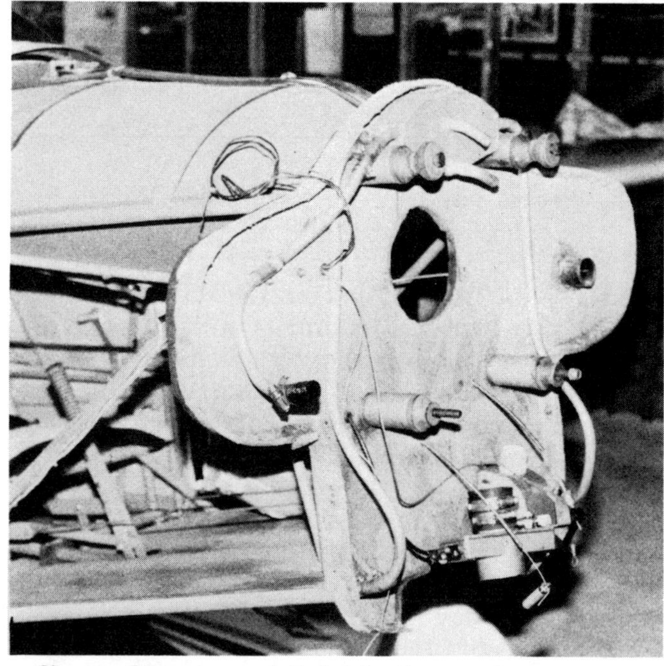

Short, Simple and light is the engine mount on a Sonerai II.

NOTES

FoKD I

exhausts & mufflers

5

Shhhh.... Mufflers at Work

People are disturbed and angered by aircraft noise. And, even though the most irritating frequency and level of such noise is generated by large jets approaching and departing metropolitan areas, all aircraft are circumstantially adjudged guilty by association.

To your ears, the purr or powerful bark of your aircraft engine may be beautiful music; to the ears of the man in the street, or the woman in the garden, however, it may be noise. (Remember, noise is merely unwanted sound, so even a baby's giggling would be considered noise by some people.) Even if the average light aircraft is quieter than the neighbor's motorcycle or lawn mower in action, it is pointless to debate this emotional issue. Airplanes are noise makers and, as such, are subject to the noise pollution laws laid down by the Environmental Protection Agency and the FAA.

The Federal Aviation Regulations (FARs) specifically those detailed in Part 36, attempt to deal with the problem of aircraft noise. The FAA developed these rules with airworthiness and safety as the first concern. At the same time — it must have been about 1974 — - the FAA attempted to abide by the spirit of the original EPA edict...to cure ALL environmental ailments, and to do it pronto. We were quite fortunate to have the FAA act as a buffer between the light aircraft users and the EPA. As a matter of fact, I believe the FAA is doing an excellent job in coping sensibly with the overall noise problems.

As stated, these standards are now issued and are the law of the land. So far, however, amateur built aircraft are excluded from the regulations, while antique aircraft and standard category aircraft are not. But, because of the FAA's interest in airworthiness and safety, I am sure the Agency is, even now, taking a hard look at the effect mufflers might have on low-powered homebuilts, and I feel that there will inevitably be a requirement for ALL aircraft to meet noise level standards tailored to their horsepower classification.

This is not all bad. Excess noise is harmful, especially to pilots. I know that the hearing of many pilots has deteriorated because of prolonged exposure to noise above normal levels.

The Air Force has long required its personnel to wear ear protectors for any sound levels above 85 decibels (dB), regardless of exposure time. Quite some time later, as the EPA was expanding its influence, a governmental agency called OSHA (from Occupational Safety and Health Act) started forcing industry to protect its employees from excess noise.

Take a look at the levels listed below to get a better handle on what our aircraft situation is in relation to other noise-generating activities (sound levels are in decibels and are very conservative):

150	*Jet aircraft taking off*
140	*Threshhold of pain*
120	*Airport terminal boarding ramp*
110	*HiFi played by average teenager*
105	*Light aircraft, outboard motor, unmuffled snowmobile at a distance of 50 feet*
95	*Busy streetcorner, motorcycle at 50 feet*
85	*Passenger car interior (subcompact)*
80	*Dividing point between safe and unsafe sound range*
70	*Business office, electric hand drill, air conditioner, homebuilder's workshop*
60	*Normal conversation (male voices)*
40	*Refrigerator*
30	*Bedroom at night (average, excluding indoor sports)*
10	*normal breathing*
0	*You can't hear anything, but maybe your dog can.*

Now, let's check the current levels established for standard category aircraft, including antiques and aerobatic production models. The noise level must not exceed 68 dB for aircraft weighing up to and including 1,320 pounds (600kg). For aircraft weighing more than that, but not exceeding 3,630 pounds (1,650kg), the limit increases at the rate of 1 dB/165 pounds (1dB/75kg) to 82 dB at 3,630 pounds, after which it remains constant for larger aircraft up to and including 12,500 pounds.

Where do we stand? Many unmuffled homebuilts will emit an acoustical power level of 100 to 130 dB at 20 feet....more to the gent in the cockpit.

You will also be hearing of acoustical changes. Even now, limitations on acoustical changes apply to antique aircraft and standard category aerobatic aircraft. For that matter, they apply to the several resuscitated older designs like the Great Lakes biplane, the Taylorcraft, etc.

As I understand it, the regulation on acoustical changes states, in effect, that the noise standards will not permit anyone to modify an aircraft so that it is noisier than it was under its original certification This means, of course, that antiquers and owners of aerobatic type aircraft will not be able to re-engine their craft with larger engines unless the resulting noise level is equal to or less than the original version. The FAA, apparently with tongue in cheek, states: "If the antique aircraft is quieter than those limits prior to the change in type design, this amendment permits noise increases up to that limit".

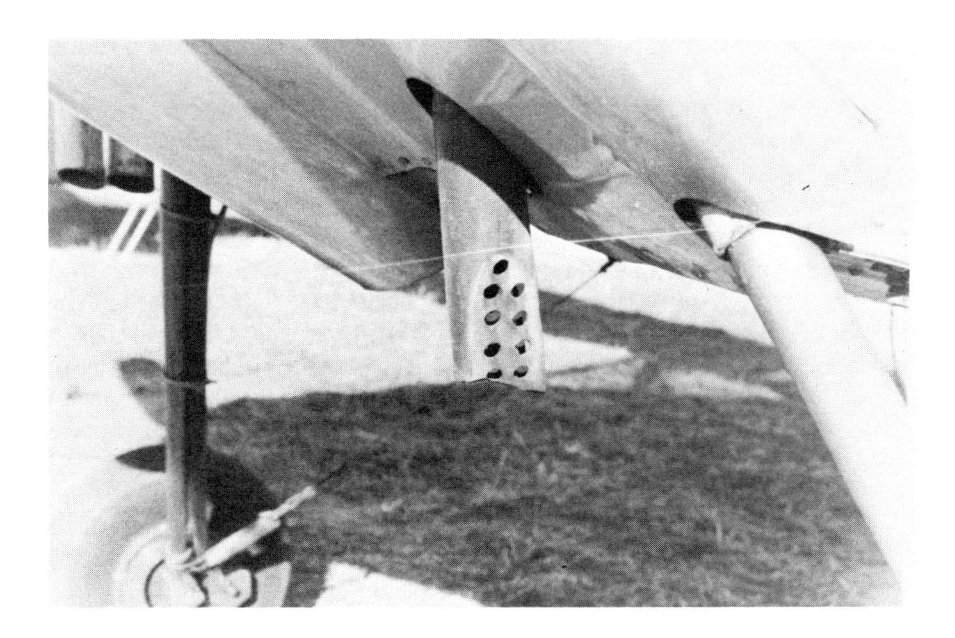

Various tail pipe treatments are tried in an attempt to alter exhaust noise patterns.

Basic Considerations

An aircraft exhaust system should be designed and built to carry the exhaust gases and heat away from the engine and do it without burning up the airplane or asphyxiating the pilot. An equally commendable goal in this age of environmental awareness is to accomplish this feat as quietly as possible. However, while you are contemplating the implications of all this, you should also be aware that there are other factors to be considered in planning your exhaust system.

The FAA offers some guidance in this respect. Its Federal Aviation Regulation (FAR), Part 23.1123 states, in effect, that exhaust manifolds are to be made of fireproof and fire resistant materials, that the assembly should allow for expansion due to temperature changes, and that it be supported against vibration and inertia loads. In addition, the FAA requires a means for inspecting the critical internal parts of heat exchangers. Even though these provisions are not mandated as far as homebuilts are concerned, they should be heeded.

But, where do you begin? Well, if the engine you have acquired came with a serviceable exhaust system, study that installation to see how you can convert it to fit under the cowling of your airplane. Modifying a good used exhaust system is much easier than building one from raw material. It might sound a bit fundamental but, remember this: if you build and install the exhaust system, your cowling must be made to fit around it. Naturally, if you already have a cowling, then the exhaust system must be made to fit inside without touching any part of it. At any rate, don't be surprised if you find that a cowling you may have ordered won't fit around your existing exhaust system.

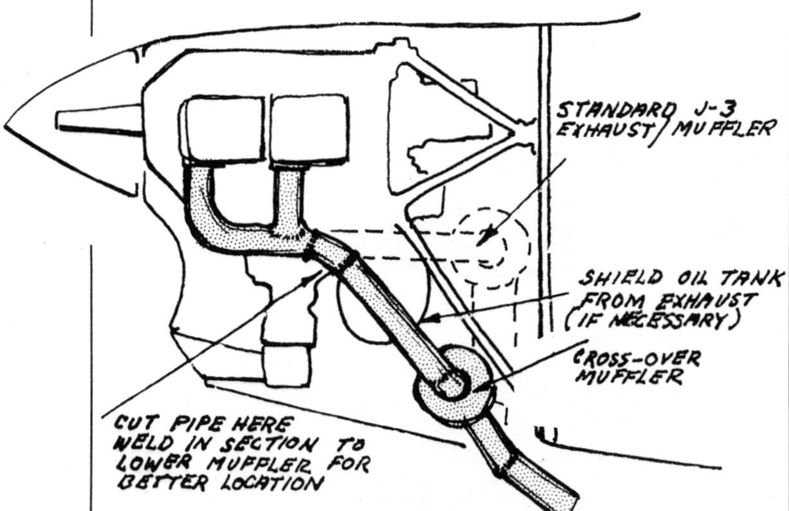

What about cowling clearances? In general, try to keep exhaust pipes away from any structure within the engine compartment and allow at least one inch clearance between your pipes and the cowling. Sometimes, with just a little cutting and welding here and there, you can adapt a standard production type exhaust system to fit your airplane. If at all feasible, this option should be your first choice. A typical example of this sort of modification is shown in Fig. 1. It illustrates how a standard exhaust for a Continental engine may be modified to serve a specific need. Such modifications are relatively easy to make and provide you with an economical and quick "fix".

Figure 1

Standard J-3 Exhaust System Modification

A standard aircraft stainless steel exhaust system in good shape should never be discarded without due consideration. Most modifications will only require a few small wedge cuts at appropriate points and, possibly, the addition of a couple of welded tailpipe extensions.

All this is fine if you happen to chance onto a good used exhaust system, but some of the old patched-up exhaust installations one often sees aren't fit to be in service in the first place. You probably wouldn't even allow some of those rigs in your shop. Well, if salvaging an old exhaust system doesn't appeal to you, what other options do you have?

You could, of course, build a complete exhaust system using new materials, making it either of stainless steel or of automotive pipe. An easier alternative would be to purchase basic manifold units to fit your engine. These can be added to as necessary to route the exhaust outside the cowling and overboard. If construction of an exhaust system is to be undertaken in its entirety, you might as well build and install exactly what you want. The chances are good that it wouldn't be any harder to build than something that is only marginally satisfactory.

SHORT STACKS

How about the installation of short stacks to get rid of the exhaust gases safely and without unprogrammed pyrotechnic surprises? The short stack term applies to any installation that consists of comparatively short sections of individual exhaust pipes bolted directly to the cylinder exhaust ports. Before I try to shoot short stacks down as unsuitable for you, it is only fair that I present the advantages of this type of installation.

Ordinarily, the exhaust pipe that is bolted to each exhaust port is no longer than necessary for it to route the gases from the engine ports to the outside of the cowling. The formula racers rely almost totally on this type of exhaust installation. If you examine photos of these fast little planes you will see that the pipes are trimmed off even with the cowl....no more, no less.

Without the longer pipes and mufflers that most of us are accustomed to in the aircraft we fly, the perky naked bark of the engine might be a delight at first, but not for long.

In defense of the short stack installations, however, I should point out that they are light, economical, easy to install, and easy to maintain and inspect. But, perhaps even more important, they produce no significant power-robbing back pressures. Furthermore, short stack installations help hold exhaust valve temperatures to a minimum. All these plusses add up to greater available horsepower....and that is good, performance-wise. On the other hand, the drawbacks to the short stack system, at least for the average sportplane, are considerable.

Who needs the noisy, unrestrained, staccato bark of a mighty four lunger on a long cross country hop? Besides, noisy aircraft are no longer welcome at many airports. In fact, it's beginning to look as if the time has come to bury the short stack along with the bones of the Dodo Bird. We must try to make our aircraft quieter for our own sakes. As added incentive, if we don't show concern for our

Individual short stacks sticking out of the cowling look nice, but where do you install a heat muff for carburetor heat?

own health, hearing and well being, "big brother", in one guise or another, will see to it that we do it his way.

Well, if you were not moved by the preceding impassioned plea and still persist in thinking to build a short stack system, consider this, too.

Short pipes do not do a good job of conducting the exhaust gases away from the cockpit area. They also permit too rapid a temperature change at the exhaust ports (drastic cooling) whenever power is reduced suddenly. This can lead to warped valves and seats. Wait, there's more! Short stacks leave no space for heat muffs or heat exchangers, and that can make it quite difficult for you to arrange for carburetor and cabin heat. So? So, with stacks, if you don't end up on the environmentalists' "hit list", or lose your hearing, or be overcome by carbon monoxide fumes, or freeze to death, you may wind up in the poorhouse buying new valves. (Wow! Talk about overkill!)

MANIFOLD SYSTEMS

A system which joins the exhaust pipes on one side of the engine is a manifold improvement over the short, individual stack arrangement. Maybe that's why it's called a manifold.

Using a common manifold to gather the

individual exhausts and then directing them overboard through a single pipe provides a quieter running engine even when no muffler is installed. Another important factor affecting the noise output of an engine is the length of the tailpipe. The longer the tailpipe, the quieter the engine. Often the manifold from each side of the engine is interconnected and only a single tail pipe exhausts the gases somewhere under the aircraft. More often, though, the homebuilder will find it easier to use a separate tail pipe for each side of the engine.

CROSS-OVER SYSTEM

A cross-over exhaust system is one that ports the cylinders in their correct firing order, thereby minimizing back pressure by uniformly spacing the exhaust pulses. To do this, the front exhaust ports are joined together with a single exhaust pipe that routes the gases overboard through a single stack. The rear cylinders are likewise connected by a cross-over pipe in the same manner, making the exhaust system look like a cowl full of spaghetti oozing out of the engine.

As complex looking as it is, there's no denying that it results in increased engine efficiency by reducing back pressures. A well constructed cross-over system is reputed to net about a 100-rpm increase....this means more available power and greater fuel economy. (Figure 2)

The usual practice is to have both sets of pipes crossing in front of the engine, although there are installations where the builder routes the rear manifold around the rear of the engine. However, to do so may complicate the installation of heat muffs and mufflers.

Unfortunately, a cross-over system is heavier, harder to build, and requires a lot more piping. The long pipes will, of course, be conducting heat through a long route inside the engine compartment, so be sure that this does not cause problems. It is also difficult to make a cross-over system compact enough in a small homebuilt that it can be cowled easily without resorting to unusual bulges.

Because of the long pipes required in a cross-over system, it is imperative that you incorporate both slip joints and ball joints to absorb the expansion and contraction of the pipes, and to minimize the effects of vibration on the installation.

Remember, since the engine floats and shakes in its engine mount, you cannot usually brace the exhaust pipes against any part of the aircraft structure. Brace them instead, against some part of the engine.

ROUTING, EXITING THE EXHAUST

Avoid developing excessive back pressures in your exhaust manifold by providing an easy out for the exhaust gases. That is, do not build in sharp angles or changes of direction or use too small a discharge pipe diameter in your tail pipes. (Figure 3)

Exhaust outlets should be located so that the slipstream will not retard the exhaust gases. This would have the same effect as an obstructed opening. Also, because of the high drag created, the exhaust pipes should not jut out into the slipstream at a 90-degree angle. It is better to exit the stacks at a swept angle (looks better too). A reduction in drag is possible if you can arrange to have the tail pipes one behind the other....outside of the cowling.

Many aircraft exhausts exit out the bottom of the cowling. This is as true for those utilizing a single large exhaust pipe as it is for those having multiple tail pipes. You have probably noticed, too, that the single stack outlets seem to be located more frequently on the lower right hand side. As a matter of fact, it would seem sensible not to outlet ex-

A J-3 Cub-type exhaust system utilizes a cross-over muffler. The long tail pipe jutting downward is enclosed by a muff for carburetor heat. The installation is simple but not particularly efficient due to back pressure.

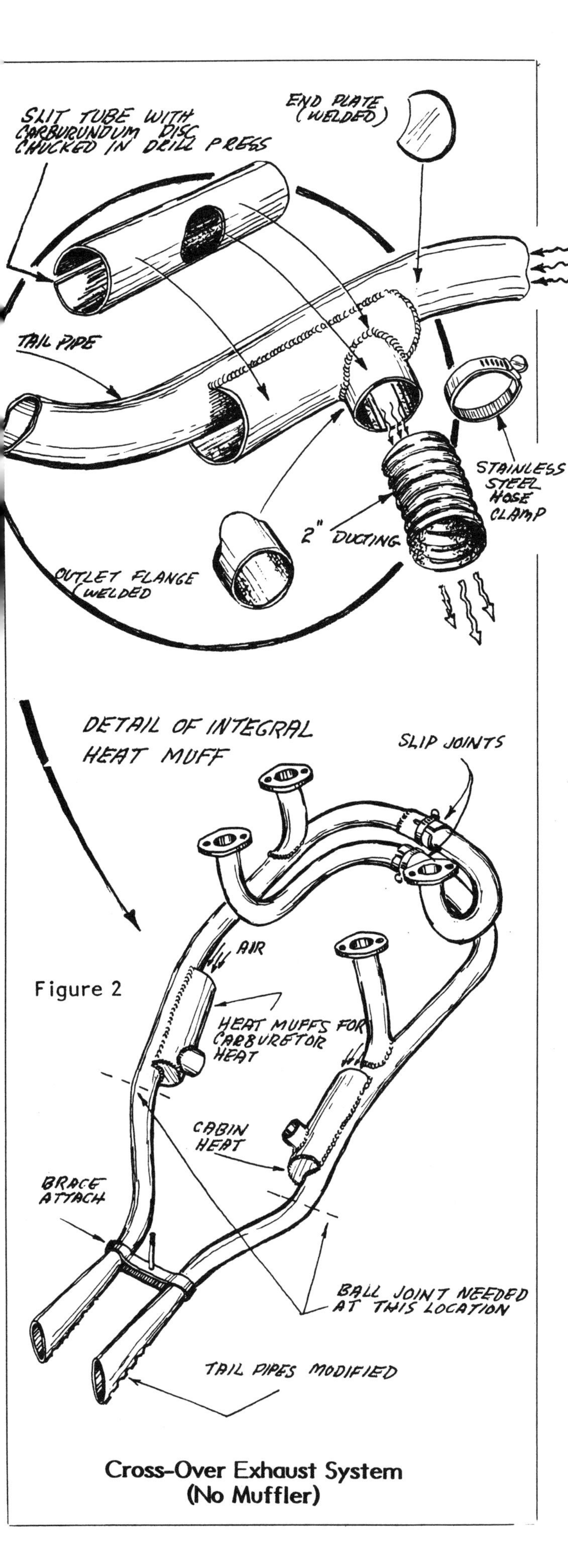

Figure 2

Cross-Over Exhaust System
(No Muffler)

haust stacks on the left hand side (as viewed from the cockpit) at any point higher than the lower longerons. The propeller, because of its direction of rotation, would tend to swirl the exhaust gases up toward the cockpit area (VW engines rotate in the opposite direction, ordinarily, and would have an opposite tendency).

If cockpit vents are to be located near any exhaust outlet, try to determine if any fumes would find their way into the cockpit. Remember, it is possible for fumes to get in even though the cockpit or cabin is enclosed.

The pipes should be long enough to clear the aircraft structure so as not to impose excessive heat on it. Even so, certain areas of the structure, like the cabin floor, may have to be insulated as well as protected with an underlying blanket of asbestos or Fiberfrax. Don't forget that parts of the exhaust system operate at temperatures as high as 1,400-degrees F — that's hot, pardner.

Maybe the best reason for running the exhaust pipes under the airplane is so that you can't see the flames leaping from those cherry-red pipes while flying at night.

BACK PRESSURE

With the introduction of manifold, mufflers, longer exhaust pipes and pressure cowls, the pipes have to be curved and intersected with each other through a very crowded engine

Be sure to brace your tailpipe to the engine and not to some part of the fuselage. Otherwise vibration might induce cracking and early failure.

compartment. All this increases the risk of causing unwanted back pressure.

Back pressure is a condition in which greater than normal atmospheric pressure is created at the engine's exhaust ports. That is to say, the exhaust gases aren't scavenged as rapidly as they are expelled from the engine's exhaust ports. Things get crowded there and the pressures build up. This build up of pressure is sometimes aggravated by pipes that are too small in certain areas. Sometimes the intersection of two pipes creates turbulence because of a poor weld joint and, very often, because of a poorly designed or defective muffler. The causes of back pressure are many and cumulative.

One thing is certain, excessive back pressure always results in a loss of horsepower. Fortunately, for the users of low-powered engines, the back pressure in a simple exhaust system is quite low. Outside of avoiding sharp bends in the exhaust system or using pipes that are too small or restricted most of us have very little else going for us in designing our own exhaust installation.

Most of us are not in a position to determine what effect our exhaust system will have on the power output of the engine. After all, how many of us have access to a dynamometer or can arrange to run torque-meter tests to obtain the required data?

I guess the best approach to planning and building an exhaust system is to pattern it after proven design practices. The results should then be predictable and reasonably effective.

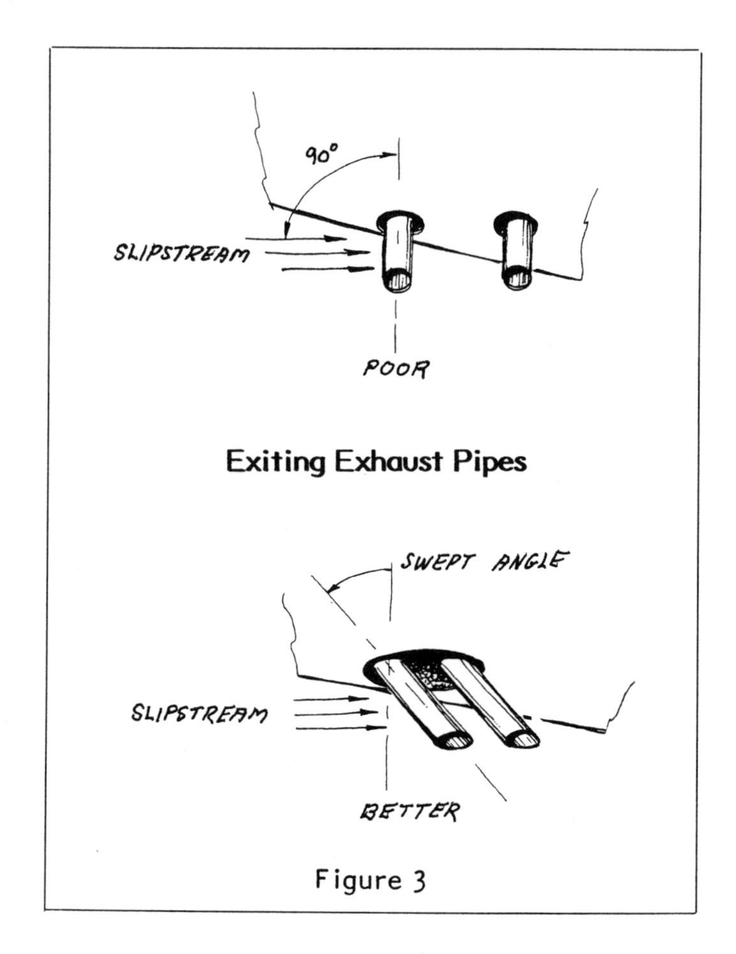

Figure 3

This twin-muffler installation dumps the exhaust gasses almost straight down.

Design Considerations

Once you have visualized the general arrangement of your exhaust system, give some thought to these design considerations:

* Should it be made of automotive mild steel or stainless steel?

* Either the exhaust system must be made to fit your cowling or your cowling must be tailored to fit the exhaust installation. It all depends on which you prefer to acquire or to build first.

* The exhaust system should be constructed so that it will be easy to inspect and maintain without obstructing access to other engine components. Furthermore, the entire system must be adequately supported and free to expand and contract with operational temperature changes.

MILD STEEL AUTOMOTIVE PIPES

The question most often asked about exhausts is "what about using automotive pipes on my airplane?" (Nobody asks questions that can be answered with a simple YES or NO anymore.)

I, for one, am losing my long standing reluctance to the use of automotive pipes in sport aircraft. Automotive pipes, when properly fabricated, are every bit as effective as the stainless steel types. They may not be as light and they may not have as long an operational life because of their susceptibility to corrosion but they should prove adequate under the conditions we encounter in recreational flying.

Many builders are using the economical automotive pipes simply because the source of supply is plentiful, locally available and, most important, curved sections such as 45-degree bends, 90-degree bends, and "U" sections can be obtained from practically any automotive supply store. These steel pipes are very easy to weld (another attribute) and the finished exhaust system, as a result, looks very good.

There are drawbacks to using automotive pipes, of course. Weight is one of them. In building a complex cross-over system the weight differential could be considerable as the automotive pipe walls are approximately .055-inch thick while the usual aircraft stainless pipes are made of .035-inch stock. Assuming that both varieties of steel are approximately equal in weight, the automotive pipe installation would be almost twice as heavy.

Nevertheless, automotive-type exhaust systems are certainly working out very well on VW-powered aircraft and more builders seem to be installing them on larger engines. Only time will tell how long these installations will perform adequately under differing climatic conditions.

The ease with which automotive pipes can be welded makes it easier to accept the handicap of their heavier weight. Still, make sure that you don't inadvertently use some pipes where the gauge is heavier than .055-inch wall thickness as it would be too high a weight penalty to pay.

STAINLESS STEEL EXHAUST PIPES

For comparative purposes you can figure that a stainless short stack exhaust installation would weigh about three pounds. On the other hand, a simple stainless two-manifold installation is not likely to weigh more than eight pounds with a muffler and heat muffs. A complete stainless system with the works, muffler, heat muffs and braces will weigh in at somewhat less than 12 pounds, a little more if it's a cross-over installation. It really depends on the system's complexity and the length of the tail pipes. As these estimates are for typical stainless systems, you can judge the weight of an equivalent automotive pipe system accordingly.

The general acceptance of stainless steel exhaust systems speaks for itself. Stainless is great material. It is very strong, with a tensile strength of approximately 90,000 psi in its annealed condition, yet it can be cut with tinsnips. It's quite ductile and can be bent and formed easily. It is also nonmagnetic in the annealed state and very corrosion resistant.

On the other hand, it's quite difficult to weld and tricky to drill holes into. Even so, it is almost a perfect material for aircraft exhaust systems.

Type 321 is the stainless steel of which aircraft quality exhaust pipes and manifolds are most frequently fabricated. If you are going to construct your own system, I suggest that the welded seam variety may be used, as it is a good bit cheaper than the seamless variety. The price differential per foot is significant. Stock aircraft pipe diameters are 1 1/2-inch and 1 3/4-inch O.D. with the wall thickness ranging from .035-inch to .049-inch for 4-cylinder engines. The tendency will be for the homebuilder to select the heavier gauge because it is somewhat easier to weld. If weight control is important, and it should be, then you should be advised that the .035-inch thickness is proving quite adequate in certificated stainless steel exhaust installations.

Exhaust pipe flanges are also available in pre-welded short stainless steel stack sections and are well worth the extra cost for the time and labor saved. You can bolt these short stacks into place at each cylinder exhaust port and then cut, fit and tack-weld the remainder of the exhaust manifold or pipes into automatic alignment. This method of building up your exhaust system will ensure a good fit.

IT LOOKS LIKE STAINLESS BUT.....

Those of you who haunt the salvage yards in your search for aircraft project materials probably have already found some 1 3/4-inch stainless pipe....or is it Inconel? Both look alike to me but Inconel is a nickel-chromium-iron alloy while stainless is a chromium base alloy.

Inconel, like stainless is a corrosion resistant steel used in exhaust systems. However, most of us will probably have enough of a problem locating and working with stainless steel stock, much less something more exotic like Inconel. Furthermore, the welding of Inconel does take a different type of welding rod than that for stainless. The point of all this is that you might accidentally acquire Inconel thinking you have the ordinary stainless steel. BE CAREFUL! It's not a bargain if you can't use it.

DRILLING INTO STAINLESS STEEL

During the fabrication process, you may want to drill some holes in the stainless pipes. If you don't use the right technique, you could learn to your surprise that it is almost impossible to do. It may help to first grind the drill bit to a slightly flatter angle than a regular bit, however, if you handle it right the ordinary bit will do the job. Make a very light punch mark. Do not try to get the drill bit started without the help of one, it may be useless effort. Use a slow speed for the bit — about half what you would use in drilling 4130 steel. Back up the work with a solid support while drilling. KEEP THE BIT CUTTING or else the steel will get so hard after the first few non-cutting revolutions that you will dull or burn the drill bit and get nowhere at all! The only recourse once you start the useless spinning is to stop and regrind the drill bit to a different angle, or maybe to punch out a hole and file it to the size you want with a rattail file. A file used to sharpen chain saw blades works nicely, too.

MODIFYING STOCK EXHAUST PIPES

A common problem encountered in adapting a stock exhaust system is the lack of clearance within the cowling or an insufficient clearance at the firewall due to the use of a shorter engine mount. In spite of these difficulties, it is easier to modify a stock exhaust/muffler system than it is to build a new one from raw material. Ordinarily, a simple modification has to be made to alter the direction of the pipes to the desired point or angle of exit. Sometimes, it may be necessary to weld in short pipe sections in order for the stacks to clear the cowling. More often, however, the modification may only consist of altering the location of the muffler. This can be accomplished by cutting a small wedge out of the pipe and rewelding that particular joint in its new deflected position.

In the example illustrated in Fig. 1 the stock J-3 muffler was too close to the firewall and the builder had no space for the generator. By welding in the short exhaust pipe section, he was able to lower the muffler location.

It also had the advantage of permitting him to reduce the length of the tail pipe and to allow it to exit at a more favorable angle. The original shroud on the tail pipe was removed and a wrap-around muff installed on the muffler.

When making such modifications, however, the builder should be aware that the muffler in its new position could add unwelcome heat to the oil if located too close to the oil tank.

Should that happen, the oil tank must be shielded from the muffler with an asbestos or Fiberfrax-backed metal baffle.

A typical installation, with its comparatively heavy muffler at the end of a long exhaust pipe, exerts quite a load on the exhaust flanges and studs. An installation like that is prone to develop cracks and weld failures. It is a good idea to install a couple of small straps to help take up some of the weight and to dampen the vibration. One end of the support strap is ordinarily attached to some point on the engine so that the vibration is contained within the shock mount frequency of the engine. Admittedly, finding a place on the engine to secure one of the straps or braces is not simple. On the other hand, if the straps were attached to some point on the engine mount or aircraft structure they might serve instead to aggravate the vibration problem. A possible exception could be in installations of the type shown in Figure 4 . Here the ball joint connection in the system is spring loaded and permits movement of the exhaust pipes without imposing stresses on the flanges and the welds. bracing the exhaust tail pipes against the aircraft structure in a rigid manner normally is not a practical solution unless that section of the exhaust system is free to move independently of the shock-mounted engine. To do this successfully, you would have to use both slip joints and ball joints in each tail pipe. Their location, too, would be very important.

One of those expensive ball joints referred to in the text. Note the spring loading and the downstream heat muff. If you look closely you will see that the edges of the heat muff are secured by a removable piano hinge wire.

Cross-over systems are particularly prone to developing cracks unless they are constructed with a similar high degree of flexibility.

Before undertaking the modification of a stock system, you might first assure yourself that the unit is in good condition, otherwise

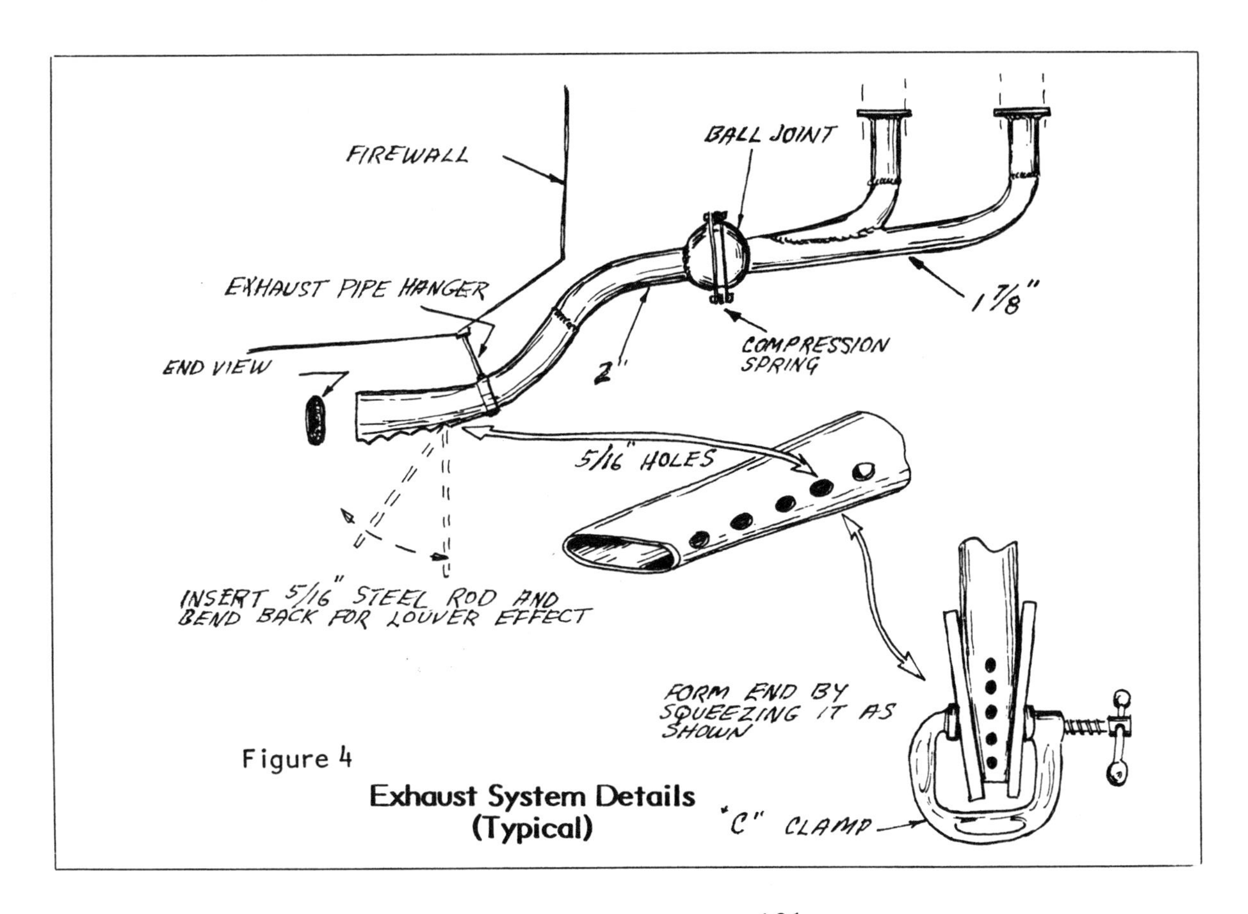

Figure 4
Exhaust System Details (Typical)

The intestine-like appearance of these exhaust pipes is an indication that the builder has worked hard to get each pipe the same length in order to obtain a tuned (balanced) exhaust system. The pay-off is increased efficiency and minimum back pressure.

there would be little purpose in going to that trouble and expense.

BUILDING WITH STOCK COMPONENTS

If you cannot acquire a good complete stock system to adapt, you might consider building up an installation using a couple of standard exhaust manifold components as your second easiest option. There are still plenty of Luscombe, Aeronca, Cub, and T-Craft exhaust stacks and mufflers to be had at competitive prices. Stock manifold units make a good starting point as they can be cut, or simply added to, to form a complete custom-built installation. Using this method will give you the advantage of a new system with a minimum of work and welding. The flanges would have already been properly fitted and welded and you would need only to add to the downstream portions to fit your airplane. Be sure, though, that the manifold components you purchase will fit your model engine.

EXHAUST PIPE FLANGES

Flanges provide a starting point or a foundation for the construction of the exhaust system and, typically, the entire weight of the exhaust pipes often hangs from them. It is no wonder, then, that the exhaust flange areas develop cracks and fail at a much higher rate than other less stressed areas of the exhaust system.

Some builders make their own exhaust flanges and weld up the entire system from new raw materials, but this is the hard way. Instead, I suggest that short stacks or stub exhaust stacks be purchased for a start simply because they already have flanges fitted and welded. If you prefer you can purchase the flanges alone.

If you are so inclined, you can, of course, make your own flanges from 3/16-inch or 1/4-inch mild steel plate but this is quite a chore unless you have access to machine shop equipment. I wouldn't recommend the exercise of cutting 1/4-inch steel stock with a hacksaw any more than I would suggest that you drill the holes in the flanges with a 1/2-inch bit and enlarge them with a rattail file. Of course, a good heavy duty fly-cutter and a slow drill press speed could provide the right size hole, but really, ready made flanges are not that expensive. You would have to buy the metal anyway....

The flange size is different for the Continental engines and the Lycomings so be sure to order what you need (normally 1 1/2-inch stacks are used for Continental A and C series engines, while 1 3/4-inch stacks are generally mated to most AVCO Lycoming engines). Actually, exhaust stack tubing can be had in diameters ranging up to 2 1/2-inch.

If you are building your exhaust system from the flanges on up (down?) you have two courses of action. You will need to decide whether you want to go to the trouble of laying out and making a welding asssembly jig, or whether you prefer to build the exhaust system directly on the engine.

If you elect to make and use a jig, you can obtain the layout position and spacing for the flanges directly from the engine manufacturer's specification sheet for the make and model engine you have.

Here is something to remember regarding laying out a jig. When viewing the specifications and drawings, you will be looking down on the engine and flange layout. But, when making your jig, the angles, as viewed for the flanges will appear reversed as the exhaust pipes really come out of the bottom of the engine....right?

If you are working with a jig (preferably a steel plate jig), bolt the exhaust flanges to the jig plate and tack-weld the stack sections together....occasionally removing your creation from the jig for trial fits on the engine.

Many builders will naturally think of using the engine as a jig for making their manifold sections. Although this is a common practice, the accepted rule is that the engine is used only to make the initial tack welded fit of the pipes. The main welding should be done somewhere else....in the clear. The flame from a welding torch can damage just about any part of the engine or its accessories if the builder is momentarily distracted.

EXHAUST PIPE BENDS

Exhaust pipe sections, both stainless and mild steel can be obtained in 45-degree and 90-degree bends. Automotive mild steel pipes may even be purchased in "U" bend angles. There is really no use in trying to bend either the autmotive pipes or the stainless steel exhaust pipes, at least not by heating them with a torch or using some other home-devised method.

Some builders might have access to a shop with a hydraulic tube bender that utilizes an I.D. mandrel. That's great! For the rest of us, though, purchasing the services of such a tube bender (if you can find one) will also be quite a pocket book bender. It may cost as much for each bend as it does for the pipe. In most instances, it is much cheaper for you to buy a ready made set of exhaust pipe sections or manifolds of aircraft quality than to obtain the raw material and pay for all the necessary bends (and welding).

Bent sections should have a generous radius, about three pipe diameters. Unless you can get pipes with the correct bend angles in them, you may have to resort to a few wedge cuts to obtain the proper overall angle. This, of course, requires more welded joints, unfortunately. As you know, the fewer welded joints, the more reliable the unit (and the less likely it will be to develop hot spots).

An exhaust problem peculiar to the VW engines is the necessity to curve the pipe rather sharply from its flange connection at the front exhaust port to provide safe clearance between it and the propeller. A sharp curve at this point may make it impossible to put a wrench on the inside nut for torquing it. The solution is to weld the exhaust pipe perpendicular to the flange even if or just a short distance ahead of the pipe curvature. (Figure 5)

OTHER DETAILS

After your exhaust system has been welded, check its flanges for flatness to forestall leaks at the gaskets. If the flanges aren't

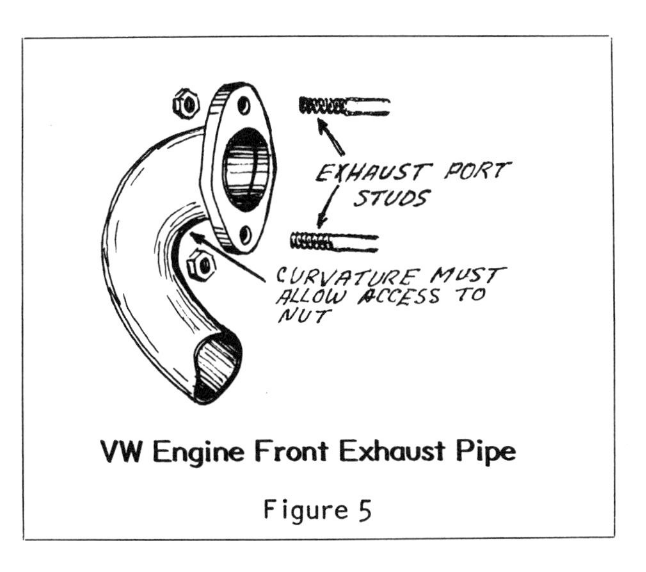

VW Engine Front Exhaust Pipe

Figure 5

smooth and flat, correct the difficulty on a bench mounted disc sander or by careful hand-filing.

Rubbing the flange over a piece of emery cloth backed by a smooth, hard surface will give you more accurate results than you can obtain by filing, provided the flange distoration isn't too severe. Sandpaper doesn't remove material very quickly and is suitable only for minor corrections.

The completed unit, if made of mild steel pipes, should be sandblasted to remove the welding scale and rust. As soon as possible after the sandblasting, the exhaust manifolds should be sprayed with a good grade of high temperature (1,200-degrees F) spray paint.

Incidentally, during the construction of your exhaust system, use a felt pen rather than a lead pencil. Lead pencil marks apparently are absorbed after the pipes get hot and cause a change in the metal, tending

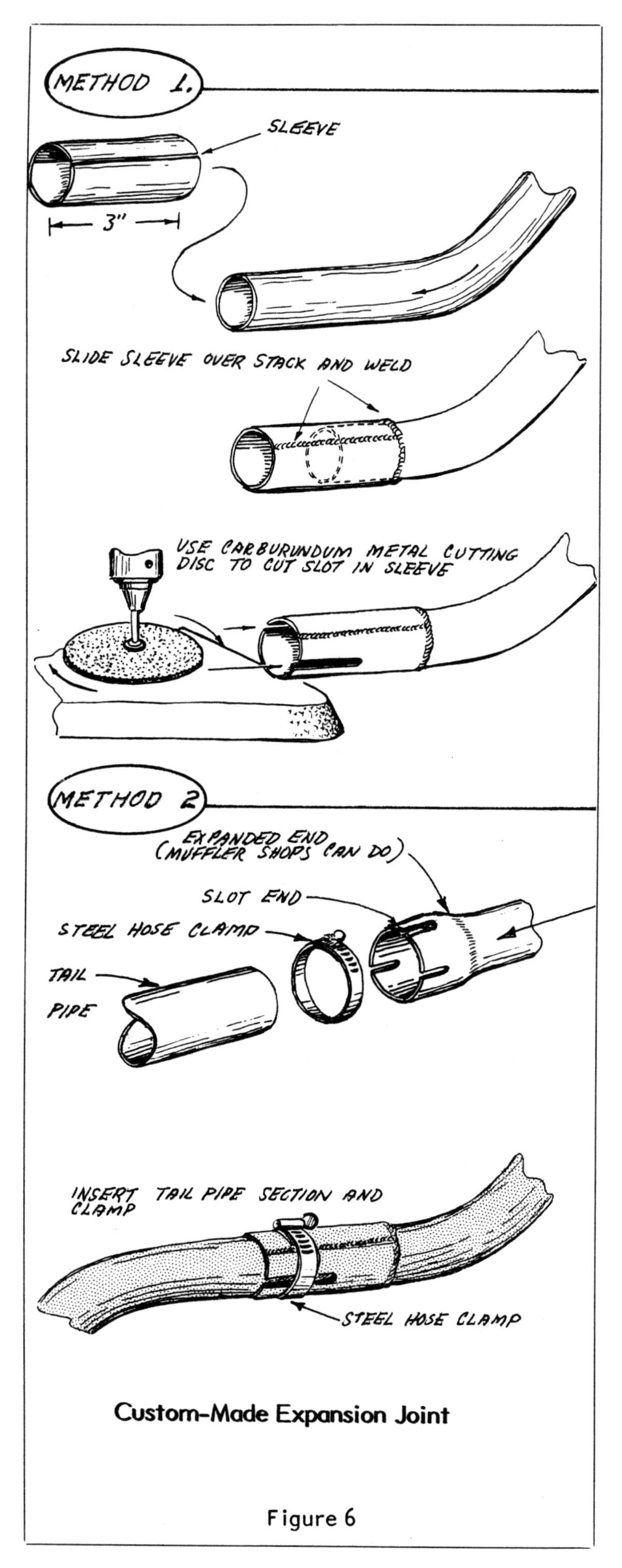

Custom-Made Expansion Joint

Figure 6

to soften it in the areas marked. This sort of thing reportedly can lead to cracks and failures. I understand that the use of zinc plated or galvanized tools could create the same effect wherever the plating of the tool is rubbed off onto the pipes. Although this appears to be a remote possiblity, it does make a good conversation piece and is interesting enough to note.

You should, of course, always use exhaust flange gaskets between the exhaust pipe flanges and the exhaust ports. Do not, however, install new copper gaskets repeatedly during the fabrication of your pipes as they will take on a permanent set. Because of slight differences between flanges, the final fit may not be as good as you would like to have it. Save them until you are all through with the fitting process.

BALL AND SLIP JOINTS

Long lengths of unsupported exhaust pipe invite severe stress, cracking and, possibly, complete failure. Uneven heating and cooling, coupled with expansion and contraction are the culprits in this instance. A good, long-lived exhaust system will usually be one that is fitted with strategically located ball joints and slip, or expansion joints. A study of existing exhaust systems in aircraft and in the photos throughout this book will give you a good idea where these assemblies might be located.

Standard ball joints are extremely expensive so builders tend to favor slip joints. If you are constructing your own exhaust system, you might be interested in a flexible ball-like joint available through Ken Brock Manufacturing, 11852 Western Ave., Stanton, CA 90680. The unit is both simple in design and inexpensive.

A couple of easy-to-make slip joints that merit your consideraton are illustrated in figures 6 and 7.

MANIFOLD NUTS

Use the correct type of exhaust manifold nuts for the installation of the stacks. These nuts, 5/16-24's (#22022) for the small Continental engines are made of brass. The Lycoming engines use the steel type nut, size 5/16-18 (#STD 1410). (NOTE: The use of improper type nuts could result in their seizure, making removal difficult or impossible.

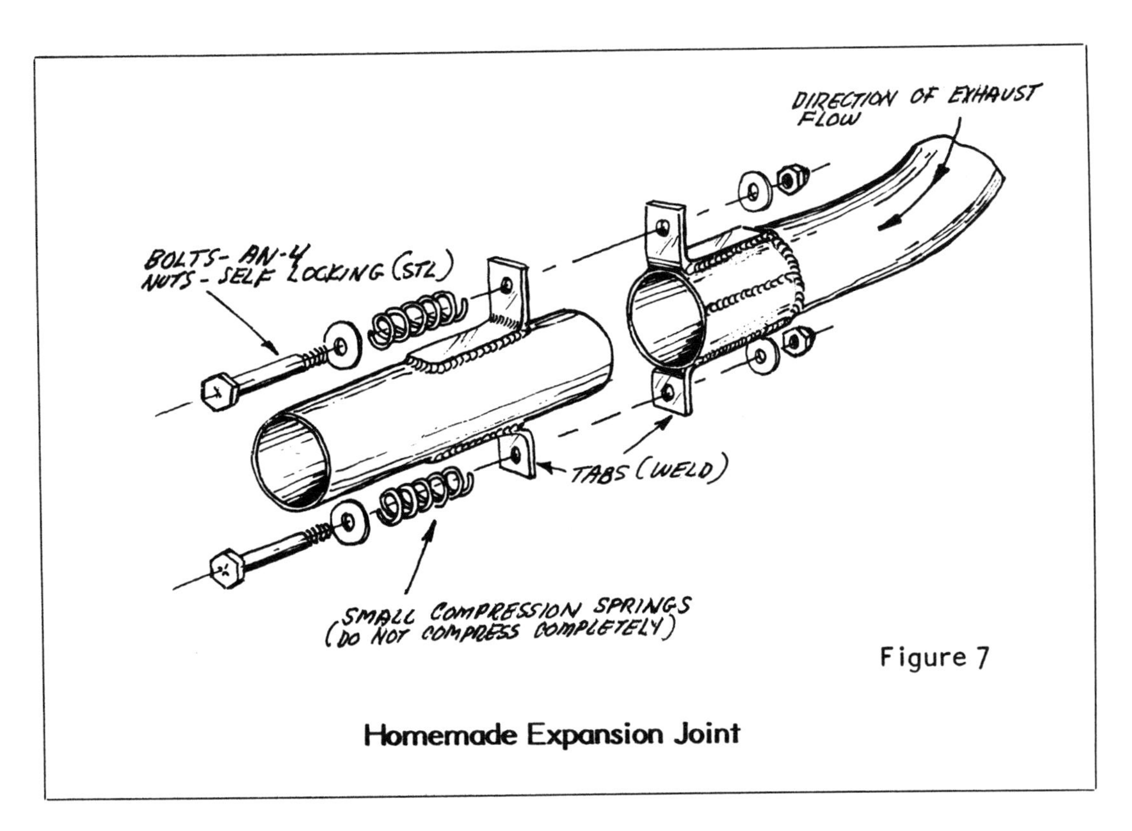

Figure 7

Homemade Expansion Joint

DUCTING

In the engine compartment, hot and cool air is usually conveyed to various units (carburetor heat, cabin heat, ventilation, cooling air on magnetos and generator, etc.) by "CAT" or "SCAT" Aeroduct flexible ducting. These ducting hoses are made of neoprene-impregnated fiberglass with internal reinforcing copper-steel wire. They also have an outside wrapping of fiberglass cord. CAT is black and SCAT is red. The difference is that the red is used where exposure to very high temperatures will be encountered, perticularly in ducting hot air to the carburetor. However, the black CAT ducting is good for temperatures up to 300 degrees F, which is plenty hot. Still, most FAA inspectors prefer that you use the red ducting because it can take temperatures up to 450 degrees F. Let that be your guide. Guess which type is the most expensive .... right, the SCAT stuff.

Ducting is easy to install. Cut it to length, allowing a little for flexing. The wire will have to be clipped with a pair of diagonals. Bend about 1/2-inch of the wire's end 90 degrees to relieve tension and to ensure that it can be trapped under the stainless steel hose clamp on the installation. This keeps the ducting from "unravelling" or getting a hole punched through it.

In a pinch, you can screw a couple of short pieces of ducting together and apply a silicon bead around the external seam.

Whever necessary, tie-wrap the ducting to a solid support to keep it from rubbing on the cowling or other parts in the engine compartment.

This large muffler and heat muff unit is matched by one just like it on the other side. These units are from a Cessna 150 and are very effective. The problem is that you need a deep, wide cowling to enclose the installation.

Exhaust Jet Effect

Getting rid of the exhaust gases represents a considerable waste of energy so it is not surprising that attempts are frequently made to utilize some of this energy. In its most elementary application, the exhaust pipe is constructed to provide a smooth, unrestricted path for the exhaust gases with its outlet pointed downstream to the slipstream. This will definitely give a little jet thrust to your airplane, however. A refinement of this jet reaction is possible by encircling the exhaust pipe with a shroud. A section of thin walled (.015-inch) stainless steel about three or four inches in diameter may be used. The ends are flared (don't ask me how you do it at home) somewhat as shown in Fig. 8. Stainless is very ductile and flaring should be rather easy for an imaginative guy to do (ha!). The idea behind this feature is to construct the shroud so as to cause the air to be accelerated past its opening using the principle of the venturi to increase both the velocity of the engine compartment air, and the exhaust gases leaving the stack.

Such an augmented system serves a dual function by accelerating the cooling air passing through the engine compartment and reducing the overall cooling drag, while, at the same time, taking advantage of the jet-like reaction of the exhaust gases.

You can increase the jet effect by constricting the exhaust tail pipe portion somewhat. You cannot, however, get something for nothing, so if you squeeze the pipe down too much, you may be creating excessive back pressures and will only be reducing the available power instead.

As I have previously suggested, an exhaust pipe can be squeezed to a fineness ratio of 5:3 with reasonable assurance that it will not cause adverse back pressure build up. In those installations where you plan to drill a series of holes in the side of the tail pipe, it may be possible to squeeze the pipes a bit flatter without a loss in efficiency.

Fig. 4 and the photo show how holes drilled in the ends of exhaust pipes can be made to have a louver-type of effect. The effective hole sizes used for this appear to vary from 5/16-inch to 1/2-inch in most installations inspected. It seems as though the builder's preference prevails.

In any event, the procedure for modifying the tail pipe in this manner is to insert a steel rod of the proper diameter and then sweep it backward toward the end of the pipe. Do this for each hole and the metal will be forced outward forming a very nice looking louver.

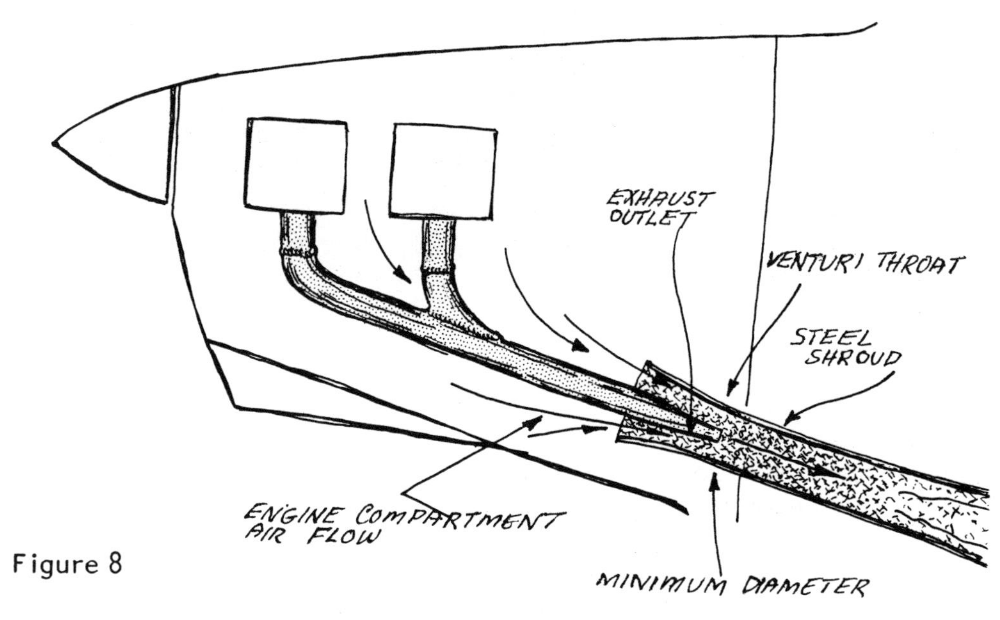

Figure 8

Exhaust Jet/Cooling System

An excellent job of making louvered tail pipe extensions. The installation could be cleaned up a bit by shortening the pipes and using standard stainless steel hose clamps to secure the extensions.

HEAT MUFFS (SHROUDS)

Heat muffs exist for two basic purposes. One is to provide hot air for cabin comfort and the other is to produce hot air for the carburetor intake when needed to prevent or eliminate carburetor icing. Normally, two muffs are installed in the typical engine installation.

Warm air for heating the cabin, or for warming your feet and ears in an open cockpit job, is obtained by encasing a muffler or a portion of an exhaust pipe within a metal shroud or muff, and causing air to circulate around that hot pipe or muffler. (Figure 9)

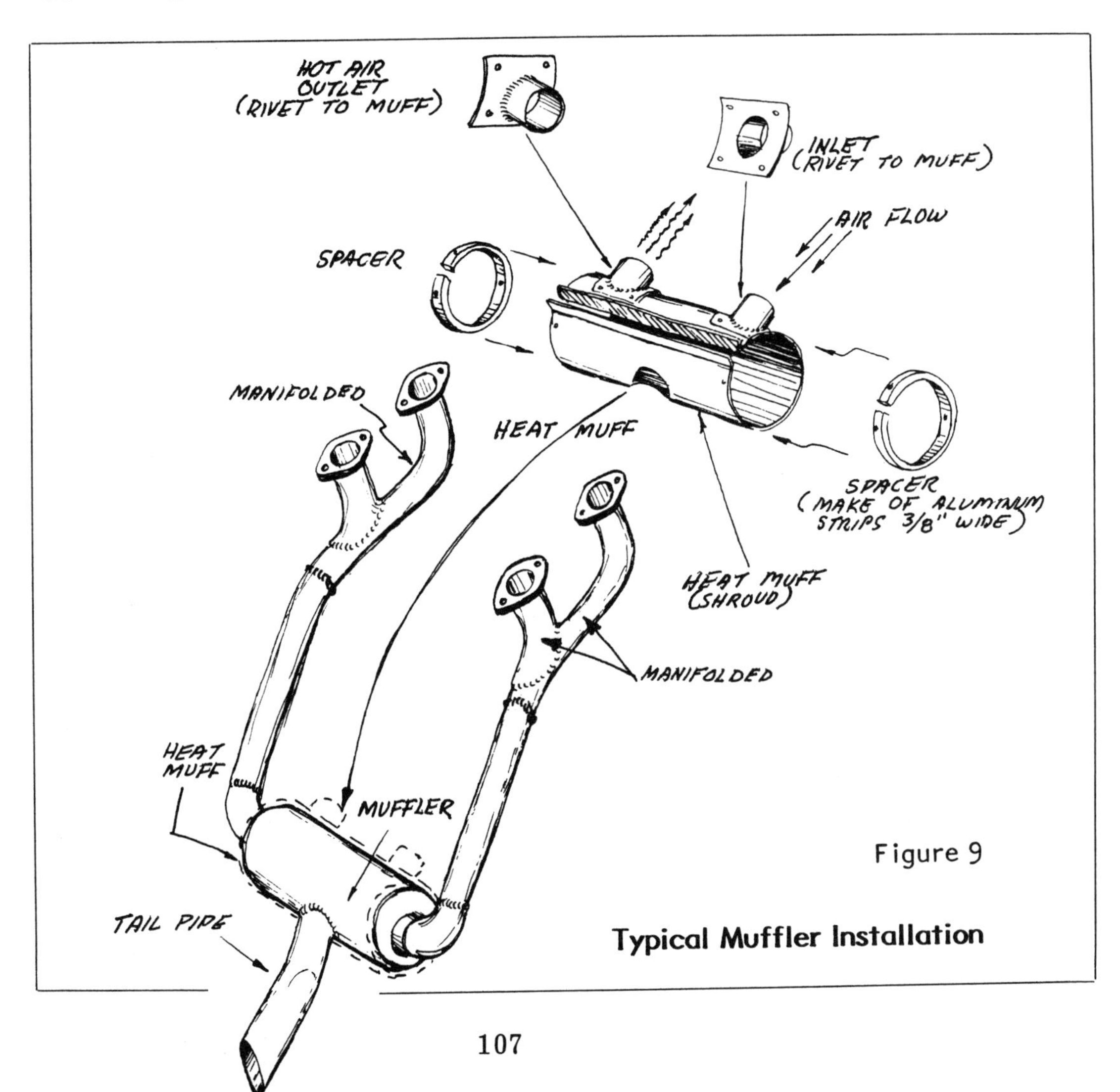

Figure 9

Typical Muffler Installation

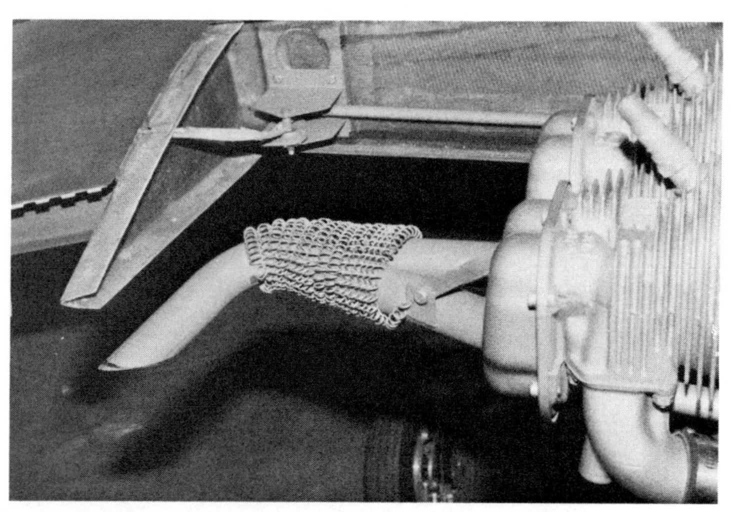

To increase the heat radiating area of a heat muff, secure one end of a screen door spring and stretch and wrap it around the exhaust pipe.

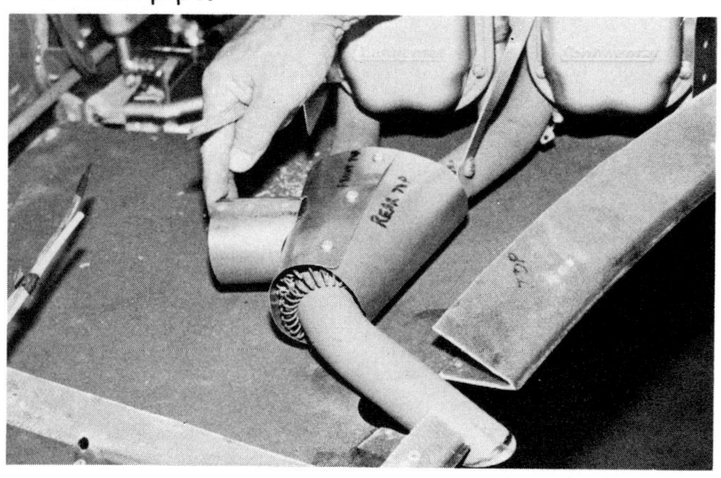

An aluminum wrap-around cover can be cut to fit and secured with sheet metal screws. A 2-inch diameter heat outlet flange can be fitted and pop riveted to the wrap-around cover.

The carburetor heat muff for this VariEze was connected to a remote carburetor heat "box" through a flexible 2-inch duct and was secured with steel clamps.

The shroud should be located as close as you can get it to the exhaust port where the temperatures are the highest. Allow a gap or a space of 1/2-inch to 5/8-inch between the exhaust pipe and its shroud covering.

You can make these wrap-around heat exchangers of .015-inch stainless steel or of .025-inch 2024 T3 aluminum. Leave one end of the muff open to provide an air inlet connection so that the air may be directed into the opening and around the hot pipe. Connect the downstream end of the muff to a flexible ducting (red) and the far end of it to the carburetor heat control box or to the cabin heat control box, depending on which hook up you are making.

The ducting to use is the Aeroproduct flexible wire supported duct impregnated with silicone rubber and capable of withstanding temperatures up to 450-degrees F. Although it is made for both industrial and aircraft use, your chances of finding it are much better in the catalogs catering to the homebuilt trade. I warn you that the stuff is very expensive in diameters under 1 1/2-inch, and very nearly out of reach for diameters larger than that. You should note that AeroDuct is also available in a black, neoprene-impregnated fiberglass capable of withstanding temperatures of from 65-degrees F up to 350-degrees F. It is acceptable for cabin and carburetor heat and is quite a bit less expensive. Occasionally you might run into an FAA inspector who believes the only ducting acceptable for cabin heat and carburetor heat is the red variety. If any doubt exists, you might check it out with your inspector. However, it is most unlikely that your hot air system will produce temperatures in excess of 350 degrees F.

The standard size air inlet and air outlet openings are two inches in diameter for most small aircraft but they can also be made to accept either a 2 1/2-inch or a 3 1/2-inch ducting. Actually, the ducting is available in a number of sizes ranging from one inch up to 4 inches in diameter. Smaller engines, such as the VW engine will require the smaller diameter ducts as the size of the ducting used should match the diameter of the carburetor heat box inlet. These are usually around 1 1/4-inch to 1 1/2-inch in diameter. And, last but not least, make your muffs removable for inspection purposes.

Although mufflers on VW engines are a rarity, these engines still require some sort of a muff or shroud to provide standby carburetor heat (or cabin heat) if a float type carburetor is installed.

With a little ingenuity it is even possible to fit a heat muff to a short exhaust stack.

Mounting a heat muff up front often poses a a problem in cowling some homebuilts because a small profile is desirable.

Those tiny muffs seen on some VW engines must be strictly for show as they are bound to be ineffective. The relatively short span of exhaust pipe available leaves very little space for a muff large enough to be effective without augmentation of some sort. It is for this reason, therefore, that VW engine users attempt to boost the amount of available carburetor heat by increasing the heat radiation area within the muff.

They do this by wrapping a coiled spring (similar to those used to close the screen door, at least in the olden days they did) around that portion of the exhaust pipe enclosed by the heat muff. Such a coiled spring transfers quite a bit of heat and improves the heat output to the carburetor (or cabin) considerably.

To provide even greater heat output, it is possible to use two springs tightly wrapped around the exhaust pipe.

Attachment of the springs is by means of a small tab welded to one end of that portion of the exhaust pipe that will be enclosed by the wrap-around shroud. If you prefer, you could also weld a tab to the far end. Each end of the wrapped spring is hooked to one of the tabs using the small hole drilled for this purpose. When only a single tab is used to retain the spring, the far end of the spring can be hooked to itself....a simpler method than providing a second tab. (Figure 10).

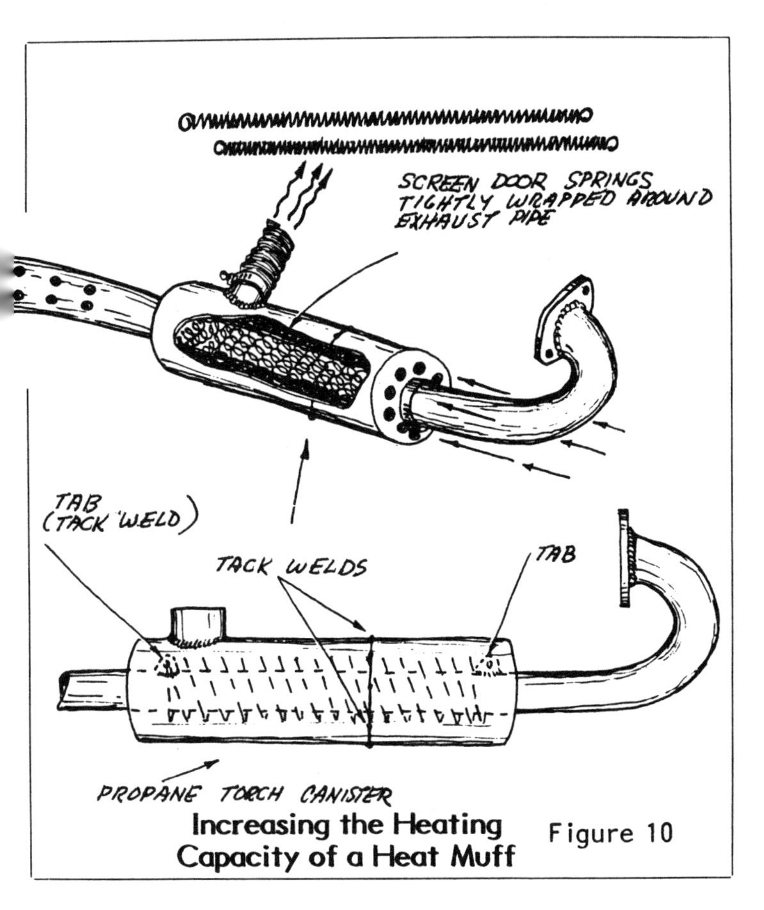

Increasing the Heating Capacity of a Heat Muff

Figure 10

Although most wrap-around heat muffs are made of .025-inch aluminum, one of the neatest looking muffs I've ever seen was one made from empty propane torch cylinders. You know, the small gas torch cylinders sold through hardware stores and used by the do-it-yourself handy man the world over. Needless to add, I suppose, is a reminder that anyone intending to use such a container must exercise caution and be sure that the cylinder is indeed empty before any welding is done on it.

Two such containers can be joined together to obtain whatever length muff you need, simply by tack welding them together. The completed muff makes a professional looking installation with a minimum of work. It will be fairly light, and when properly installed, will really clean off the VW intake manifold quickly when moisture and frost appear.

The Sound and the Fury

The noise you associate with exhaust is due to the rapid expansion of air during combustion and to the turbulence of its high speed exit from the engine. Smooth out the combustion pulsations, reduce the turbulence of the exiting gases, and you will have succeeded in quieting the exhaust noises. The commonly accepted means for reducing the noise of engine exhaust is through the installation of one or more mufflers.

A muffler works by providing a large chamber in which the sound of combustion and turbulence are reduced. The quality of the muffling, therefore, depends on the volume of this chamber and on the downstream resistance the gases encounter. Unfortunately, the muffler chamber increases the downstream resistance, resulting in an undesirable increase in combustion chamber back pressure and a reduction in power availability.

The addition of a muffling system on your aircraft, therefore, is likely to result in a loss of performance and an increase in cost, weight and maintenance. Unfortunately, while it is often a negligible factor in larger, more powerful homebuilts, it has been an unacceptable burden for low or underpowered aircraft. It is unfortunate but true, any significant loss of available power in lightweight aircraft powered by a small engine is considered as unacceptable. For this reason, you will only rarely find a muffler installation in a homebuilt powered by an engine of less than 65 hp. Mufflers are particularly rare in VW engine installations.

The situation is changing, however. Ultralight aircraft are appearing with a variety of clever lightweight muffling devices for their angry, hornet-like, high revving engines. Much to their credit, Ultralight aircraft builders are trying hard to muffle the shrill whine of their low-powered engines.

As mentioned, in order to muffle engine noise you must somehow smooth out the flow of the turbulent gases, or at least, modify the ordinary exhaust pulsations. Longer pipes do, of course, help a bit. So do some of the modifications of the tail pipe. Some builders add unusual treatments to the ends of the tail pipes in order to effect some alteration in the exhaust pulsations.

One such variation is to flatten the last six to eight inches of the tail pipe and then drill random holes into it. The effect of the flattened pipe is to help accelerate the flow of exhaust gases. This type of tail pipe modification is standard on the VW-powered Fournier (a French-built motor sailplane). Furthermore,

The popular Cessna 150 muffler and shroud (heat muff) used by many homebuilders. One mounts on each side of the engine. Unfortunately, such an installation often prevents the installation of a close-fitting cowling.

with the pipes pointed straight back, a little bit of jet assist is also realized.

Although these tail pipe designs do help to alter the noise patterns, they are not true mufflers that significantly suppress or reduce engine combustion noises. It is interesting to note that the tail pipe modification is most effective when both pipes exit at the same point and are as close together as possible.

HOW ARE THEY PUT TOGETHER?

The standard aircraft muffler consists of a tube of some specific length and diameter. It is usually a rather large stainless steel tube and may or may not have built-in internal baffles. The practice has long been to construct aircraft mufflers with baffles and with the inlet and outlet tubes staggered. In some mufflers the exhaust gases are forced to change direction drastically in their pell-mell trip to the free atmosphere.

To construct your own muffler with similar baffles built in, is a chore of questionable value unless you have some relentless urge to experiment and to induce forced labor. The practical thing to do would be to omit the baffles completely as they usually are the first area of failure in the exhaust system. Burned, disintegrated or collapsed baffles in standard category aircraft often obstruct the flow of exhaust gases and have been known to cause serious loss of power or engine failure.

Mufflers are most effective when they are located as close to the exhaust outlet of the cylinder as it is practical to place them. The Cessna 150 provides an example where this concept is effectively practiced. Its individual mufflers are hung on each side of the engine on short exhaust risers. In some Piper models, too, the cross-over system mufflers are integrated as far upstream as possible.

As for muffler size, the Cessna muffler is about 4-inches by 10-inches. (It looks bigger with the shroud around it.) These are really the minimum size for the muffling job at hand. Actually, a muffler measuring 6-inches by 10-inches would be much more effective. With due regard for the extra weight and lack of space under the cowling, your own muffler or mufflers should be as large as possible....something on the order of at least 270 cubic inches if at all possible.

Forget about internal baffles as their usefulness is questionable when everything is considered. Instead, make the diameter as large as you can manage. A good length of tail pipe downstream from the muffler will increase its effectiveness noticeably and will assist in smoothing out the pulsations of the exhaust gases.

The expense of good aircraft mufflers tends to keep them off most homebuilts. However, any builder can incorporate a standard aircraft muffler into his exhaust system with very little difficulty.

Ultralight builders are developing a variety of muffler-like tail pipes. These are proving quite effective and perhaps more homebuilders will find them suitable installations for conventional aircraft. One thing that most homebuilders will not do is stick the tail pipes straight up like a farm tractor even if that does direct the exhaust noise away from the populace below. It may be all right for an ultra-slow, ultralight, but a bit out of character for the racy little homebuilts.

BUILDING YOUR OWN

I think you must be aware that attempting to build your own open chamber muffler would be more difficult than fabricating the basic exhaust pipe portion of the system. And most of us simply would not undertake the construction of a muffler for that very reason. The FAA, in its Inspection Aids has regularly reported frequent incidents arising from failures occurring in and around mufflers....at the point where the tail pipe attaches to them. Thus, the construction of a hand-made muffler would have to be of high quality in order to be free of such difficulties.

THE SWISS MUFFLER

As you may know, Sweden and Switzerland were among the first two countries in the world to impose strict — nay drastic —limitations against all noise pollution. Switzerland, not only set low levels, they have enforced them since January, 1974. As a result of this action, a number of light aircraft, including some type-certificated models, were grounded, (I repeat, GROUNDED!) until their owners did something about noise! Oh, the officials weren't prejudiced about it because the action affected motor bikes and trucks as well.

Needless to say, the homebuilders were motivated and quickly came up with an unusual muffler which seems to have become the standard Swiss homebuilt muffler.(Figure 11)

Essentially, it consists of a length of stainless steel mesh rolled into a tubular shape. The inside diameter of this tube must

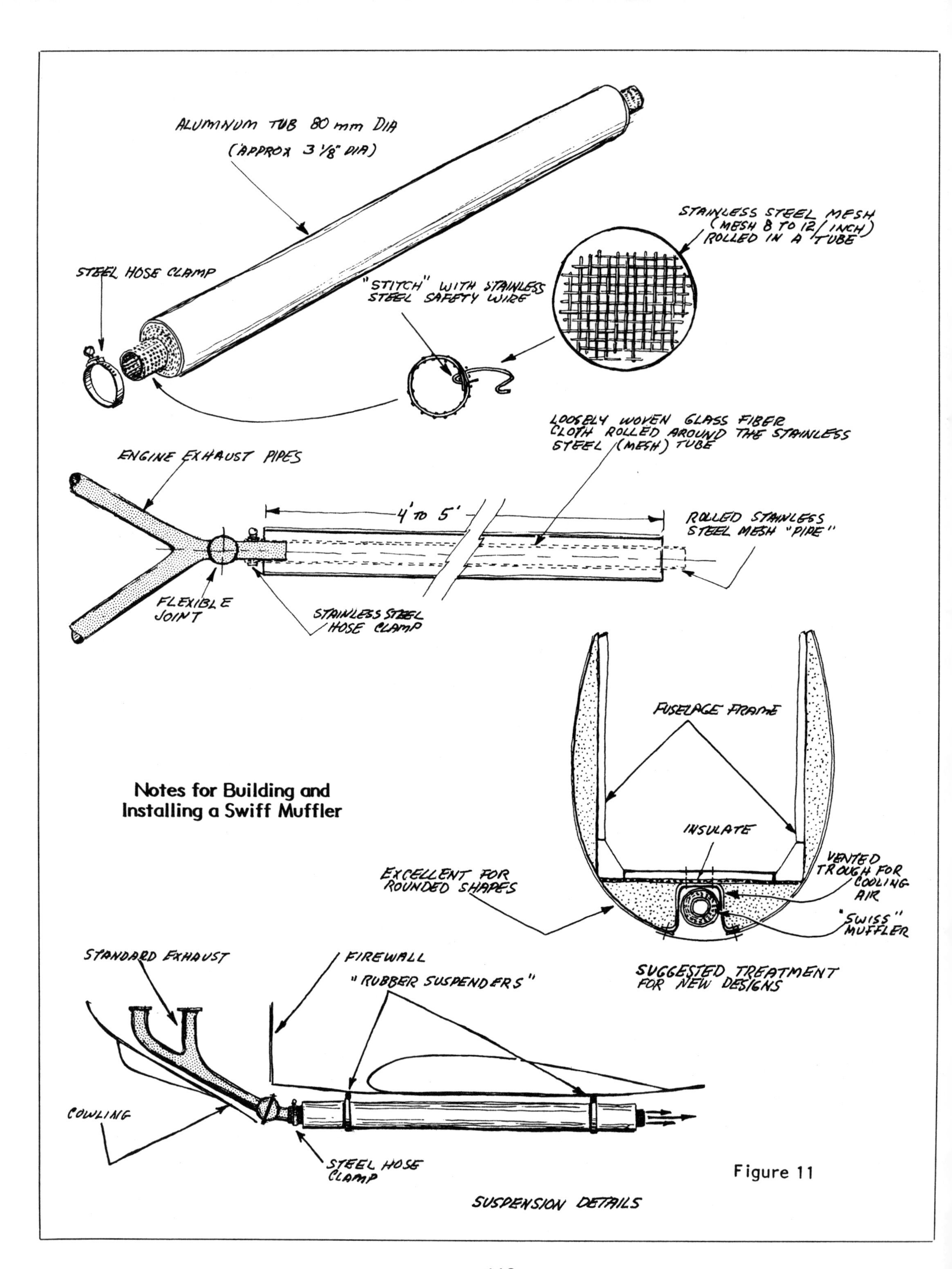
ALUMINUM TUB 80 mm DIA
(APPROX 3 1/8" DIA)
STAINLESS STEEL MESH
(MESH 8 TO 12/INCH)
ROLLED IN A TUBE
STEEL HOSE CLAMP
"STITCH" WITH STAINLESS
STEEL SAFETY WIRE
LOOSELY WOVEN GLASS FIBER
CLOTH ROLLED AROUND THE STAINLESS
STEEL (MESH) TUBE
ENGINE EXHAUST PIPES
4' TO 5'
ROLLED STAINLESS
STEEL MESH "PIPE"
FLEXIBLE
JOINT
STAINLESS STEEL
HOSE CLAMP
FUSELAGE FRAME
Notes for Building and
Installing a Swiff Muffler
INSULATE
VENTED
TROUGH FOR
COOLING
AIR
EXCELLENT FOR
ROUNDED SHAPES
"SWISS"
MUFFLER
STANDARD EXHAUST
FIREWALL
"RUBBER SUSPENDERS"
SUGGESTED TREATMENT
FOR NEW DESIGNS
COWLING
STEEL HOSE
CLAMP
SUSPENSION DETAILS

Figure 11

be large enough to slip over the outside diameter of the engine's exhaust pipe. The length of the hand-made stainless "mesh" tube varies in length with the builder and, perhaps, the engine used. It seems, however, that most of them are around four to five feet in length. That's right, four to five feet long! This is for 65 to 85 hp aircraft like most of the European two-seaters, the Jodels, Emeraudes, Minicabs, etc.

Next, loosely woven cloth, or what we call roving, is rolled around the 4-to-five-foot-long stainless steel mesh tube and its diameter is built up until it becomes a tight push-fit into a seamless, 80mm diameter aluminum tube. (This would be approximately 3 1/8-inches in diameter, if you don't have a metric scale). Do not use fiberglass mat. It will probably shred away and be reduced to dust by vibration after a very limited use. The aluminum tube is apparently readily available in Europe as this sort of tube just happens to be the standard vertical rainwater downspout found on houses. It is about 0.5mm thick and can be obtained from building supply stores in Europe. Any similar light-wall, large-diameter tube could be used. Here in the U.S. about the only source for such a large diameter aluminum would be companies producing telescopes and telescope kits.

The whole assembly is then slung underneath the fuselage and clamped onto the single exhaust outlet with a steel clamp. The muffler is supported in at least two places along its length with steel clamps and rubber suspenders of a sort.

Way out, you say? You may think this is ungainly, heavy and draggy, but perhaps we can accept the word of our European reporter who says the results are outstanding. You no longer hear the exhaust....just the prop and the slipstream. It is reported that flying takes on a new dimension and taxiing on crowded ramps becomes a dangerous adventure unless you have a wing walker to 'shoo' people away.

This muffler is certainly inexpensive and easy to make and to maintain. Although its useful life is as yet unknown, inspection and replacement, even if a bit more frequent than for standard aircraft stainless steel mufflers, would certainly be no deterrent to its acceptance. This muffler system should not generate any power-robbing back pressure. It can be added to any existing design, although the aesthetic qualities might be subject to a few quaint remarks. Owners of replica warbirds could make theirs larger in diameter and pass them off as bombs or torpedos. Most important, however, these mufflers do the job simply and permit the Swiss builders to continue flying in the face of restrictive governmental limitations.

The original Swiss muffler. It looks impressive and the muffling results deserve merit. The entire unit could be hidden from view (and drag) in a tunnel under the fuselage.

On further reflection, there is no reason why the muffler could not be concealed in a sort of built-in trough underneath the fuselage ....at least it's an idea worth trying.

The typical exhaust system is rather easy to construct and install if you do not have to install mufflers and heat muffs. A system complete with a muffler and heat exchangers will be a bit more difficult to design and to construct properly. It may even be more troublesome to maintain, but the results are worth it....a quiet airplane.

A few known sources of materials are: CPI Sales, P.O. Box 14149, Houston, Texas 77021; Ludlow Saylor, 8474 Delport Drive, St. Louis, Missouri 63114 — for stainless mesh (cloth); and A. Jaegers Optics, Lynebrook, New York, New York 11563 —for the aluminum tubes. Anyone reading this book years after it is published may have to find other sources as these folks may have retired or moved off to one of the planets.

Welding Stainless Steel

Your stainless steel exhaust components may be welded with either an electric arc or an oxy-acetylene flame, but of course, a heliarc does the finest job.

Anyone using an electric arc welder should obtain flux coated rods to use for the arc surrounding the hot metal in the weld arc as the rod is being deposited.

As for oxy-acetylene welding, I would assume that the majority of builders will have, at one time or another, tried their hand at it. However, as proficient as they may be in welding mild steel or 4130 steel, the welding of stainless is something quite different. Still, if you have ever done any welding at all, you should try welding stainless at least once.

Serviceable (not pretty, but serviceable) welds can be made in stainless, believe it or not, by almost anyone who can use an oxy-acetylene outfit. It is not as difficult or as tricky as welding aluminum. Nevertheless, do not start on your stacks and expect to produce nice looking beads if you have never tried welding thin wall stainless pipe before. Usually one's first beads look like a disorganized string of abandoned raisins. Don't despair. Once you have a better understanding of the characteristics and behavior of stainless during welding, you will be able to achieve acceptable results with a little practice.

Welding stainless requires a certain preparation of the work before you can start welding and it requires a certain technique during welding.

First, obtain some 1/16-inch stainless steel rods and a small jar of flux for welding 18-8 stainless steel. The numbers have reference to the chemical composition of the steel, which is made up primarily of chromium, nickel, carbon, silicon and manganese. The stainless melts at about 2,500 degrees F to 2,679 degrees F, about the same range as that of 4130 steel, yet there is something different about stainless steel.

Stainless dissipates heat only 40% as rapidly as 4130. This means that not as much heat is required to do the welding and it is necessary to use a smaller flame and tip than you would expect. Because the heat is not diffused by the surrounding metal there is a good chance that you might inadvertently burn a hole in the thin exhaust pipe material before you realize it. So, be warned, use a small tip, about one size smaller than you would use in welding 4130.

Stainless steel or any metal containing a lot of chromium will immediately start to oxidize if heated with a flame. An excess mixture of oxygen just makes things worse. To avoid oxidizing the metal it is advisable to use a neutral flame. Since most welding regulators will not hold a precise gas mix, it is better to adjust the torch so that a slight excess of acetylene is visible. This would be when a fine feather of acetylene shows about 1/16-inch around the inner flame cone. This is just about right because an unwanted change to an oxidizing flame can be easily detected and readjusted. Here are two types of improper flame adjustment and their effects: (Figure 12)

1. Too much oxygen — oxidizes the molten metal, making the weld porous and interfering with adhesion.

2. Too much acetylene — reduces corrosion characteristics of the stainless and tends to make the weld brittle as the acetylene takes up the excess free carbon.

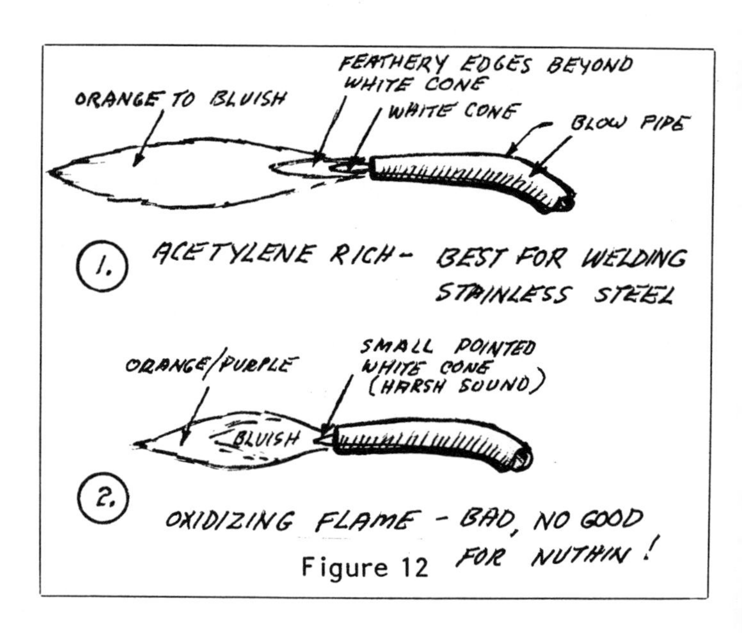

Figure 12

During the welding process, stainless steel must be protected from the air, otherwise, the oxygen and nitrogen present in the atmosphere will combine with the hot metal and adversely affect the adhesion of the weld. You can obtain this protection by using that flux especially formulated for welding stainless to dissolve the chromium oxide which forms on the molten metal. This flux is a white powder which has to be mixed with water (or alcohol, depending on its type) to a paste-like consistency.

Always brush the flux to the underside of the joint and allow it to dry before starting to weld. It helps also, to apply flux to the welding rod. Applying flux to the top side of the metal is not particularly important as the flame will protect it. Actually, not using flux on the top surface of the metal being welded may even make it easier for you to see the color of the hot metal by eliminating some of the glare caused by melted flux.

Excessive heat and the inadvertent development of a large molten puddle could combine to intensify oxidation to a point where even the flux cannot counteract the oxide. So, while the primary purpose of the flux is to conteract the tendency for the stainless to oxidize under the heat of welding, it is not a panacea.

WELDING PROCEDURE

Welding a decent bead around an exhaust stack requires a good initial fit of the two parts. Try for a fit that permits the parts to match within 1/16-inch of each other all the way around.

The two parts should be brightened with emery cloth. Get the edges and weld area clean because this is more important when welding stainless than when welding 4130 steel.

Apply the welding flux to the freshly cleaned area and, perhaps, to the welding rod, too, if you are so inclined. Jig up or clamp your work so that you can weld downhill. This permits the flux to dissolve and flow along with your weld, providing greater protection against oxidation during the welding ritual.

The classical procedure is to hold the filler rod just ahead of the flame so that it melts at the same time as the metal at the seam becomes molten. The texts had me believing that the technique to use was the forehand method but this "screwdriver and hammer mechanic" can't seem to do anything with that method except burn holes (not very good ones at that). I would suggest that the reverse technique might also be more successful for you, if you are not an experienced stainless welder. And, while you are developing your own technique, you might try tilting the torch more than the recommended 80 degrees and pull it back slightly at frequent intervals to keep a close check on the degree of heat evident at the weld.

Directing the flame back over the completed portion may help you keep from burning through unexpectedly. Thin wall pipes melt through quickly so vigilance is in order. Try slipping a smaller tube (as a back-up) into the pipe being welded. This will act as a leveler for the weld bead and will help prevent welding sags on the underside of the weld. Although it will also serve as a sort of heat sink, it is unlikely that it will interfere with obtaining good penetration in the relatively thin exhaust pipe material.

Well, so much for background information. Why not try your hand at welding some thin wall stainless for practice?

To get started, begin by tacking two pieces together with very small tack welds in about three places around the pipes. Check your welding tip to see that it is in good condition and that the flame doesn't come out cockeyed or forked. It is hard enough to make an acceptable weld when everything is just right anyway.

Use a somewhat smaller flame than you would for mild steel, but one big enough to bring the metal up to welding temperatures without too much waiting. Fill the joint completely and do not move on until the deposited filler from the rod diffuses. Once you start a seam, finish the entire bead without interruption. If for any reason you find you have to stop, reheat the entire weld area (within two inches) to a red hot color before beginning to proceed with the weld again.

The inner cone flame should be directed so that it just about touches the molten area. If you don't hold the flame until the edges of the filler rod are feathered (melted) there will be a tendency for your weld bead to be piled high and you will have a narrow bead with poor penetration. Apply the rod sparingly.

While some good welders say it isn't necessary to anneal a weldment in stainless, the heat of welding does reduce the corrosion resistance in the weld areas. This deficiency can only be corrected by uniformly heating the weld area to 1,900 degrees F to 2,000 degrees F (sort of a lemon color) and allowing it to cool rapidly. Actually, the thin walled pipes are cooled rapidly enough in still air without quenching.

There really is not too much welding required in the construction of an exhaust system....perhaps six joints or so. This might raise the question of how big an effort you should make in learning to cope with the welding peculiarities of stainless. However, since our motivation is 50% educational, most of us are willing and determined to try to do as much as possible ourselves.

You could always take the exhaust pipes to have them heliarc welded commercially. One problem here is that you will have to jig the whole system and will not be able to check alignment and clearances until after you get the pipes back. Then what if they don't fit?

MILD STEEL ROD FOR STAINLESS?

How about using a mild steel rod for welding stainless? It is being done you know. One problem with using a mild steel rod, however, is that it is not corrosion resistant. Another is that it has a markedly different coefficient of expansion. Stainless conducts heat only one-third as quickly as ordinary steel, while it will expand about one and one-half times as much as 4130. This could create some severe stresses in a weld. However, this effect may be minimized in an exhaust pipe weld by making the weld in one continuous run completely around the pipe. Another drawback, according to some sources, is the reduced corrosion resistance of the mild steel weldment. Funny thing though, it doesn't seem to be a serious matter in service and the welding of stainless with a mild steel rod has worked well for others. Besides, the appearance of the welds is much more appealing to one's aesthetic senses.

CHECKING YOUR WELDS

After all welding is completed, you should check your seams for tiny holes and leaks that you may not have noticed while welding. Do this by directing the torch flame inside the pipe and moving it around the weld

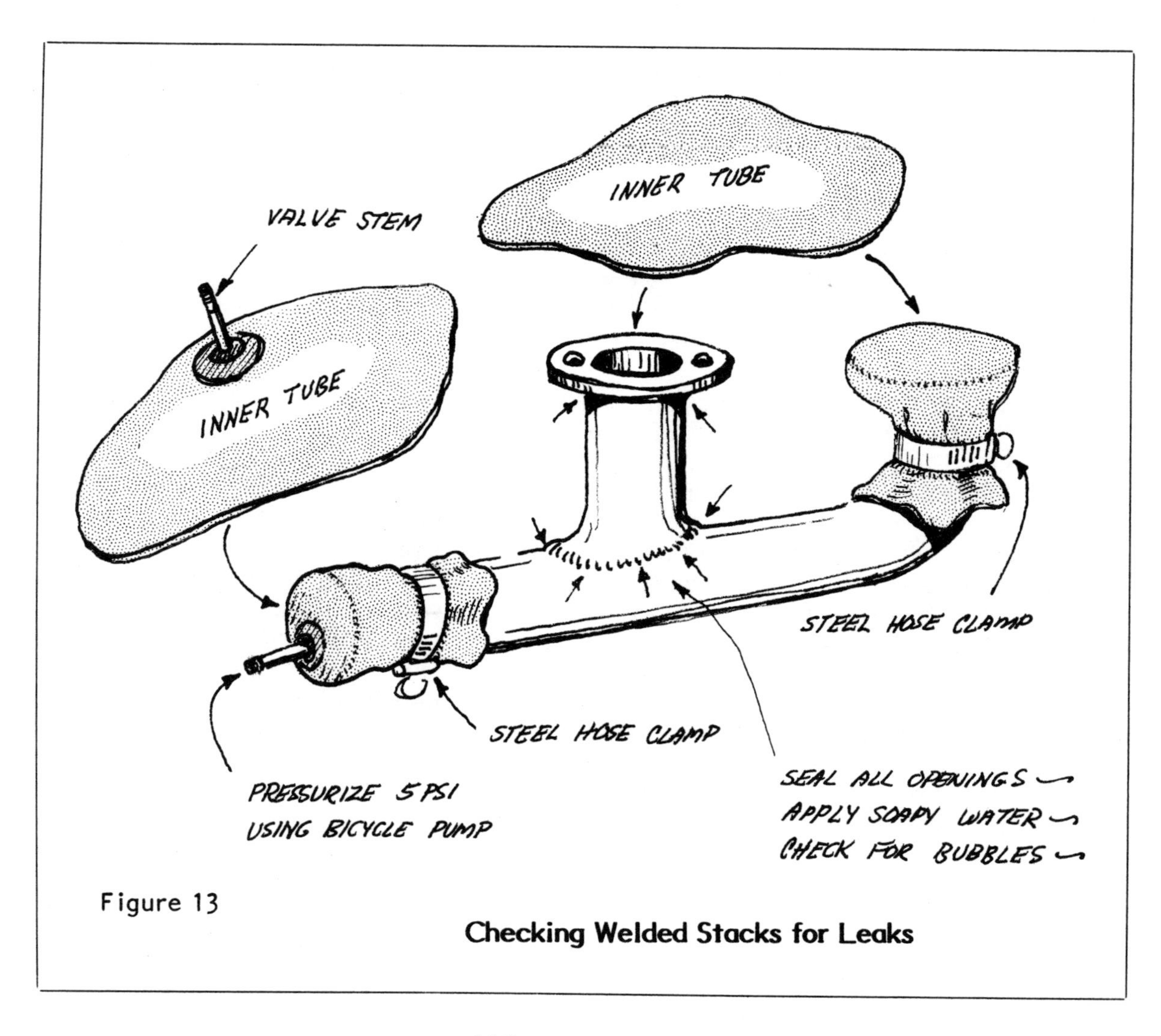

Figure 13

Checking Welded Stacks for Leaks

areas while you watch for sparks or signs of light around the outside of the welded bead. I suppose you could do it on the outside and look in, too. However, both of these methods would be difficult to do effectively if the exhaust pipe section were long.

You could always give it the 'pressure test'. This will certainly show up all poor welds and fits, but it is a pesky, time-consuming process. To perform a pressure test, you will have to make and install cover plates with gaskets of some sort to block off all openings. Then you will need to pressurize this sealed exhaust system with no more than 5-psi of air. Next, apply a mixture of liquid detergent and water (two tablespoons to a cup of water) to all joints and watch for bubbles. (Figure 13)

The louvered, squeezed-end treatment is also being used with VW engines. Exiting both exhausts close to each other has a modifying effect on exhaust noise.

cowls & cooling

6

Cooling by Cowling

Many people still think a cowling is used only to streamline an airplane and to make it look good. Years ago, this is just about all the cowling did for an airplane. It rounded off the nose and hid the clutter of the engine compartment.

Although the early builders using air-cooled radials and small 4-cylinder air-cooled aircraft engines cowled them, they usually allowed the finned cylinders to stick out into the slipstream where they would be, in theory at least, properly cooled. Actually, this type of cowling, while imparting a quaint look to the airplane, did very little for drag reduction or efficient cooling.

In time, what may best be described as an internal-flow cooling concept (also called pressure cowl cooling, ram air cooling, etc.) emerged. It changed the face of airplanes and the familiar protruding cylinders disappeared from view....along with a lot of parasite drag.

With the so called ram air or pressure cowl type of cooling system, a little air is made to do a lot of cooling rather than a lot of air cooling but little, and doing it unevenly at that.

The ram air internal-flow cooling installation commonly used in light aircraft consists of several identifiable areas. They are:

1. The cowling
2. The air inlet areas
3. The pressure chamber or plenum
4. The baffles
5. The expansion chamber
6. The outlet area

These cooling system areas (in most installations) function as follows: Cooling air inlet openings up front in the nose of the cowling allow air to enter into the upper portion (pressure chamber) of the engine. Internal metal baffles and the roof of the cowling, which together form sort of a four-sided barrier around the upper engine area, intervene and the cool air entering the compartment is directed down through the cooling fins of the engine cylinders. Additional pieces of metal baffling force the air to curve closely around all the cylinders, further enhancing the effectiveness

If streamlining is no problem, you can let it all hang out. A cowling as simple as this one can be made of sheet aluminum and it has no compound curves to complicate matters.

Most homebuilts have air inlet openings that are far larger than necessary for efficient, low-drag cooling. This one is not guilty of such excess.

The great potential for drag reduction is illustrated in this cowling designed for updraft cooling.

of the cooling flow.

By the time the cooling air has found its way down through the engine it has picked up heat given off by the hot cylinders and the hot exhaust pipes and has expanded. It has also lost much of its velocity in the large cavern-like area (expansion chamber) beneath the engine and, therefore, needs a well designed outlet to expedite its departure. If the opening for the exiting air is no larger than the inlet opening, the departure of the air will most likely be slowed and so will the flow of fresh air through the engine, adversely affecting its cooling.

COWLING INLETS

The lack of data regarding the design and the cooling effectivenes of various kinds of cowling inlets is obvious. A look around any flight line will convince you that there is a lack of standardization among the aircraft parked there.

The different treatments given each of their inlets can only imply that any kind of inlet will do, or that since a nice big opening up front will let plenty of cooling air inside, nothing else matters. Well, that's wrong. Some experts say that fully 10% of the total power produced by the engine is absorbed by drag generated in getting the air through the engine compartment. Others quote even higher percentages. If all of that entering air (and drag) is not actually cooling the engine, some of it is being wasted, and, so is the engine power that is used to push that excess air through the engine compartment.

The size and location of the cowling inlets depend upon the amount of air required to cool the engine at a given power setting, the velocity of the air, and the resistance imposed by the internal baffles to the passage of that cooling air. These factors vary so much with differing conditions of flight that a simple calculation of inlet location and design by the individual builder is impractical. In fact, it is the kind of design problem that can keep an aeronautical engineer and his interactive graphics computer occupied for days.

Obviously, the average homebuilder does not approach this matter with the same mathematical compulsion as the engineer. If he is designing and building a cowling for his own airplane, he will probably be concerned with its overall appearance long before he even gives a thought to the location, shape or size of the air inlets.

Any builder who intends to alter the shape of a ready-made cowling designed by somebody else should at least attempt to stay with the location and, possibly the size, of the inlet openings originally provided for the cowling. The same thought should apply to the size and shape of the air outlet area.

A few specifics regarding air inlets may be helpful....not as design criteria but rather as aids to better understanding of air inlets in general.

Air inlet flow is most likely to be disrupted during high angle of attack climbs when the air tends to accelerate over the upper cowl as though it were an airfoil. During this flight condition more air tends to flow across the

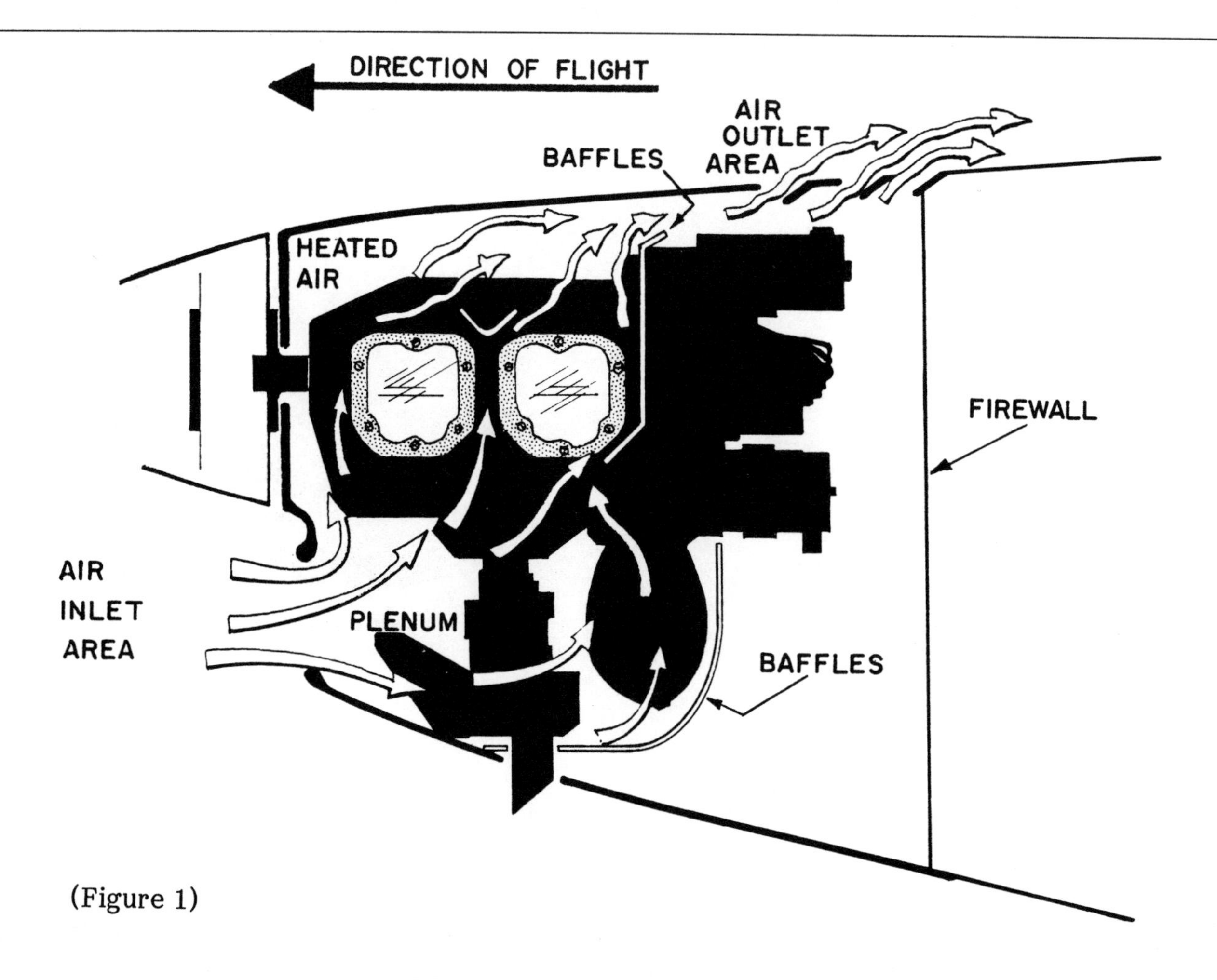

(Figure 1)

Up-Flow Cooling

Every once in awhile a builder will get the idea that it would be clever to cool his engine with cooling air that enters at some low point in the cowling, flows up through the engine and exits at the top of the cowling through louvers just ahead of the windshield. After all, he reasons, hot air rises so wouldn't it be easier to allow it to flow in that direction? Then too, isn't there a relatively high pressure area along the bottom of the fuselage which makes it more difficult to expel low velocity hot air there?

In general, all that is true, but....the aircraft engines we use, Continentals and Lycomings mostly, were designed for down-flow cooling. With these engines, the builder will find that, if cooling air is routed upward, it will have already passed around the hot exhaust pipes, which heat the air considerably before it ever reaches the engine cylinder barrels and cooling fins. This premature warming of the air is a severe handicap to overcome. Also, outlets for the air in the top of the cowling introduce the potential for an oil-spattered windshield anytime the slightest oil leak occurs. This could be dangerous unless the pilot is real good at 'blind flying'.

A few notable examples of up-flow cooling may occasionally be seen in the old Navions and Swifts still flying, but the manufacturers, it seems, are not particularly interested in the arrangement. (Piper does have one in production now, however.)

A few one-of-a-kind homebuilts, too, have had this type of system installed with varying degrees of success. An up-flow cooling system may be identified by the large single opening in the nose of the cowling just under the propeller. For the most part, however, the up-flow cooling concept seems to be more unique than popular.

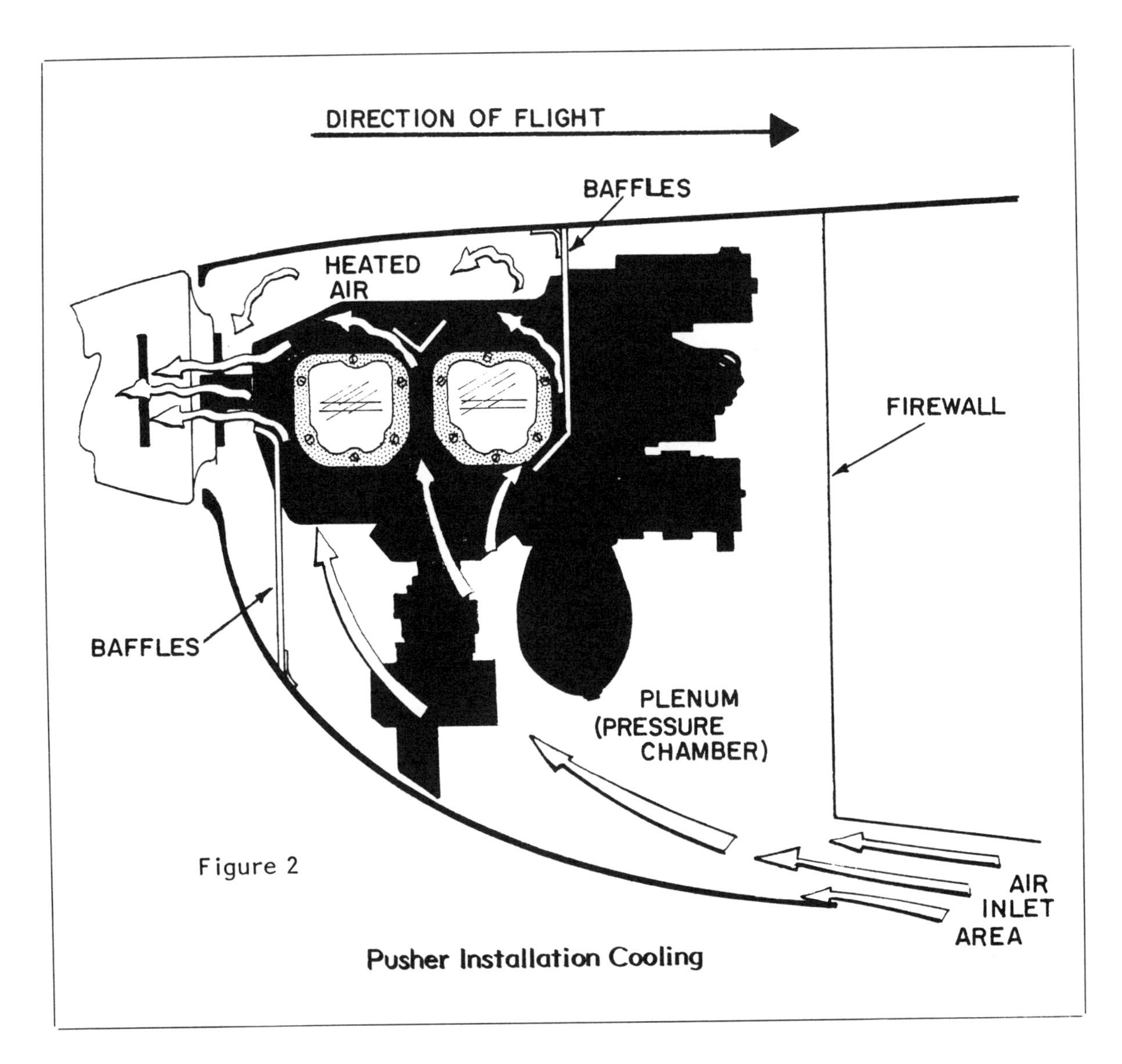

Figure 2

Pusher Installation Cooling

inlet opening rather than into it. Compounding the problem is the propeller with its thick hub whipping the air into a turbulent frenzy. This, combined with inlet openings that are not the proper size and shape, may cause the air to pile up and could adversely affect its entry into the cowling.

Inducing and sustaining a mass flow of cooling air into the cowling inlets and around the engine is helped by properly shaped inlet lips. Your inlet ducts (openings) should be as well rounded or contoured inside the cowling as they are externally at the point of entry.

Note: It is often essential to draw off some of the cooling air to other parts of the engine, the magnetos for instance. Also, inlet area cooling air is often bled off for the oil cooler rather than to provide a separate external scoop.

One thing you will notice once you start comparing the inlets of your favorite aircraft types is that the faster the aircraft, the more likely that its cowling inlets are small. This is great for higher speeds but such an installation could produce cooling problems during slow speed climbs (high angles of attack) with high power settings. Prolonged ground operation, too, is sure to cause engine overheating unless some way is found to increase the flow of cooling air through the engine at low forward speeds. One way is through efficient design of the outlet openings. Another is with the installation of cowl flaps.

OUTLET AREA

Ordinarily, slower aircraft have fixed cowling outlets located somewhere under the aircraft or on either side of the fuselage (gills). These openings are often too large or too small. It is a rare installation that features a cowling outlet opening that achieves a perfect dynamic balance with the inlet opening. Remember, the outlet area must be somewhat larger than the inlet area.

An additional complication is the need to maintain a pressure differential between the incoming air in the pressure chamber above the engine and the expansion chamber below the engine where the 'used up' heated air has lost most of its velocity. You need to make it easy for that air to exit. Thus, it should leave the aircraft parallel to the slipstream rather than perpendicularly or at a high angle to that high pressure slipstream. Usually, this requires that the bottom of the cowling project below the fuselage profile slightly. When side located outlets (gills) are used, the aft edges of the cowling, similarly, have a greater width than the fuselage profile. Both arrangements permit the air to exit parallel to the slipstream. Some builders might not like the disrupted appearance of such a cowling but it does look normal when installed....and it works, too. To obtain a better flow of exiting air, you should also avoid placing turbulence-producing units or obstructions in the outlet area. In fact, you want to try to produce a low pressure area at the outlet.

Sometimes the bottom lip of the cowling is deflected downward severely, creating a stalled low pressure area at the outlet. Although this might help produce a sort of low pressure area for the exiting air, it is at best a repugnant, eye-sore kind of solution. It is better, perhaps, to make the opening larger and to alter the shape of the bottom cowling area. Examine other aircraft and talk to other builders to see what works best.

The small inlet opening, close-fitting spinner to propeller installation, and the closely cropped exhaust stack are all worthy of note on this sleek homebuilt.

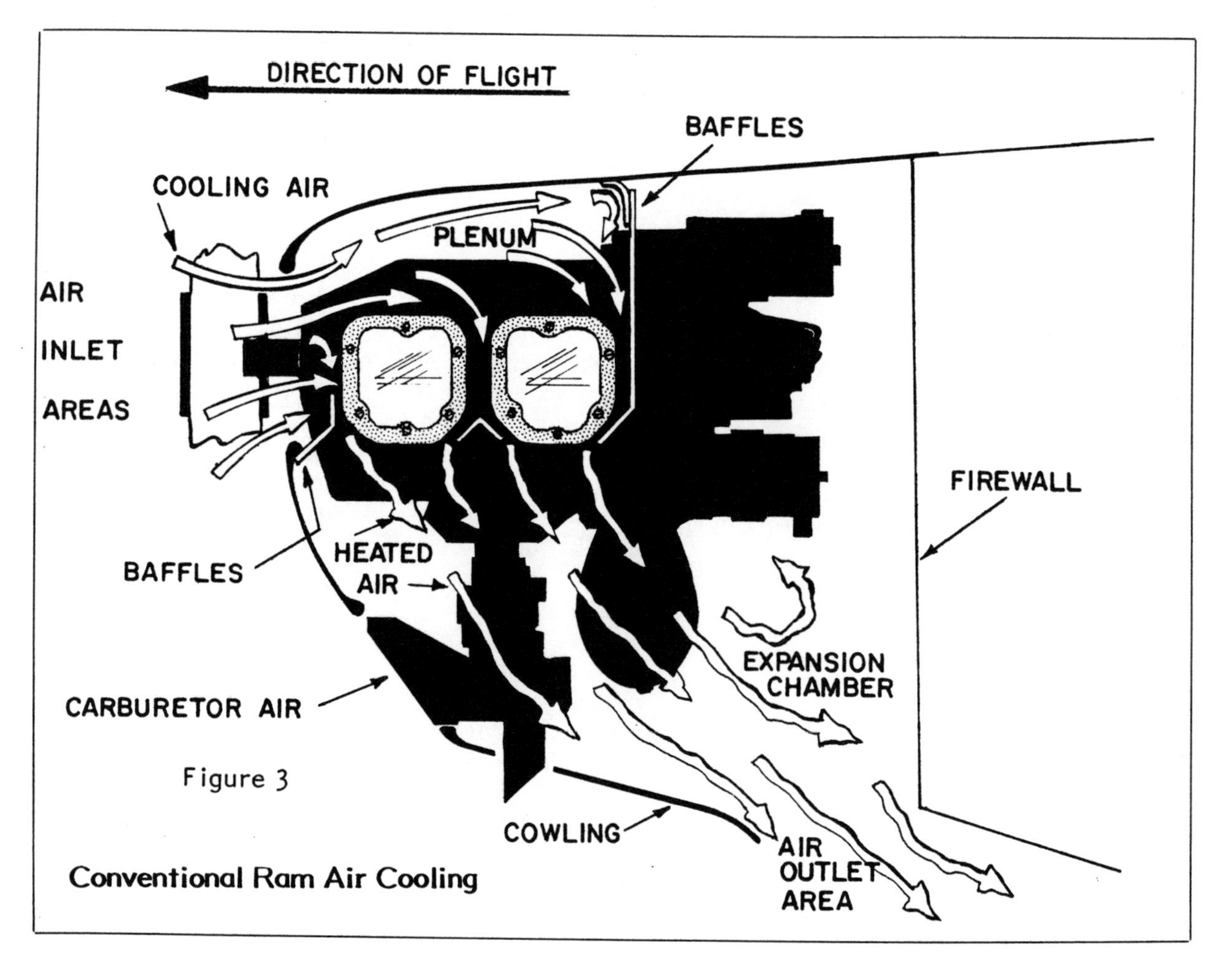

Figure 3

Conventional Ram Air Cooling

In this south end view of a VariEze pointing north, you can see the fit of the baffles around the engine mounts and the neat installation of the seal strips that ensure a snug fitting cowling and no lost cooling air.

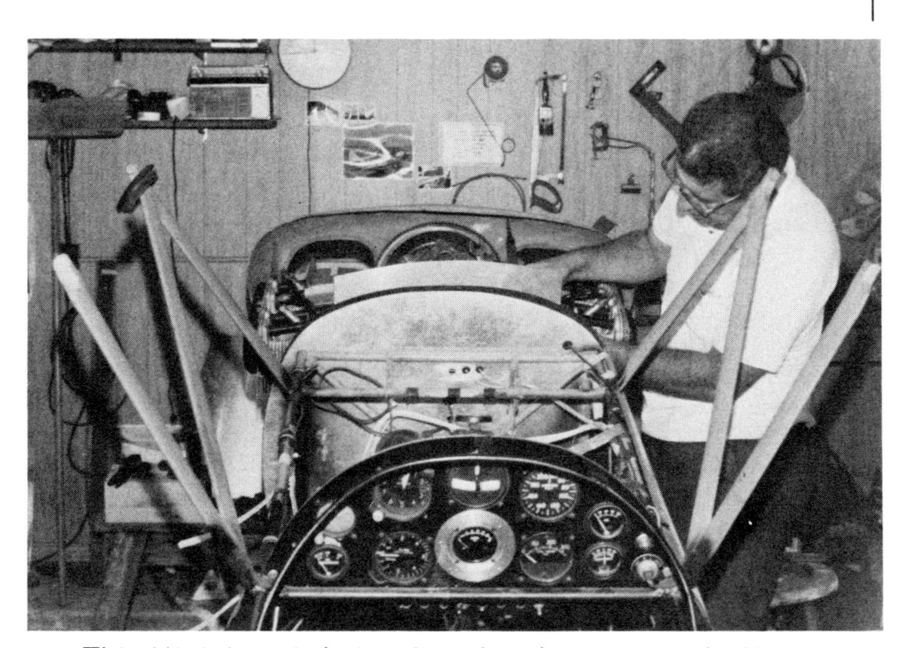

This Miniplane is being fitted with its engine baffles. It can be a very time-consuming task if the builder doesn't have patterns.

Baffling Business

Builders laughingly refer to baffles as 'baffling'. This is because the function of these bits of metal can be puzzling until most of the pieces have been fitted in place.

Baffles will not do a good job of cooling the engine unless each is properly cut and fitted. Moreover, each is dependent upon the other baffles in order to perform efficiently. For example, the entry baffles must not impede the entry air flow. Rather they must be designed and installed to cause the air to enter smoothly and direct it efficiently into the pressure chamber above the engine. The internal baffles, in turn, must force the cooling air down through the engine cylinders. Thus, the function, design and installation of baffles are inextricably connected.

Baffles can only perform their function effectively if they are tightly fitted to the engine. Leaving gaps of 1/16-inch or 1/8-inch here and there is not good form!

Fortunate, indeed, is the builder who acquires his engine complete from the firewall forward. Such a find consists of many expensive parts and assemblies including a complete baffle system. While I would not classify baffles as expensive items, they are difficult to make without the benefit of patterns.

If a fairly good set of baffles came with your engine but do not quite fit your cowling, don't be hasty in assuming that you'll have to make a complete new set. You may be able to modify the original baffles by riveting extension metal strips along the top edges and then trimming them to fit the shape of the cowling. However, aircraft building being what it is, you will probably not only have to fabricate your own baffles, but you will also have to puzzle out where to locate them, how to make them, and then how to attach them.

Good materials for making templates include sheet aluminum, formica, or a light gauge galvanized sheet. All bend lines should be marked or scribed on the template (NEVER ON THE BAFFLES) and all hole locations should be accurately spotted and drilled. Notes that would be useful to you later in making and installing the baffles should be added to each one with a black felt-tip pen. The drawings here will save you time and can be helpful in establishing the size and shape of your engine baffles.

MAKE PAPER PATTERNS

If you have no templates, use the illus-

This pressure chamber-type aircraft cooling system is very effective. It forces the cooling air through the engine cylinder fins, giving maximum cooling with minimum drag.

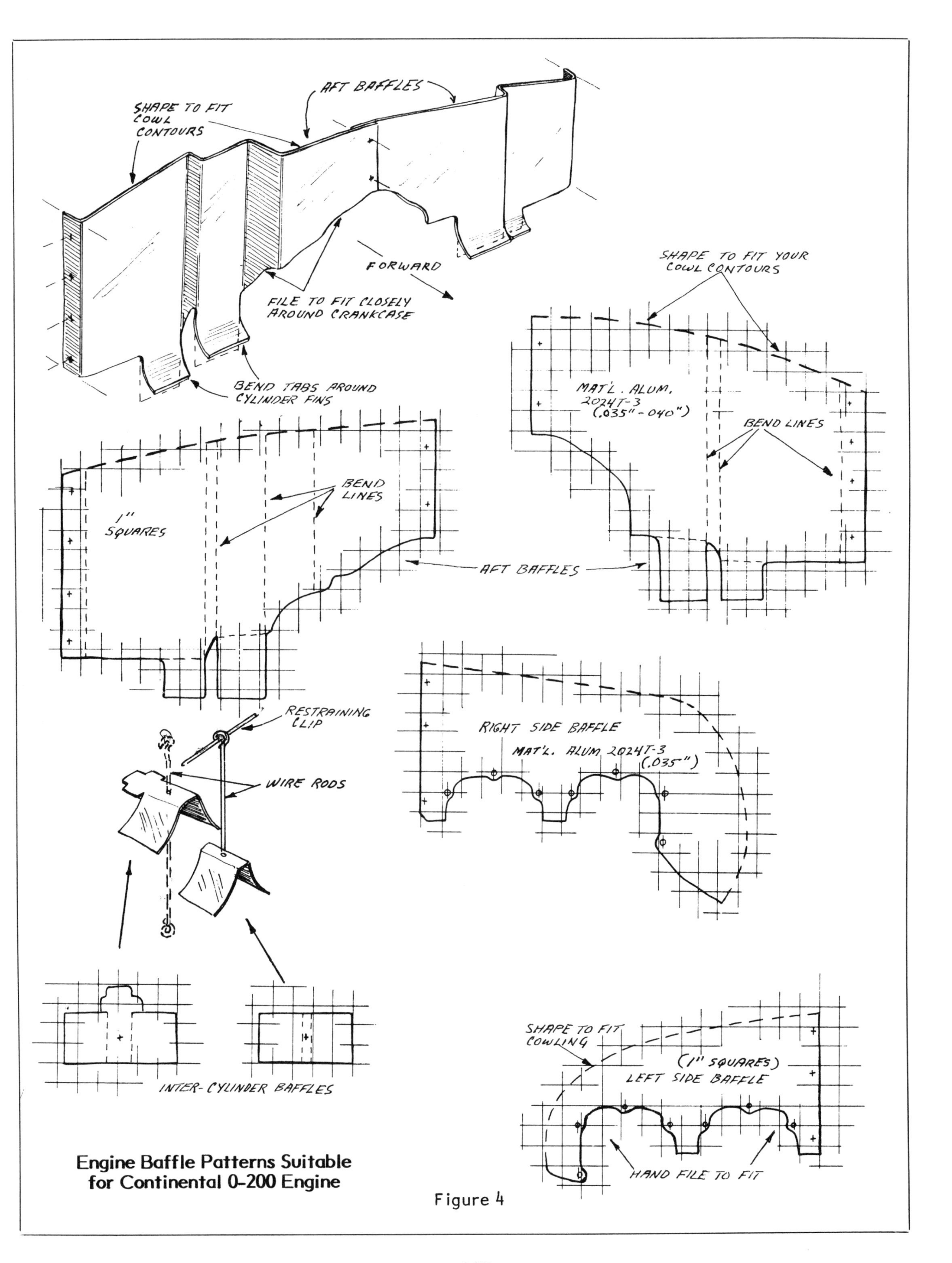

Engine Baffle Patterns Suitable for Continental 0-200 Engine

Figure 4

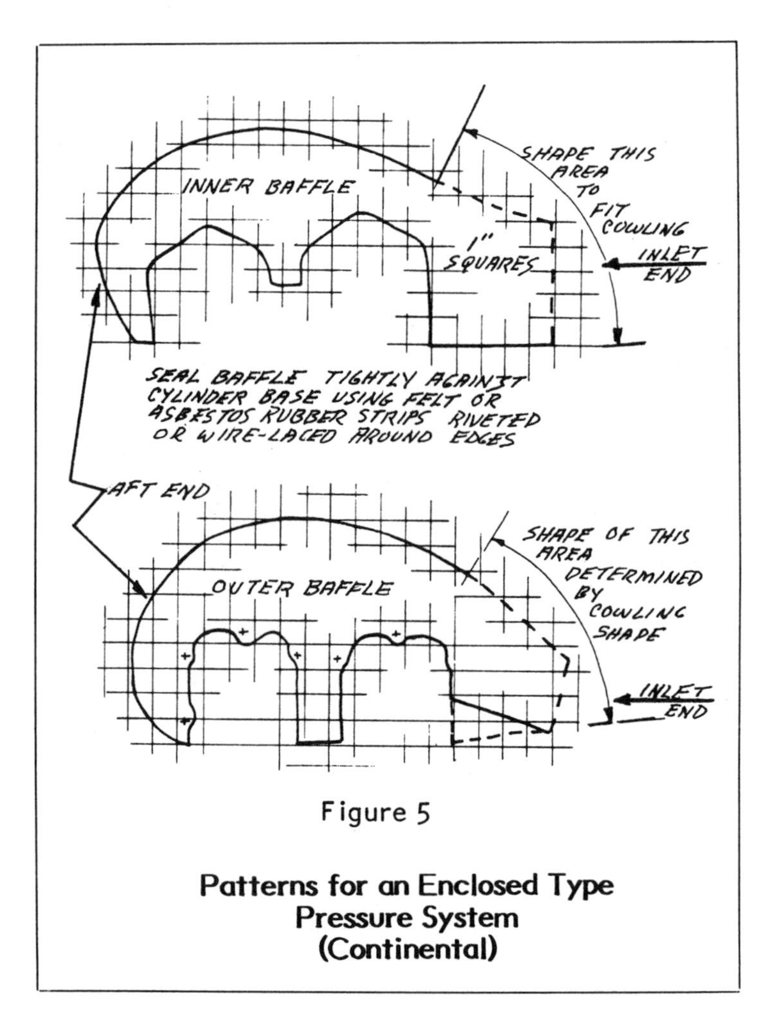

Figure 5

Patterns for an Enclosed Type Pressure System (Continental)

trated patterns drawn to full size on stiff poster paper or cardboard (not corrugated).

After cutting out the patterns, you will probably find that there are some areas that fit like round pegs in square holes. Don't get unduly vexed as a poor fit can be easily corrected. To do this you will need to place your patterns on the engine and tape them in place temporarily.

Now would also be a good time to locate all the attachment holes. You can find where the holes are by rubbing some small hard object such as a dowel over the general area of the holes. The sharp edges of the holes will be outlined on the paper (sort of like cutting a gasket in place on an engine).

The next step is to correct your paper pattern, which is still taped in place on the engine. Using small pieces of scotch tape or masking tape, stick these small pieces onto your paper pattern to fill the gaps of all poorly fitting areas. Remove the pattern. The bits of tape will remain stuck to the pattern, outlining the real shape accurately. There you have it, an accurate pattern that can be traced directly to the baffle material or onto your permanent template material, if you care to make such. Remember to include extra width wherever a flange is to be made.

BAFFLE MATERIALS & HARDWARE

Your baffles may be made of .025-inch or .032-inch aluminum sheet. Anything thicker would be needlessly heavy. Either 2024 T3, or 6061 T6 aluminum, which is less prone to cracking and may be more economical, can be used.

Cut the baffles out on a bandsaw. Use a metal cutting, fine-tooth blade and run it at the regular (wood cutting) speed. If tin snips are used, leave a little extra margin around the cuts and file them down to the line. Cutting with most tin snips leaves tiny crimp marks so allow a little extra metal for dressing these crimps out. To lessen the likelyhood of future cracks developing in stressed areas, take the time to file the edges smooth and eliminate sharp corners and nicks. The patterns shown indicate that the top portion of the baffles must be made to fit the curvature of your cowling. This also means, of course, that having your cowling completed beforehand will help guarantee a better fitting installation, otherwise leave plenty of excess material for later fitting the baffles to the cowl. Shaping of this portion of the baffles could be a bit tricky even if you do have the cowling. However since you need only to trim the baffles within 1/2-inch of the cowling, they can be trimmed down a bit at a time until you obtain the fit you want. Do not try to fit the bare metal baffle edges too snugly against the cowling as

Here's a neat asbestos rubber seal between the cowling and the baffles. Note, too, the attachment brackets on the firewall for securing the cowling.

the vibration of the engine in flight will cause the baffle to wear through the cowling in a short time. Rather, you must rely on the asbestos or felt seal strips for ensuring a close fit.

Baffle sections ordinarily are assembled with 1/8-inch dome head pop rivets (Monel, Cherry MSP 42, MD 424BS or similar). Use pop rivets in areas that do not have to be separated for removal, and machine screws and nut plates (anchor nuts) where disassembly may be required later. Cleco fasteners or small sheet metal screws may be used during trial fitting but these should be replaced with permanent fasteners later for rigidity. Baffles may be secured to most any convenient attachment point on the engine (threaded holes, studs, bolts, etc.)

Note: Never use the engine cylinder hold-down studs and nuts for the attachment of baffles or braces for baffles. That is, placing soft metal under the cylinder hold-down nuts could cause them to loosen up and the cylinders to fail.

MAKE THEM TO LAST

Have you ever noticed how vibration has caused the front baffles on some airplanes to take on the appearance of a cheap comb? A little heavier material could have reduced this tendency.

Even with a proven baffle installation, it is quite possible that you may find a cracked baffle in the first 50 hours of your test program. Such a small crack can usually be stopped cold by drilling a small 3/16-inch hole at the very end of the crack. Nevertheless, in some cases it might also be wise to rivet on a reinforcement plate there as the crack may be an indication of severe vibrational stress at that point.

The secret to long lasting baffles is in careful design. A baffle made to minimize flexing and vibration will last indefinitely. Often it is necessary to rivet small reinforcement plates on in areas of greatest stress. Baffles that vibrate at any point due to the lack of support or inherent stiffness will break or crack after a few hours of operation.

Sometimes the cracking and breaking of the side baffles may be aggravated by tight fitting attachment bolt holes. This is one place where you do not want a snug fitting bolt hole in aluminum baffles.

SEALING BAFFLES

Seal all along the bottom of the baffles and at any place where the fit is not perfect. Many builders use silicone rubber for this purpose (it comes in a tube and is commercially available at hardware stores).

Although the bottom edges of the baffling are fitted closely to the engine contours, the top edges need to end about 1/2-inch short of the cowling. Otherwise, the shaking of the engine on its mount may eventually cause baffle edges that touch the cowling to wear through.

The space between the top of the baffling and the cowling surface has to be fitted with a strip of anti-chafing material that is also capable of sealing off all air leaks. This installation is best done after the cowling has been fitted.

The material used most often is an asbestos cloth treated with a rubber-like neoprene substance (cowl gasket material). Another frequently used sealing material is felt strip. It is available at virtually any building and

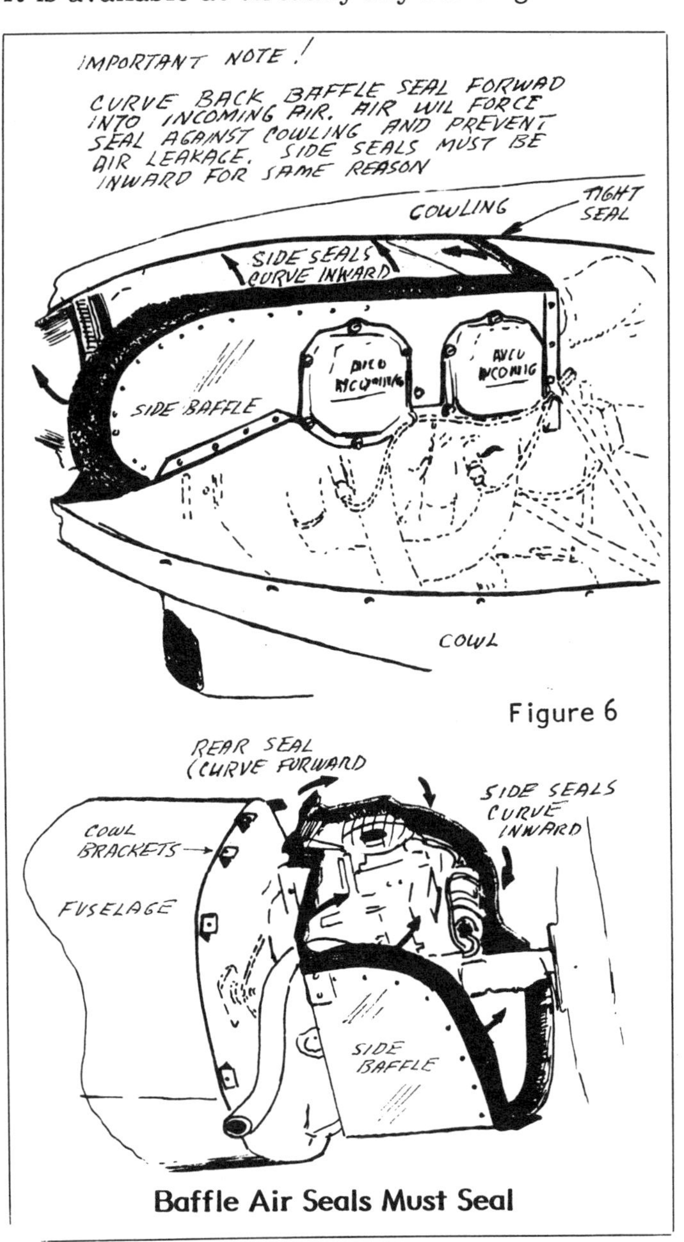

Figure 6

Baffle Air Seals Must Seal

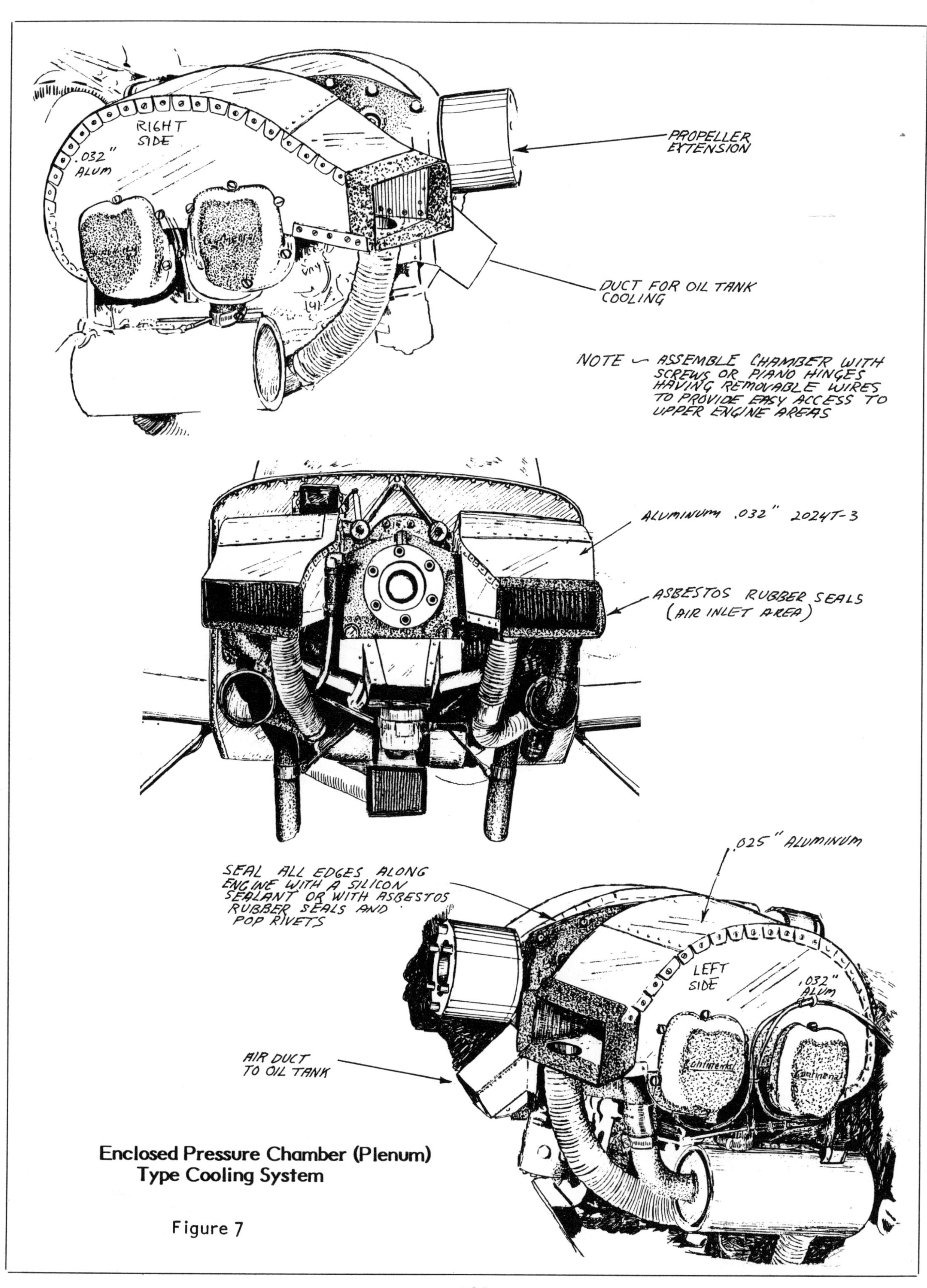

Enclosed Pressure Chamber (Plenum) Type Cooling System

Figure 7

supply outlet in rolls 1/8-inch thick and up to 1 1/2-inch wide as an insulating weather stripping material. Felt strip is generally inferior to the wear-resistant asbestos rubber and might soak up oil, but it would be a rare aircraft engine that leaked oil from the top of the engine. Using felt for baffles seals is quite effective and a fairly common practice among homebuilders.

Attach the seals along the edges of the baffles by using pop rivets with small washers or by lacing them on with safety wire. If you decide to wire lace them, it will be necessary to drill small holes along the edges of the baffle to accept the wiring. Keep the holes at least 1/4-inch inboard. The holes needn't be closer than one inch apart in most places. Install the seals on the pressure side of the engine compartment with the seals bent in towards the center of the engine so that the cooling air tends to force the seal against the cowling.

Install your seal strips so that they will be forced up snugly against the cowling by the entering air.

The rubberized asbestos cowl gasket or baffle seal material is rather heavy, about 1/2 pound per square foot, so, you shouldn't make the strips any wider than necessary for the location being sealed. Install them in such a manner that they will form a positive seal against the cowling no matter how often you remove and reinstall the cowling.

All gaps in the baffle areas must be sealed effectively no matter how small they might be. Don't overlook protecting ignition harness wires passing through baffles with grommet seals.

Remember, any cooling air that escapes without flowing through the engine cooling fins is power-wasting cooling drag.

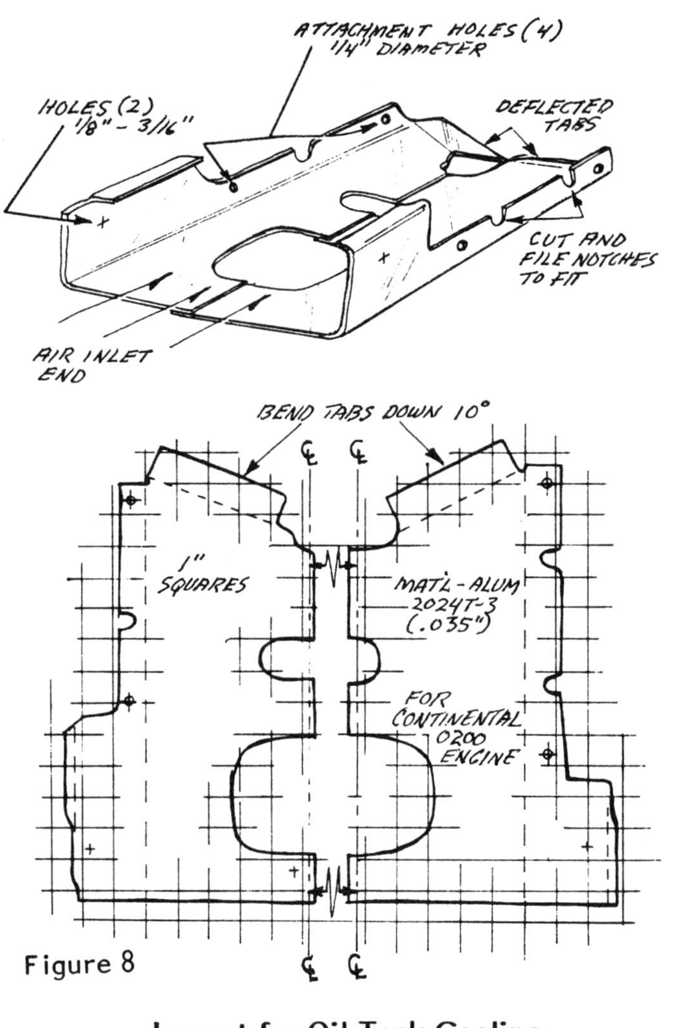

THE INLET AREA DEFLECTORS

It is difficult to show useful drawings for the front deflectors or inlet baffles. So much depends on the shape of the cowling and the cowl openings that, undoubtedly, it would be just as easy for you to custom fit your own patterns. There are only a few common options available to you for installing the front baffles. Inspect a few aircraft on any flight line and you will see for yourself how it is done. These front deflectors can be seen without removing the cowling. Notice their attachment points and, if possible, how these brackets are made and secured to the engine. Note that the front deflectors usually cover only the bottom half of the cylinders. Don't be surprised if you see some installations where the baffles are installed perpendicular to the airflow and are not sloped.

INTER-CYLINDER DEFLECTORS

Whatever baffle system you build, don't forget to install the inter-cylinder deflectors. Some of these baffles are very simple; others, such as those found on some of the larger Lycomings, are a bit fancier. Usually these inter-cylinder baffles are nothing more than a couple pieces of bent aluminum held in place by a wire poked up between the two cylinders and secured to a short wire straddled between the top fins of the two cylinders. A more sophisticated threaded 1/8-inch rod and a special cross piece is also commonly used to hold the small inter-cylinder plates snugly against the bottom of the cylinders. Most

Figure 8

Layout for Oil Tank Cooling Optional (Continental 0-200)

That slot-like air inlet under the spinner is characteristic of a Volkswagen engine installation. Air is conducted under the ribbed crankcase to aid in cooling.

This view of the cowling from the bottom shows the carburetor air inlet, the round fuel overflow outlet, and the rather large cooling outlet.

engines will already have these inter-cylinder deflectors in place. When you install your own engine, just make sure that you have a set in it.

OIL TANK COOLING

It is necessary to direct some additional cooling air under the crankcase (VWs) and sometimes onto the oil tank (Continentals) to help keep the oil temperature down. It seems like a shame to have to do so but to me this makes more sense than directing a blast of cooling air directly onto the housing in which the oil temperature bulb is housed. The drawing shows this baffle pattern for Continental installations in two parts to make it easier for you to trim fit the patterns to the exact shape of your engine.

After you have checked them, join the two pieces together. It should be possible to make this cooling channel out of a single piece of .035-inch aluminum. It could be that your particular engine installation may not need this extra help for cooling, so try your cooling system without it. You can always add it later if the oil temperature continues to read on the high side.

HOW ABOUT SOMETHING DIFFERENT?

If you have sufficient space between the engine and the cowling and you are building a high-performance aircraft, you might want to try an enclosed pressure baffle system. This is a low-drag type pressure baffle system, as shown for a Continental O-200 engine.

The pattern drawings give you the major dimensions for the two most important parts. The wrap-around top portion can be easily measured and made after the inner and outer vertical baffles are secured temporarily in place on the engine.

The two long curved top pieces on this baffle system curve around the rear of the engine and are bent tightly around the bottom of the cylinders, leaving, uncovered, about 2 1/2 inches at the very bottom of the cylinder to permit the air to escape.

In cutting and fitting all of these baffles keep in mind that the whole thing must fit under the cowling, that the frontal inlets must be tailored to the cowling profile, and that you must make some provision for access to the upper spark plugs. This means that you cannot rivet the whole assembly into one nice neat permanent enclosure. Extensive use may be made of sheet metal screws or machine screws and nut plates to aid in easy removal and to satisfy the need for access to the plugs.

KEEP YOUR BONNET ON

When most any type of cowling is removed, the engine will run hot, as the cowling shape and openings are designed to enhance cooling. You should, therefore, make it a practice not to operate your engine with its cowling removed any longer than is absolutely necessary.

What a smile! The air inlet opening in this very stream-lined cowling is adapted to upflow cooling.

Cowlings in General

It might be satisfying to build your own cowling to the exact design and shape you want but it can be just as gratifying (it's certainly quicker) if you can purchase one that is ready-made. Sometimes, the cost of making one will approach or exceed the price of a ready-made cowling, even if you discount the value of the time you spend making it.

Of course, many builders cannot purchase a cowling for their particular design as none exists for it. Any design being built in large numbers, though, is certain to offer sufficient profit potential that someone will be producing ready-made cowls and other parts for it.

Sometimes a cowling for one design can be adapted to a different kind of airplane. Such adaptations, however, often prove vexing, so if the firewall dimensions are not approximately the same, it would be best to forget that idea and build your own.

If you think most cowlings are not streamlined enough, but too blunt at the front, you can blame the engine manufacturers. They are building the engines with the cylinders so close to the propeller flange that there is not much you can do to the cowling to impart a sleeker look for it unless you install a propeller spacer (prop extension) or use a large spinner.

If you purchase a cowling be sure you understand whether it was intended for use with or without a propeller extension, and how much of an extension (3 inches, 4 inches, 6 inches, 9 inches??). Obviously, a ready-made cowling could turn out to be too short or too long for your installation. Is it better to be too long? Nope, because the nose of the cowling will probably be slimmer, having been intended for use with a prop extension. If that is the case, cutting off some of the aft edge of the cowling to shorten it will not solve the problem....it still will not fit up front unless you, too, use a prop extension.

Using a standard 4-cylinder aircraft engine in a W W-I replica introduces a real challenge in baffle and cowling design.

SPLIT YOUR COWLING, BUT....

Your cowling should be installed so that it can be removed without first having to remove the propeller and the spinner. This means that
you will have to split the cowling somewhere to get it off. The most practical way to divide a cowling (so that no joints show when viewing the airplane from either side) is along the top and bottom centerline. This, however, introduce a need to reinforce the nose area, particularly in cowlings with a single large air inlet opening up front. Some builders prefer, for this reason as well as for personal convenience, to split their cowlings along the top longeron line, making the top and bottom sections removable separately. Others prefer to split the cowling horizontally but provide a couple of large hinged hood-like sections on both sides in addition.

This discussion is probably academic if you intend to purchase you cowling. Vendors selling cowlings usually ship them already split in halves....and the split may not necessarily be as you would prefer. They do this to facilitate packing and to reduce the costs of shipping. The separation usually is as the designer has depicted it on the plans. Anyway, it is much easier to fit the cowling when it is split because you can work with each half separately.

FITTING THE COWLING IN PLACE

If the engine is offset (to compensate for torque) one side of the cowling might have to be shortened to fit it. Symmetry will be affected and it will look awful to your eyes. Accept the fact and take care that the front end is centered around the crankshaft.

It is better to fit the cowling with the engine already installed. This way you will be assured that everything clears the cowling by at least 1/2-inch. If not, you may have to relocate or modify that offending something. This problem crops up most when the first-time builder, making his own cowling, fits it too closely over the spark plugs and the front left plug fails to clear. The remedy is to cut a hole in the cowling and scarf on a fiberglass bump. However, it will always be an obvious bump no matter how cleverly you streamline it. If the exhaust system fails to clear, it would be much better to rework it, if you can, than to mess up lines of your cowling with bumps.

Before you begin your cowling installation, temporarily tape the two halves together with masking tape or grey duct tape and see how the fit will be at the firewall.

Next, make a temporary plywood disc of 1/2-inch or 3/4-inch material and bolt it with a couple of bolts to the propeller flange. The diameter of this disc should be that of the propeller spinner you intend to use or the diameter of the opening cut in the front end of the cowling to clear the propeller flange. It depends on your cowling design. The plywood disc will be used to support the nose end of the cowling during the fitting and installation process. After you have reassured yourself that the cowling is centered around the propeller flange or spinner backplate with the aid of your plywood disc, secure the cowl in that position using tape, clamps or any other means at your disposal.

FASTENING THE COWLING

Begin to secure the cowling along the firewall flange. DO NOT RUSH. Work from the middle toward the split edges of the cowling. If possible, first use Cleco fasteners, inserting one in each drilled hole. Later, you can open the holes to whatever diameter necessary for the type of fasteners you will be using.

All cowlings should be split so that they can be removed easily without the necessity of taking the propeller off first.

Cowlings are cantilevered from the firewall, that is, the cowling is not attached to or braced by the engine. Most builders use AN507 machine screws for a countersunk installation with just about any style of matching anchor nuts. The recommended size is 10-32 (thread size). If you prefer a button-head screw, the proper length AN525 machine screw and matching anchor nut would be a good choice.

Use a minimum of fasteners. Although cowling security is important, so is easy removal. Builders preferring a quick-disconnect capability can use Camlok type fasteners or Dzuz fasteners. Both types of fasteners require

A cowling with a large, wide opening for the inlet air should never be split vertically. It's too difficult to make rigid.

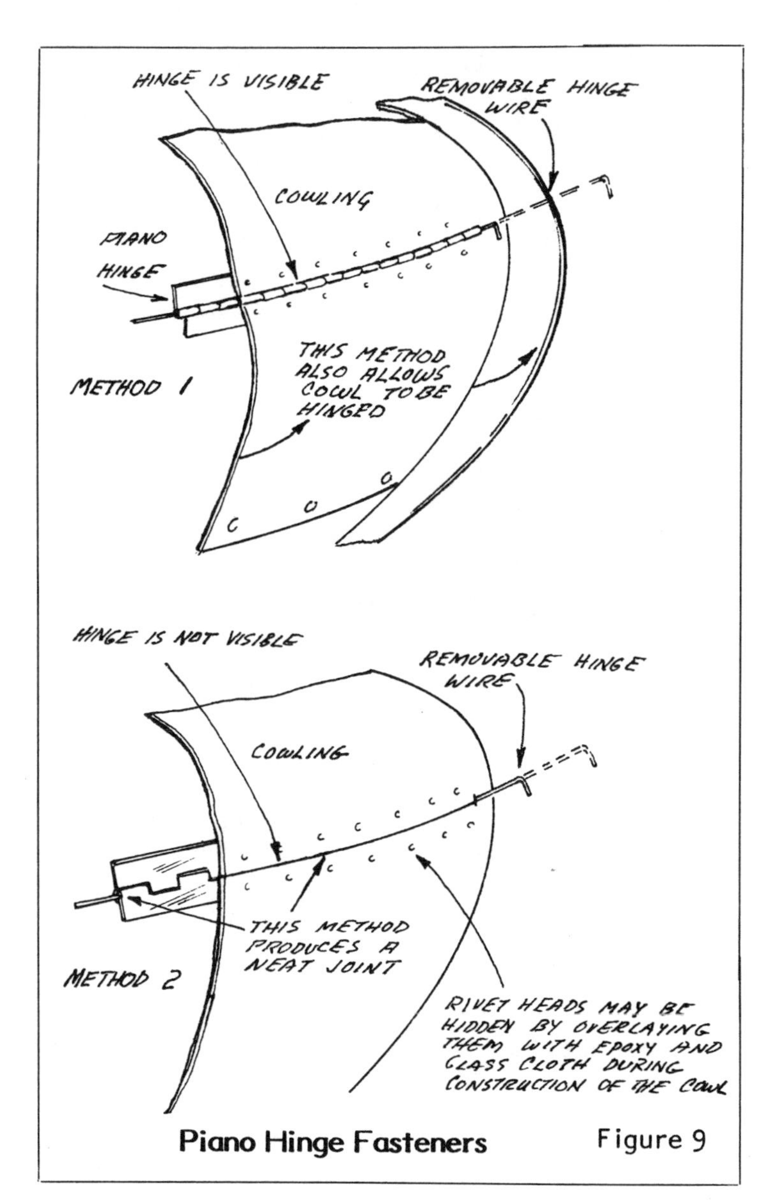

Piano Hinge Fasteners Figure 9

special installation tools. For this reason, many builders prefer to install piano hinges with removable wires to substitute for fasteners wherever practical.

During the course of fitting the cowling you will have to install and remove it countless times. If it is difficult to remove because of too many difficult-to-unscrew fasteners you will waste a lot of time, to say nothing of the time you will be wasting in years to come maintaining the airplane.

Don't lose sight of the fact that the cowling needs to be secured to the aircraft so that it won't vibrate and flutter along its downwind edges, particularly around the outlet area. If necessary, a short strut-like internal support may be installed.

CUTTING THE OPENINGS

If cowl openings are already cut out for exhaust pipes, be sure they match your installation or you will have to patch over them and cut new ones. Ordinarily, however, cowls will not have any holes or openings in them other than the nose air inlets.

As for cutting openings into the cowling yourself, don't cut them until you are positive that they are needed and that they are where you need them. Make them only as large as they have to be. The exceptions, of course, are the exhaust pipe openings, which should be large enough to clear the pipes all around by at least 1/2-inch. This amount of clearance is required to allow for engine shake in the shock mounts. Later, the excess opening can be

Wherever possible, install piano hinges with removable wire to simplify the installation and removal of the cowling.

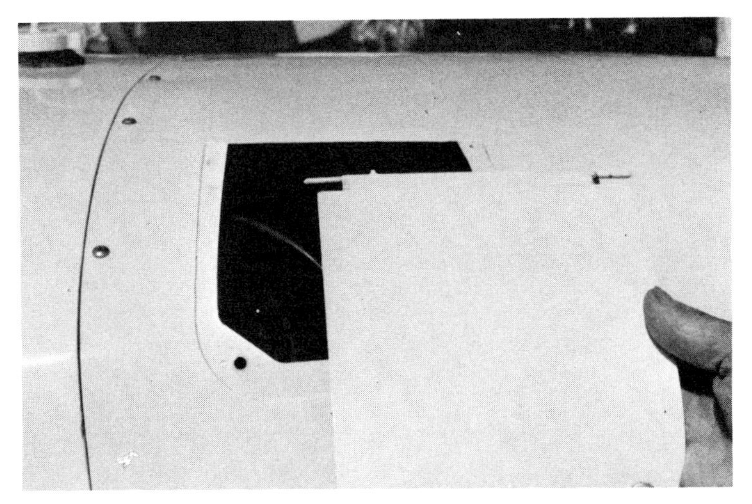

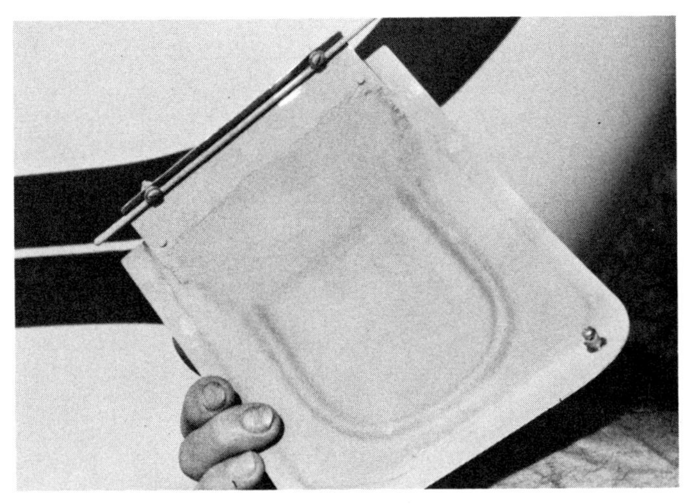

Most airplanes feature an oil inspection door mounted on a piano hinge and secured with Hartwell fasteners. This one uses only half a piano hinge and protruding wires for security. Simple to make and no hinges show.

sealed with asbestos fabric if you prefer. This is generally a good idea as it will prevent disturbing the orderly flow of the internal cooling air on its way to the cowl outlet.

Another hole you will need is the one for an oil inspection door. Determine its location and cut the opening in the cowling to provide easy access to the dip stick. Construction of the door is fairly simple if you use a piece of .032-inch or .040-inch 2024 T3 aluminum and a short section of piano hinge. For a fastener, install a Hartwell H-5000-2 quick release latch. If there is severe compound curvature in the cowling where the oil inspection door will be located, it would be easier to make the door from the fiberglass section cut out of its cowling. In that case you could break a hacksaw blade in half and use the sharp point to start the cut for removing a square from the cowling. Use the cut out for your oil door and it will have the exact contour required.

It is better to make the opening on the larger side rather than to make one that is not big enough to get your hand in to reach the oil dip stick. The door also serves as an inspection peak hole for the rear areas of the engine installation.

The completed cowling installation on a typical two-place airplane will weigh in the neighborhood of 12 to 15 pounds. Homemade cowlings may weigh more because the builder frequently overbuilds the nose piece area to a thickness of .080-inch or more when .050-inch would do. He is also likely to use an excess of resin, further adding to the weight of the cowling. Many cowling sides are built up too heavily for the job at hand. What's wrong with having the side panels a bit flexible?

Cowl flaps are important on a tightly cowled, fast airplane but are also essential for this float-equipped storebought.

This gill-like cooling air outlet is designed to help smooth the exiting air and to reduce overall cooling drag.

Cowl Flaps

The problem of establishing an efficient air outlet in the cowling becomes more acute as aircraft performance increases. A fixed outlet is not adequate for the range of the cooling flow velocity that must be accommodated in faster aircraft.

(Some aircraft have a single cowl flap; others have more than one — depends on design and builder preference. I generally discuss them in plural terms here.)

Cowl flaps provide a method of increasing or decreasing the air outlet area (air exit). For example, opening the cowl flaps gives the maximum outlet opening, and the cooling air flow velocity through the engine increases, carrying off more engine heat. As a result, engine temperatures go down. This is particularly important during climb airspeeds when the engine is running at a high power setting and giving off a lot of heat. The engine heat would not, otherwise, be dissipated because of the relatively slow airspeed and reduced velocity of cooling air through the engine. On the other hand, in cruise, because of the greater airspeed, the cooling air velocity is high while the engine power setting is comparatively low. In this case, the engine doesn't need all that cooling. Closing cowl flaps reduces the amount

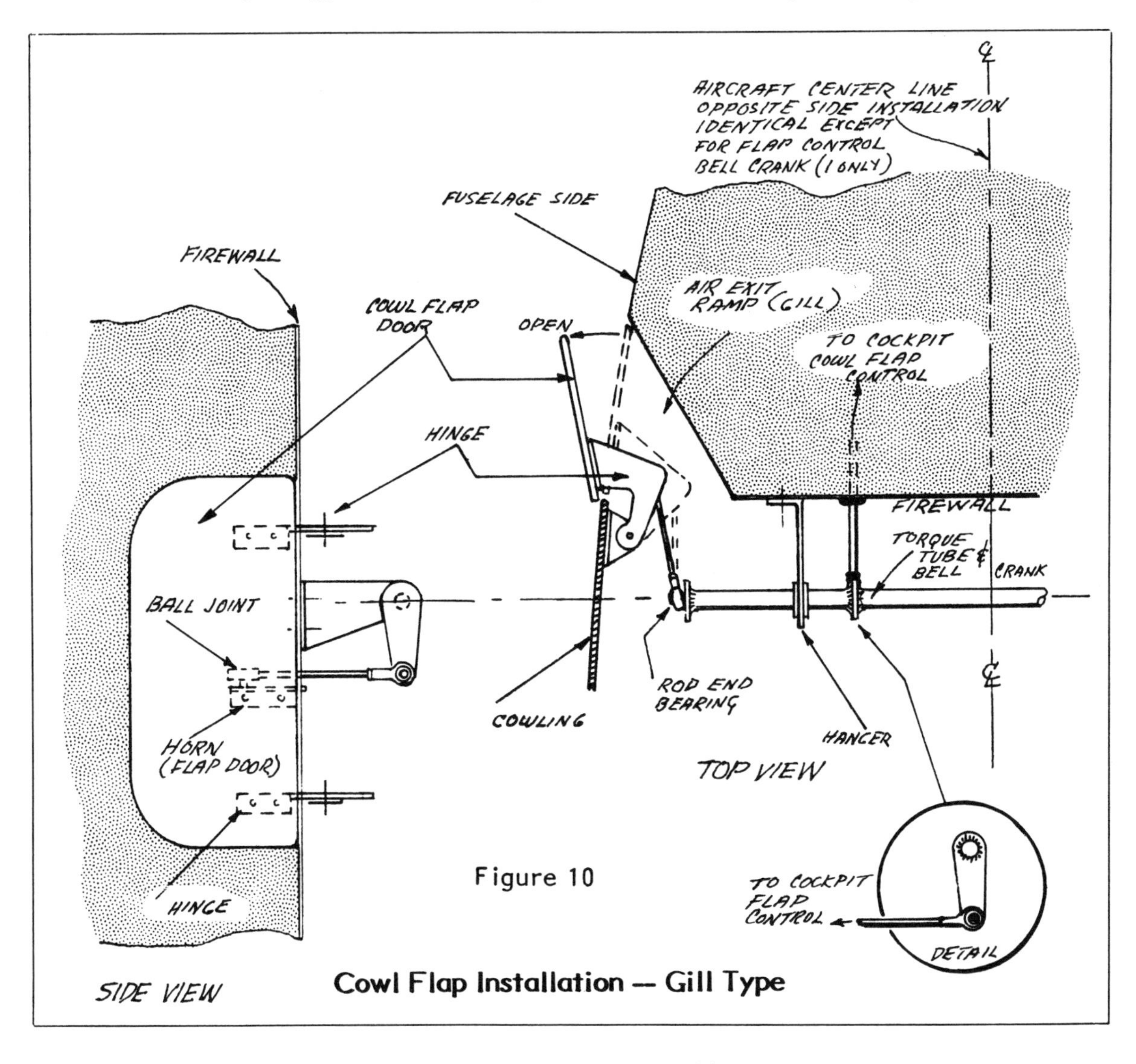

Figure 10

Cowl Flap Installation — Gill Type

of cooling air going through the engine. It also streamlines the aircraft, thereby reducing drag. This is good for the engine as well as the airspeed because, believe it or not, it is possible to operate an engine at temperatures too low for its mechanical health.

Maintaining the engine's operating temperatures within limits by controlling the flow of cooling air is as important during take-off, climb and descent as it is while the airplane is in cruise or running up on the ground. The key instruments to watch in order to monitor the effects of cowl flap operation are the oil temperature gauge and the cylinder head temperature gauge.

After you install the cowl flaps and are faced with the problem of their adjustment, you will wonder to what angles they should be permitted to open and close. Unfortunately, this determination may have to be worked out in a do-and-try process, supplemented by flight testing.

As a starting point, provide a 15-degree open angle and ensure that the cowl flaps in their fully closed, or streamline position, do not cut off all the air flowing through the engine. At least a limited outlet must remain at all times. If you initially install the control cable so that it can creep, the cowl flaps will assume their ideal position (closed angle) in cruise. The control unit can then be modified to provide the range of action required. Adjust the linkage for the cowl flaps to coincide with the cockpit control positioning. Assure yourself that when your cockpit control is in the fully open position, the cowl flaps are actually fully open. In addition to ascertaining that you have the correct response to the cowl flap control, be sure that the final installation is not subject to creep. The slipstream will exert considerable pressure on the cowl flap when it is fully deployed so that design of the control linkage, or the friction in the system, must be adequate to resist its force.

Your cowling/cooling system installation, if effective, will permit you to operate your engine safely at maximum level flight power and at its best climbing speed (even when the ground temperatures are as high as 100 degrees F), an objective that sometimes takes considerable effort and time to achieve.

Homebuilts equipped with cowl flaps generally have a single flap spanning the bottom of the aft edge of the cowling. This single flap is a rectangular section cut out of the cowling and hinged on piano hinges. A small actuating arm is fabricated and riveted or bolted centrally for raising or lowering the cowl flap. The actuating linkage usually consists of a stiff sheathed control wire. This type of control may not provide sufficient resistance in the system to prevent the creeping of the cowl flap at various speeds and a detent arrangement may have to be devised for the cockpit end of the control. A more positive system employs push-pull rods and bell cranks. However, this may needlessly complicate construction and may be too heavy for the simple job to be done.

The typical gill-type outlet on the T-18 is not so typical with this one named KONG. The outlet is, in effect, a cowl flap that is controllable by the pilot.

Another common cowl flap treatment is to make two separate cowl flaps and install them just below each of the exiting exhaust pipes. In this sort of installation both flap sections should be interconnected so that they operate simultaneously.

AUTOMATIC COWL FLAP ACTUATOR

What? Install a thermostatically controlled automatic cowl flap actuator in a homebuilt? It may sound grandiose but it's really a relatively simple installation and is certainly worthy of your consideration. It offers an improved method of controlling engine cooling in just about any aircraft equipped with cowl flaps.

Does it really Work? A prototype installation in a 150 hp Globe Swift limited cylinder head temperatures to 445 degrees F maximum on extended climb-outs and prevented rapid

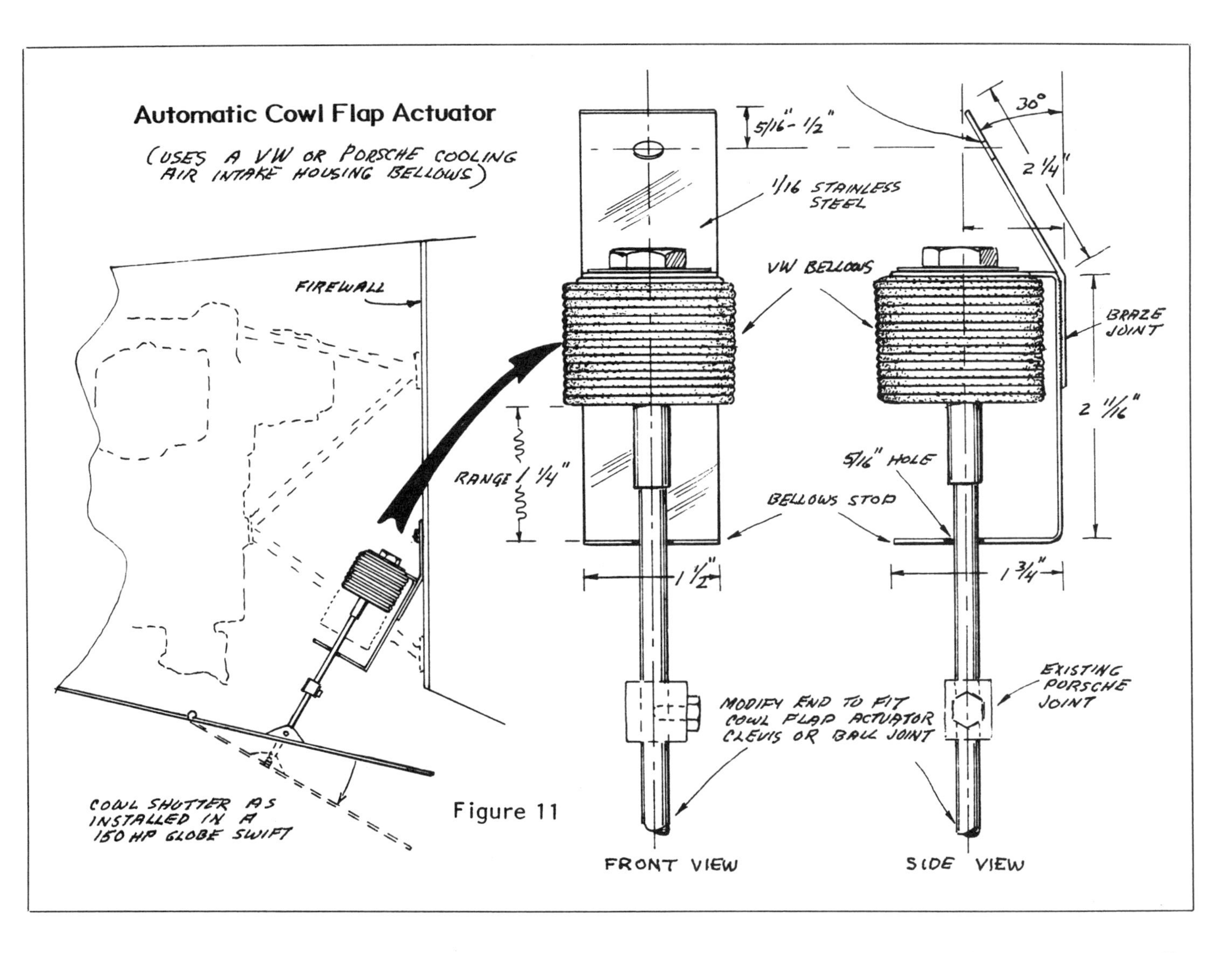

Figure 11

cooling by keeping the cylinder head temperatures between 320 degrees F and 338 degrees F during rapid, power-off descents.

As you know, during ground operations, take-off and climb, the cowl flap is usually in its OPEN position (if the pilot doesn't forget) while, during cruise conditions, the flap is CLOSED. Closing a cowl flap after reaching flight altitude reduces the amount of cooling air flowing through the engine because less is needed at higher cruise speeds and lower power settings. A side benefit from this is a reduction in drag and sometimes a satisfying increase in airspeed. However, all this cockpit activity spells WORK. Any pilot whose philosophy embraces economy of effort (not to say laziness) would undoubtedly be pleased with the luxurious utility afforded by an automatic cowl shutter control installation.

THE BELLOWS

The vital center of the automatic cowl flap actuator is an expandable bellows. When exposed to a heated air environment, the bellows expands, proportionately, in relation to the temperature. Properly utilized, this characteristic makes automatic cowl flap actuation possible.

In its normal, collapsed state (room temperatures), the bellows chamber measures approximately 1 1/8-inches in length. When the air temperature surrounding it increases, the bellows expands. You can confirm its expansion characteristics by checking it in an oven or in water at different temperatures.

The following table will give you a general idea of the bellows' behavior based on household oven laboratory results:

Temperature	Bellows Length
72 degrees F	1 1/8-inch (collapsed)
140 degrees F	1 7/8-inch
170 degrees F	3 inches
250 degrees F	4 inches

Because the bellows will expand more

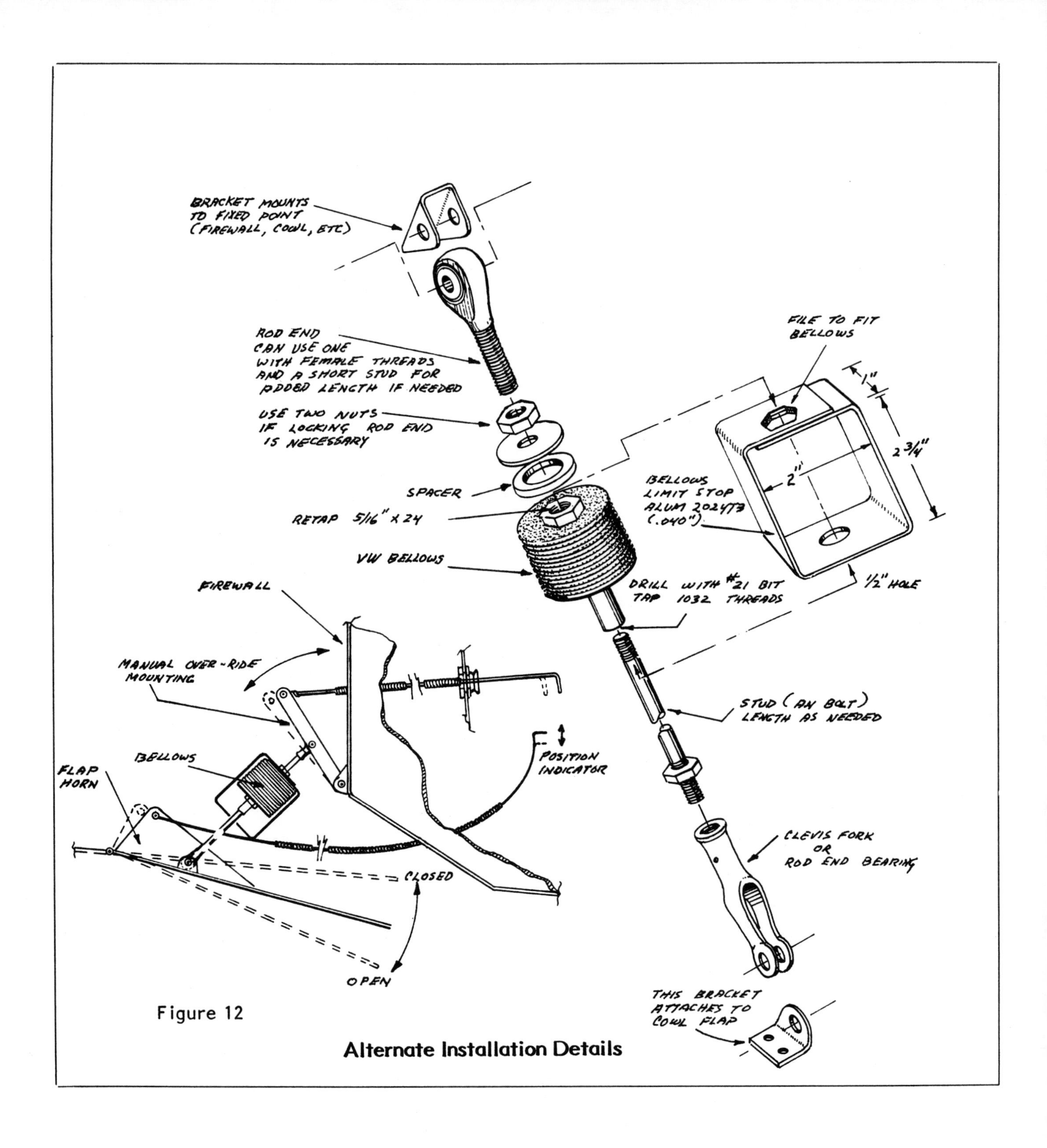

Figure 12

Alternate Installation Details

than three times its normal, closed length, it will be necessary to build in a limit stop to control its maximum expansion. The bellows, as utilized in the Swift application, was permitted a maximum expansion, or travel, of approximately 1 1/8-inches to 1 1/4-inches. In its fully expanded, restricted length it could sustain a load of more than 25 pounds without collapse.

Where can you obtain a bellows? They have been standard installation parts in VW and Porsche automobile engine air cooling systems for years. Incidentally, the proper nomenclature for these units is, "Cooling Air Intake Housing Bellows".

The little devices are virtually trouble-free and, because of this, they may not be stocked at most automotive parts houses dealing in foreign car parts. Of course, a search through a foreign car junk yard would probably yield one for you at minimal cost.

Except for the addition of an expansion limitation bracket, the installation used in the Swift required no modification to the bellows. The Swift cowl flap actuator, as installed, had one shortcoming. There was no way for the pilot to determine the position of the cowl shutter while in flight. He could really only make a somewhat belated determination of whether it was open or closed by keeping an eye on the cylinder head temperature and oil temperature gauges.

Here's another problem. What if the bellows develops a crack or a hole in it during flight? A guy could end up with one hot engine because the bellows wouldn't respond to the increase in engine temperature and open as required. A fail-safe proviso, in the event of such an unlikely malfunction would be to incorporate a manual override control for the automatic actuator. A simple mounting can provide the necessary control over the bellows if, for some reason, it ceases to function. Figure 12 illustrates.

Having the capability to manually open the cowl flap could contribute to your peace of mind during the interim when you are experimenting with the basic adjustments for the cowl flap actuator. The automatic cowl flap actuator with a built-in mechanical override would permit experimentation with much tighter control over the cowl flaps. All in all, it would be an interesting experiment and you wouldn't have to mess up your airplane to try it.

It's never too late to modify and improve your cowling. Note the carburetor air inlet modification on this bird.

lines & hoses

7

About Aircraft Plumbing

Most of us wait until our airplane is almost finished before giving any thought to the hoses and lines that will be needed to complete it. We have little or nothing in the way of tubing, hoses or fittings to use and very little idea of what we will really need or how many of each. I suppose this is inevitable because one really can't tell beforehand what sizes or lengths of hose will be needed. After all there are so many of them . . . the fuel lines, the flexible hoses to and from the oil cooler, the brake lines, pitot and static lines and, of course, the instrument lines. They are all different in some way and yet so much alike. Of all those mentioned, the fuel and oil hoses and lines will probably occupy our greatest attention. Of course, if the aircraft is a retractable that uses hydraulics, we have still another demanding system and its tubing to deal with.

Aircraft plumbing is very expensive and must be installed properly to keep expenditures under control. The term "plumbing" includes all of the hoses, tubing, fittings, connectors and the methods used to fabricate and install them. Since the plumbing is composed primarily of two types of lines, rigid (metal) and flexible (rubber or plastic), the comments here will be divided between them in the same manner for convenient reference. (Figure 1)

Rigid, or more accurately, metal fuel and hydraulic lines, are usually used behind the firewall (that is, on the side opposite that occupied by the engine). On the other hand, all fuel and oil lines inside the engine compartment are typically flexible. This is because the engine is hung on shock mounts and moves rather freely. Rigid metal lines (tubing) used under such conditions could harden, crack and fail from continued vibration, creating a serious fire hazard.

METAL LINES IN GENERAL

I guess it is accurate to say that the metal tubing used in most aircraft for fuel, oil, coolant, instrument and hydraulic lines is made of an aluminium alloy. Less frequently, one will find corrosion-resistant steel tubing but its use is usually limited to high pressure hydraulic lines. Copper lines, once the most commonly used tubing, have just about disappeared from "store-bought" aircraft. Not so from homebuilt aircraft though. Amateur builders still use a number of copper lines in their planes — primarily because these lines are readily available locally.

All things considered, however, aluminum lines are superior to copper lines because they are lighter, easier to work with and have greater resistance to corrosion and fatigue.

ALUMINUM LINES

Aluminum alloy tubing is considered the basic general purpose tubing for low or negligible fluid pressures. It is difficult to identify different types of aluminum tubing and I doubt if most builders can accurately tell the difference between 1100-0, 3003-0, and 5052-0 aluminum alloy. 6061-T6 and 2024-T3 are considerably stiffer and this would be noticeable when directly compared with the softer types mentioned, The best tubing to use for fuel and oil lines is the readily available 5052-0 aluminum tubing. It is soft enough to be easily flared and formed for routing.

You can use 5052-0 aluminum tubing for all your metal lines including pitot/static, vacuum, primer, fuel oil and low and medium pressure hydraulic lines. Here are a few common sizes:

PRIMER LINES — 1/8-inch x .035-inch.
BRAKE LINES — 1/8-, 3/16-, or 1/4-inch x .035-inch.
PITOT/STATIC LINES — 1/4-inch x .028- or .035-inch (vacuum also).
FUEL LINES — 3/8-inch x .035-inch for aircraft engines (normally).
OIL LINES — (rare with homebuilts because external oil tanks are infrequently used).
OIL PRESSURE GAUGE LINE — 1/8-inch x .035-inch.
OTHER OIL LINES — Most other oil

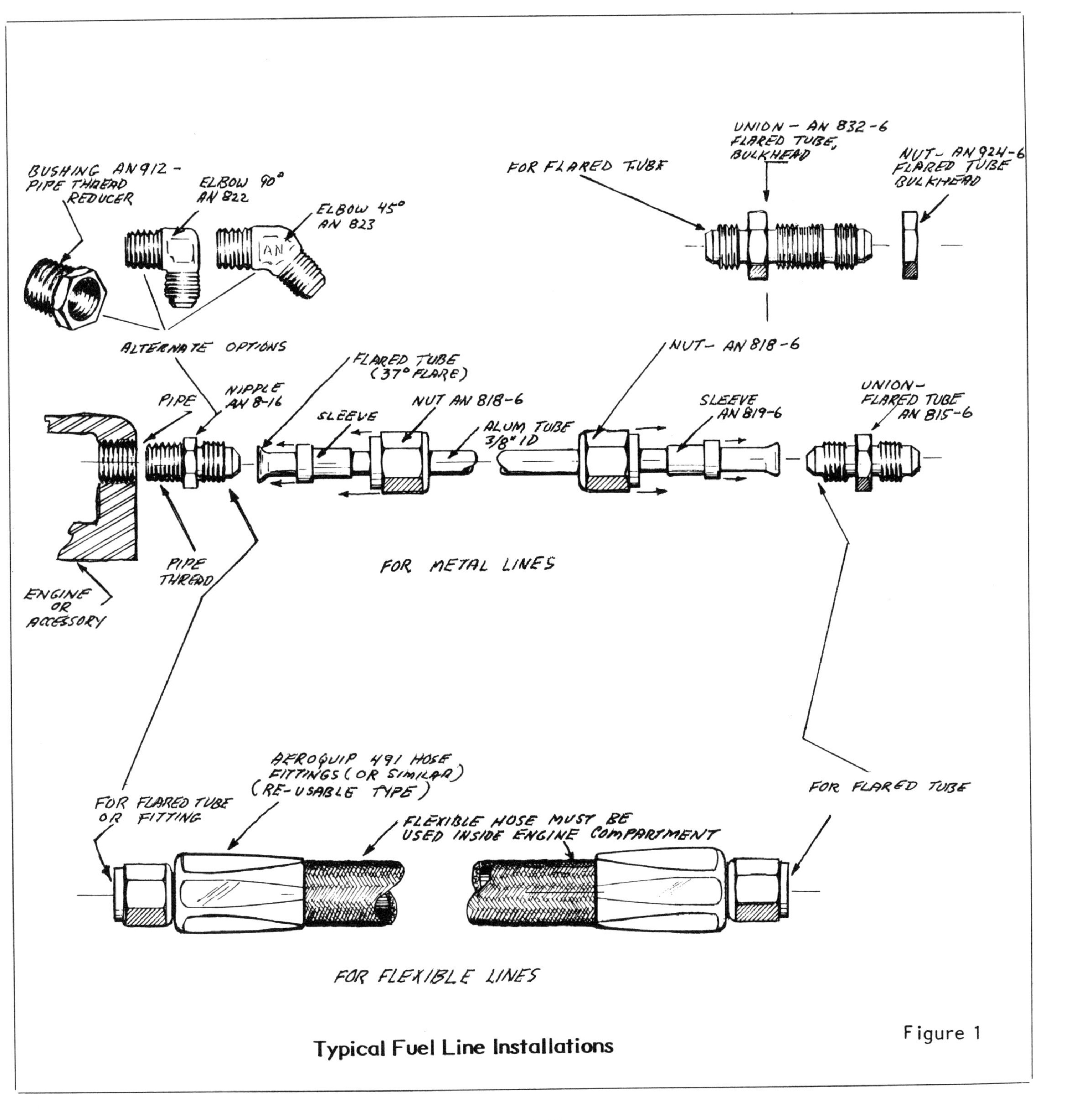

Figure 1

Typical Fuel Line Installations

lines, including those to and from oil cooler are made of -6 or -8 flexible hoses.

HYDRAULIC LINES — 1/4-inch x .035-inch.

ANNEALING ALUMINUM TUBING

Aluminum tubing also hardens under prolonged vibration and after bending and severe forming. If the tubing becomes difficult to bend it should be annealed using one of the following methods:

* *Heat the aluminum tube uniformly with a torch to the point where it will char a soft wood scrap when you touch the tube to it. Allow the tube to air cool.*

* *Another (perhaps safer) method is to heat the aluminum tube with the carbonizing flame of a welding torch until the tube is coated with carbon soot. Readjust the torch to a neutral flame and carefully play the flame over the blackened tube until all of the carbon is burned off. Allow the tubing to air cool.*

STEEL LINES

Corrosion resistant steel lines are seldom found in homebuilts. When used, the tubing is in its annealed state and, therefore, can be formed with a flaring tool. The flaring process actually work-hardens the steel and strengthens it somewhat, permitting the use of tubing with thinner walls. As a result, the final installed weight is not much greater than an aluminum line. Nevertheless, you probably will not consider steel lines unless you are building a hydraulic system in your aircraft.

COPPER LINES

Copper lines reportedly fail more frequently than aluminum alloy lines under similar operating conditions. Nevertheless, the use of copper tubing is still important in numerous light aircraft.

If you intend to use copper tubing, more than likely it will be for your oil pressure gauge. If so, you should be aware of its limitations. Copper tubing hardens with age and vibration. This is a fact you must face. Therefore, for your own safety and to increase the reliability of your installation, you should anneal your copper lines after they have been formed and before installation. You should also adopt the practice of annealing your copper tubing lines periodically — say annually. As a point of interest, both pressure indicating and fluid carrying copper lines usually fail adjacent to a line support clamp or at the flared ends in the area of the connector fittings. If your copper line has a long run with little support, form a large two-inch loop in the tubing to minimize the effects of vibration. Secure the entire length as best you can to cut down the vibration.

ANNEALING COPPER TUBING

Annealing copper tubing is almost a lost art because it is not being practiced as much as in the past. The process is simple: Heat the copper tubing with a torch to approximately 1,000 degrees F. The tube will take on a red glow in a dimly lighted work area (is there any other kind?). Rainbow colors will show up, verifying that the proper temperature has been reached. Quench the entire tubing assembly in water. That's the textbook way.

Some builders report considerable success in annealing their copper lines in the same manner they do aluminum lines. That is, coat them with the carbonizing flame of a welding torch, readjust the flame to neutral and burn off the carbon coat. One difference: the copper line is quenched in water.

TUBING JOINTS IN GENERAL

Tubing connections may be made using either a flared, flareless, or a beaded joint. Although beaded joints are used in vacuum and fuel systems in production aircraft, you don't often see them in homebuilt birds because a special tool is required to form the bead in the tube.

The flared type of joint is used with a sleeve and a coupling nut fitting. The flareless type, of course, requires the flareless

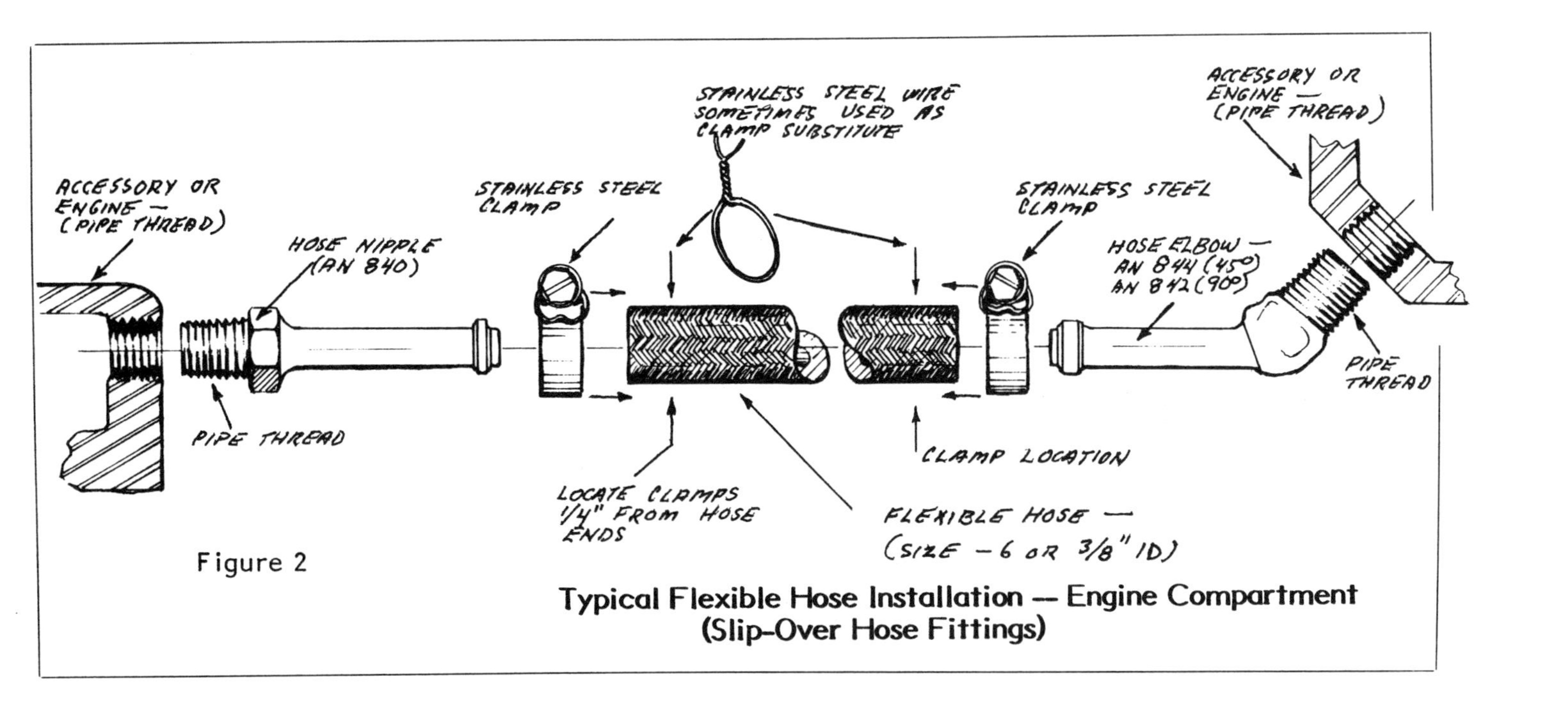

Figure 2

Typical Flexible Hose Installation — Engine Compartment (Slip-Over Hose Fittings)

fittings. Both types of joints are acceptable and provide safe, dependable tube connections.

Double flared tubing joints are superior to single flared types because they are smoother, more concentric and less prone to leak. However, not many builders have a double flaring tool, so single flare tubing joints are more prevelant among amateur-built aircraft.

FORMING FLARED TUBE ENDS

Metal lines in aircraft have traditionally been formed using flared tube fittings. In this kind of installation, the ends of the tubing are flared to fit special connections. Aircraft fittings have a 37-degree flare angle built in and the tubing must, therefore, be flared with

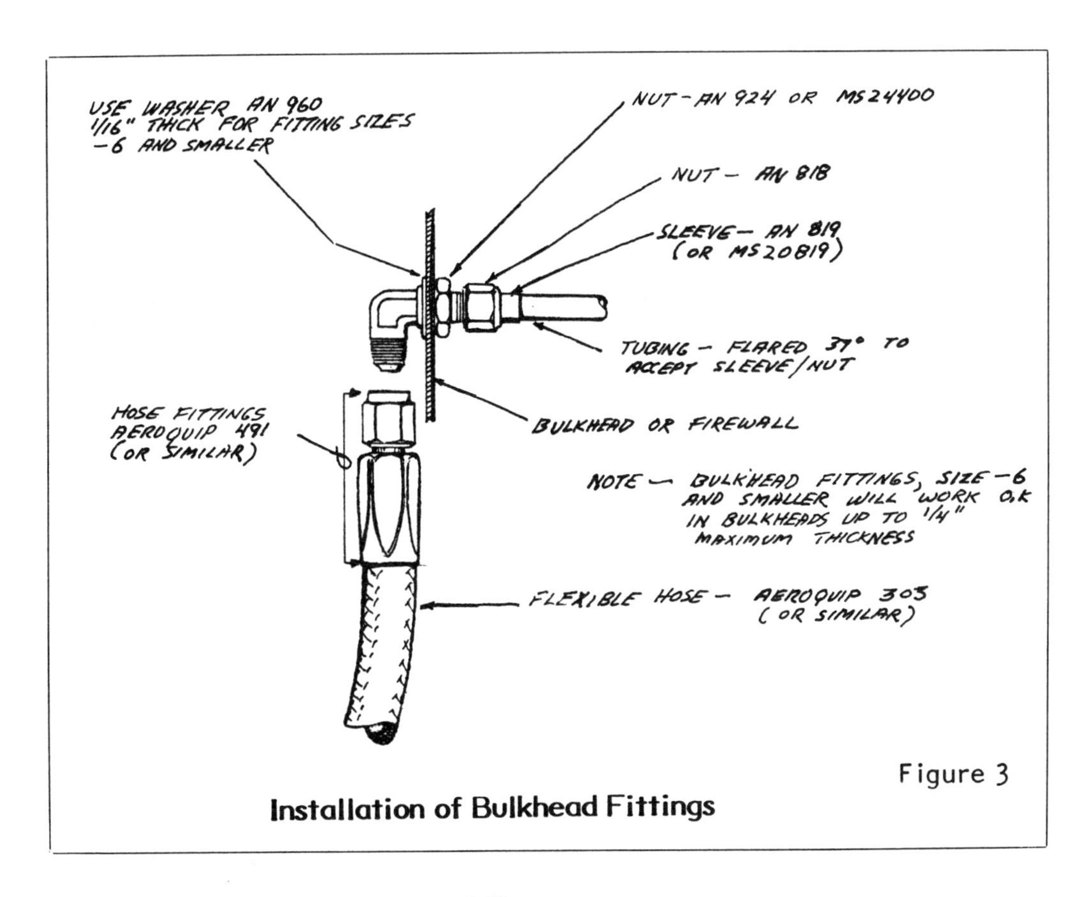

Figure 3

Installation of Bulkhead Fittings

Tables full of aircraft goodies attract the bargain-conscious builder, but beware of some of those old salvaged flexible hoses. They may cost more in trouble than they're worth in money.

a tool having the 37-degree angle forming cone. A regular automotive flaring tool must not be used to make the flared ends for aircraft fittings. Its flare angle is too severe and can ultimately cause the fitting to crack or fail completely.

A flared tube connection consists of a sleeve (AN819) and a coupling nut (AN818). Remember, the nut goes on first then the sleeve before the flared end in the tubing is formed. I'm sure you won't slip them on in the wrong sequence or completely forget to slip them on and not realize your mistake until after flaring the tube. No one ever does that!

The installation of leak-free lines is so important that a review of the technique for fabricating a flared tube connection is in order.

MAKING A FLARED TUBE CONNECTION

1. Cut tubing to the required length with a tube cutter or a fine-tooth hacksaw. File the cut end smooth.
2. De-burr the tube end, inside and out with a small knife blade or de-burring tool.
3. Clean out the tube (compressed air blast) removing all filings.
4. Slip on Coupling Nut AN818 and Sleeve AN819 correctly and make flare with 37-degree flaring tool.
5. Check the flare by sliding the sleeve up against it. Flare should extend 1/32-inch to 1/16-inch beyond it but its diameter should be no greater than the sleeve. Never re-flare an undersized flare. It will become brittle. Likewise, do not file down an oversized flare to fit .... that's SLOPPY work!
7. Examine the flare for cracks, nicks and scratches. If any are present, reject the flare. It will leak.
8. Clean out (blow out) the tubing again before installing it. If the completed unit will not be immediately installed, cap or tape over both ends to keep out dirt.

INSTALLING METAL LINES

A metal line should always be slightly longer than the straight line distance between the two connections. This is to allow a slight bend to be put in the line to take care of future expansion and contraction and to minimize the effects of vibration on the line.

Small tubing (1/8-inch to 1/4-inch) may be easily bent and shaped by hand. Larger tubing sizes should be bent with the help of a tube bender for all severe bends. Gentle bends, of course, can be hand formed even in 3/8-inch and 1/2-inch sizes.

A length of welding rod or stiff house wire can be used to form a three-dimensional template for the line before making and installing it. This will help you establish the best routing for it, determine the exact length needed and ensure a perfect fit.

Keep the number of connections and fittings to a minimum. A joint that is not there will never leak so why use another fitting when a simple bend in the tubing will serve as well? The installation will be lighter and less expensive, too.

Never apply a thread compound or lubricant directly to the flare faces of the tube or fitting. It will ruin the metal-to-metal contact that is necessary for a good seal between the two.

Always get the line assembly properly aligned before tightening the fittings. Start the nut with your fingers to ensure at least two or three turns before putting the wrench to it. Torque the fittings on 1/4-inch tubing to 40 to 65 inch pounds, and, on 3/8-inch tubing, to 75 to 125 inch pounds. Anything smaller falls somewhere in between. Steel nuts (grey) can be torqued slightly more than the aluminum (blue) nuts. Avoid over-tightening! If the connections leak, your flared tube has imperfections or some dirt is trapped somewhere. If you are using a second-hand sleeve, examine it carefully, it may be cracked.

To minimize corrosion do not mix materials. That is, for aluminum tubing stick to aluminum fittings. Steel with steel, etc.

Use clamps to support the lines and to reduce vibration. Use plastic or rubber cushioned clamps and space them approximately 9 to 10 inches apart (for 1/8-inch tubing) or 12 to 14 inches apart (for 1/4-inch lines). The larger the diameter of the line, the farther apart the clamps could be. Obviously, in a small airplane, one or two clamps may be all you can install for any particular line. Use your own judgment.

ABOUT SURPLUS, USED FLEXIBLE HOSES

You will find all sorts of flexible aircraft hoses at Fly-Ins and used aircraft parts markets around the country. You will be sorely tempted by all the 'bargains' you see and your sales resistance will be at an all time low. BEWARE, AMIGO! Some of those 'hard-to-find' parts displayed in the 'Country Store' or 'Fly Market' are items that the seller no longer wants, needs, cannot use, has to get rid of, won't work or are otherwise unusable or worthless (select one).

Consider where those hoses might have come from. For the most part, the unused surplus hoses are from military sources or from aircraft manufacturers' stores. In either case, the hoses are probably quite old and have exceeded their "authorized shelf life".

A hose's age limit is generally established to be about four years. The term refers to an arbitrary age limitation set by the manufacturer as the normal storage life for flexible hoses. It does not have anything to do with the life of a hose that is in use in an aircraft nor is it an accurate indication of the length of service of such a hose after installation. A hose that appears to be in excellent condition upon visual examination may have deteriorated in places you cannot see.

It is difficult to determine the airworthiness of any flexible hose assembly, especially those having a metal braid lining and/or covering. One test you can perform, however, on both surplus and used hoses is to flex them to see if they have taken on a permanent set. This is not a very scientific means of determining condition, but when a hose's flexibility is gone, so is much of its useful service life.

If you are still unsure about the condition of the hose, put it close to your ear and flex it. If you hear a crackling sound you can be sure the hose is becomming brittle and should not be used.

Another thing to watch out for in buying surplus hoses: the fittings may be of an oddball variety and you could have trouble obtaining matching connectors.

ABOUT BUYING NEW FLEXIBLE HOSES

It is generally true that new hoses and lines will provide a better and longer-lived installation than those obtained from used or surplus sources. I say 'generally' because sometimes even the hoses purchased from an authorized parts outlet may have been sitting on that dealer's shelf for several years. When buying a new hose, try to assure yourself that you are not getting one that has a high shelf life age.

Some hoses display the manufacturer's markings, which, in many cases, indicate age. Unfortunately, these markings vary considerably. For the most part, the markings on the hose cover (or on an attached embossed metal tag) display the manufacturer's name, part number, dash number, hose size, SAE number or rating and date of manufacture .... sometimes. (This information is collectively referred to as the "layline".) The layline may indicate the quarter year in which the hose was manufactured, thus giving you a clue as to the age of the hose. For example, the marking "4080" means the hose was manufactured in the fourth quarter of 1980. A marking of 2071 would indicate that the hose has been on the shelf a very long time indeed, having been manufactured in the second quarter of 1971.

You may, however, purchase a hose that has no such markings or has unreadable markings so that the layline may not be of much help.

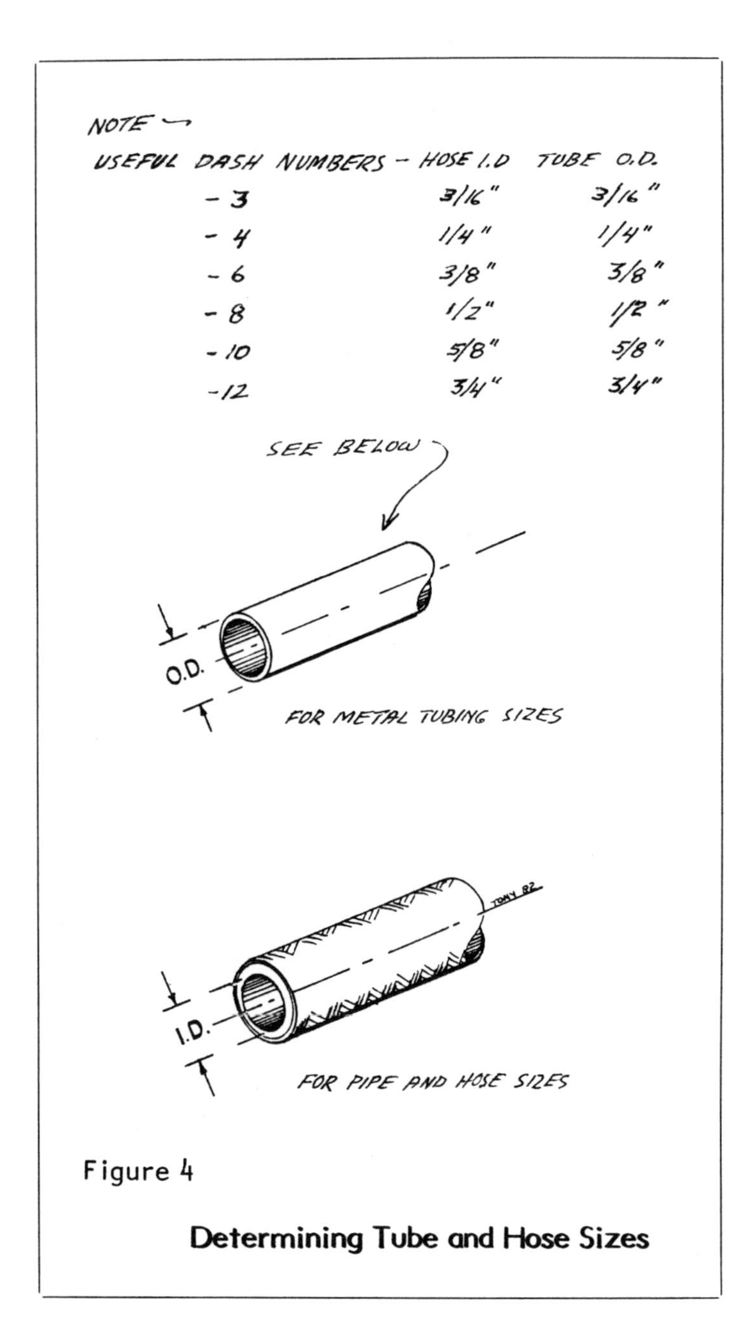

Figure 4

Determining Tube and Hose Sizes

In that case about all you have to go on are the dealer's assurances, the physical appearance of the hose and its flexibility.

IDENTIFYING TUBING AND HOSE SIZES

Tubing and flexible aircraft hoses, as well as their fittings, are identified by dash numbers assigned by the manufacturer. Metal tubing is sized according to its outside diameter (OD) and is measured in sixteenths of an inch. A 3/8-inch (OD) aluminum tube measures 6/16-inch, so the manufacturer uses a -6 as his size identification for all 3/8-inch tubing and matching fittings. A -4 size is assigned to 1/4-inch tubing (4/16-inch OD). (Figure 4)

Although a flexible hose is bulkier on the outside than an equivalent metal tube, the hose's internal diameter is closely related to the tubing size and its dash number matches that of the metal tube.

HOSE SELECTION

Hose selection is based on intended use. You need to know what gas or fluid the hose must carry and under what pressures; the internal and external temperatures to which it will be subjected; and, to a degree, the amount of flexing that will be imposed on the hose.

Some builders make the mistake of assuming that any type of rubberized hose with fittings on both ends is adequate for their fuel, oil or hydraulic brake system needs as long as the the fittings are the correct size and the hose length is adequate. This is not so. Many light rubberized hoses are intended solely for use in low pressure air and vacuum or instrument lines. Moreover, these hoses are not ordinarily compatible with the chemicals in fuel or oil, nor are they capable of withstanding high temperatures and high pressures.

Recently, preoccupation with weight control has led builders of composite aircraft to use lightweight plastic fuel lines. They are installing these lines both ahead of and behind the firewall. Although some, plastic lines are well suited to carrying fuel, they lack the fire resistant quality of standard aircraft flexible hoses.

Many FAA inspectors do not approve of such installations so you'd better check with your FAA inspector before spending good money on something you may not be allowed to use.

Although some types of plastic tubing are very resistant to detrimental effects of fuel and are relatively long lived, they are best used behind the firewall and not inside the engine compartment. The final word on the subject is the FAA inspector's.

AIRCRAFT, INDUSTRIAL FLEXIBLE HOSES

Locating the proper high quality flexible hoses for your project may be difficult if you live in a small town or city. However, one source you shouldn't overlook is the typical automotive speed shop. These auto shops stock and use goodies like 4130 steel tube, rod end bearings and Aeroquip hoses. You can have them make up whatever length hose you need complete with the proper fittings, or you can do it yourself.

AEROQUIP 303 HOSE — This is a good all purpose hose stocked by most aircraft parts departments, suppliers to the homebuilder trade, and by some automotive speed shops. It is by far the most popular of the flexible hoses being installed in homebuilts. It is a versatile standard hose that can be custom made to whatever length you need. Best of all, you can use Aeroquip 491 hose fittings to make up your own hoses easily (and at a considerable cost savings). The hose has a synthetic-impregnated, oil resistant cotton braid outer cover, while its inner tube is made of a seamless synthetic rubber reinforced with a single wire braid over a layered cotton braid. A hose made like that must be good. It has an operating temperature range of from -65 degrees F to 250 degrees F and a maximum operating pressure of 2,000 to 3,000 psi for the relatively small hose sizes you would be using. This capability also makes it acceptable for your hydraulic system (brakes, gear, etc.) needs.

AEROQUIP 601 HOSE — This hose is designed to withstand an even greater range of operating temperatures — between -40 degrees F to 300 degrees F — and is widely used for aircraft fuel and oil lines. It is recognizable by its external stainless steel braid. Its maximum operating pressure of only 1,000 psi, however, makes it unsuitable as a hydraulic line. Such a hose may be made up with Aeroquip 816 hose fittings to whatever lengths you need.

AEROQUIP FIRESLEEVE — Any of the aforementioned hoses or lines may be protected with a special fire resistant or fireproof cover. This cover is in the form of a sleeve which may be slipped over the regular hose. You must order the sleeve to fit the outside diameter of the flexible hose. It is available in long lengths. Cut it to fit the length installation you have and clamp both ends.

There are, of course, other types of aircraft and industrial hoses that would be suitable for use in you fuel, oil and hydraulic applications. Unfortunately, they are not commonly available to most of us and are not, therefore, included in this brief tabulation. Check with the manufacturer of whatever hose you wish to use. Always abide by the manufacturer's limitations.

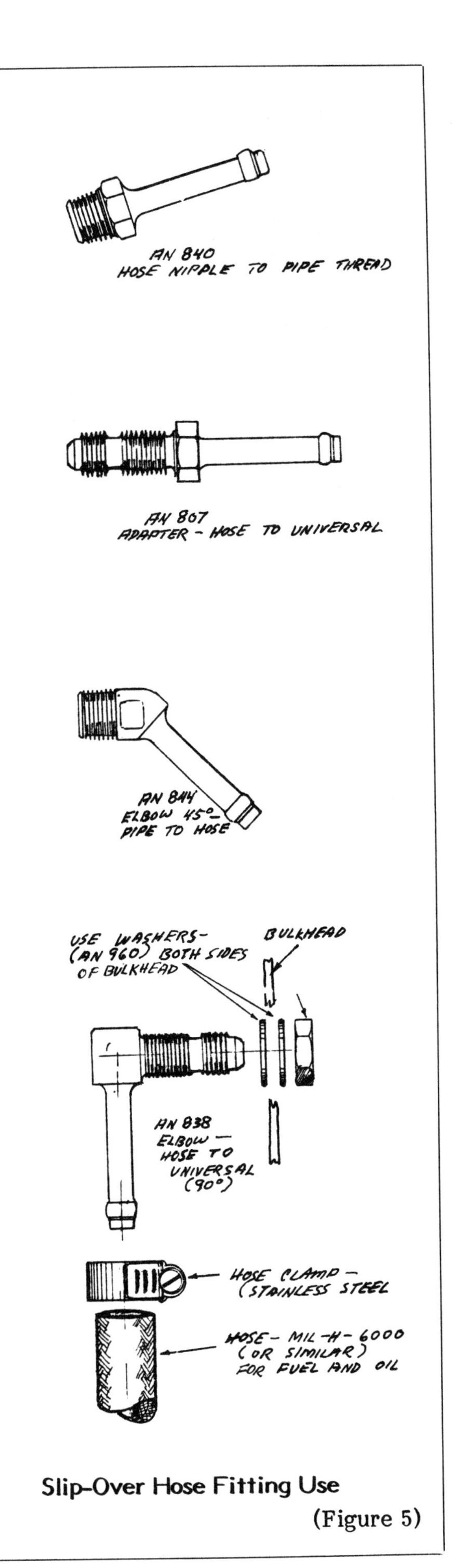

Slip-Over Hose Fitting Use

(Figure 5)

Oil coolers are always connected to the engine by flexible hoses. This up front installation is one frequently by homebuilders.

This installation is similar to the one above but there are some differences. Note different type oil cooler, different technique for baffle seal installation and ignition harness routing. Yet both installations are equally functional.

TIPS ON INSTALLING FLEXIBLE HOSES

If you will not be installing your hoses for a long time, I recommend that you coat them internally with lightweight engine oil and plug or cap the open ends. Keep them capped until you are ready to install them.

Avoid making sharp bends in a hose and route it so that the fuel will flow uphill or downhill without reversal.

Ask any experienced mechanic. The use of thread lubricant in the assembly of the fuel system fittings is not necessary. Should you deem it essential to do so, use something like plain old engine oil. Any thread sealant or lubricant used must be applied lightly and only to the threaded male portion of the fitting. Don't allow the lubricant to slop or drip across the end of the fitting.

Make your fuel system with as few connections and fittings as possible and secure each line to make it as free of vibration as possible.

Fuel lines should be separated from hot exhaust stacks by a minimum distance of 1 1/4-inch to prevent heat damage and possible fuel vaporization.

When connecting hose fittings, be sure the parts are properly aligned.

Don't try to screw the pipe thread end of a fitting into a flare fitting nut. It won't work.

Don't stretch hoses tightly between connections, but also avoid excessive slack (length) that will give the installation a messy appearance. Route all hoses carefully, keeping them clear of sharp edges, moving controls, manifolds and hot exhausts.

Avoid twisting a hose when tightening the fittings. A twisted hose will have a very short operational life.

Hose connections that are only finger-tight are as bad as those that have been overtightened. Connections that are too loose will leak, but overtightening may result in cracked or severely deformed sealing seats in 37-degree flare fittings.

You can make your flexible hoses more fire resistant by slipping a firesleeve (Aeroquip 624) over each of them and clamping them at the ends with stainless steel clamps. The firesleeves may also be helpful in reducing the fuel vaporization tendencies.

CAUTION: Do not permit residual fuel from a disconnected hose or line to drip into any area of the aircraft structure. It creates a very serious fire hazard. As you know, fuel fumes are heavier than air and will settle to the lowest levels where even the slightest spark can ignite them.

One final reminder. Installing old or substandard fuel and oil lines is no way to practice economy. A failed hose could result in a failed engine or system, and could cause you to make an unexpected landing.

N6100U

fuel systems

8

Managing the System

In an auto you turn on the key, glance at the gas gauge, hit the starter and go. There is no fuel valve to turn on, no tanks to switch and check against the gauges, no fuel pump, transfer pump or booster pump to operate, no fuel pressure gauge to monitor; in short, there is no fuel management problem. Pretty hard to mess up a system like that, isn't it?

Airplanes, on the other hand, have all sorts of these puzzling gauges, selectors and switches, each of them demanding some action before you can even taxi out.

When you think about it, it's easy to see why so many pilots are dissatisfied with the fuel management requirements of many of the airplanes they fly. They are dissatisfied with the accessibility, location, and logic of operation of fuel selector valves, which are often opposite to the pilot's instinctive actions.

Too many of the airplanes we fly require tank switching in order to properly manage the fuel supply. Why should this be? Why can't an aircraft's fuel management system be as simple as an automobile's? Granted, airplanes are more complicated than autos and they have balance problems due to the installation of fuel tanks in the wings, etc., etc. But, why should the pilot have to do everything? If complex systems are necessary, why can't they be made to work automatically without the assistance (or interference) of the pilot? After all, isn't it the pilot who switches to an empty tank, misreads fuel gauges, and does all sorts of other pilot error things?

The fuel management problem results primarily from the lack of a standardized fuel system. The kind of fuel system we have now usually consists of a fuel selector (with the "proper" markings), one or two fuel quantity indicators and perhaps a booster pump switch. With only these few elements to cope with you'd think standardization would be easy. If so, it certainly hasn't been accomplished thus far. The fuel systems now in use do not safeguard against pilot error incidents.

During March of 1978, the FAA published a report entitled "General Aviation (FAR 23) Cockpit Standardization Analysis (AD/A-052 803)". It's a good publication and should be of considerable interest to anyone who designs, builds, or flys airplanes. In essence, this report, compiled by Robert J. Ontiveros, Roman R. Spangler and Richard L. Sulzer, outlines a set of recommendations for cockpit standardization.

According to the report, many safety experts believe that certain design elements, particularly in the cockpit area, contain features that tend to lure pilots into taking improper action and making bad decisions based on erroneous impressions. This and previous studies, along with the recommendations made by GAMA (General Aviation Manufacturers Association) to the FAA, all point out the need for the standardization of fuel management functions.

I believe that cockpit fuel management remains a complex, and sometimes confusing problem because we have been too complacent to develop standardized automatic fuel systems that do not require assistance from the pilot. I am not too sure, however, that I would know a standardized fuel system if I saw one. But one thing is sure, if the demands on the pilot are kept to a miminum, there is sure to be less opportunity for his mismanaging its operation.

The consequence of fuel mismanagement is, of course, fuel starvation and engine failure. An engine dying for lack of fuel immediately commands your attention, confronting you with the necessity of making rapid and critical decisions. If you are flying an aircraft with multiple fuel tanks and a nonstandardized cockpit layout, you might well have difficulty determining which tank is presently on-line and, even determining the quantity of fuel remaining in each tank. Your need for a rapid assessment of the available information is critical. Lacking this, you may make an incorrect decision and take inappropriate action.

FUEL SYSTEM OPTIONS

If you can get by with it, the single tank gravity flow system with a simple ON/OFF valve installed just beneath the tank is the most typical and, perhaps, the best fuel system installation you can make or buy. A simple fail-proof wire and cork fuel level indicator completes the basic setup. Adding embellishments to this clever rig merely detracts from the simplicity of the system. However, an electrical fuel indicator or some other visual fuel level indicator may be substituted, if you prefer. (Figure 1)

The installation of a second tank always increases the complexity of any fuel system and the workload for the pilot. It also increases the potential for pilot error. A good installation is one where both tanks are connected to a single Y or T fitting and feed the engine simultaneously. (Figure 2)

High-wing aircraft and biplanes equipped with a tank in each wing can still provide you with a simple gravity-flow system that requires no particular fuel management attention. It is simply ON all the time and the engine will continue to run as long as you have fuel.

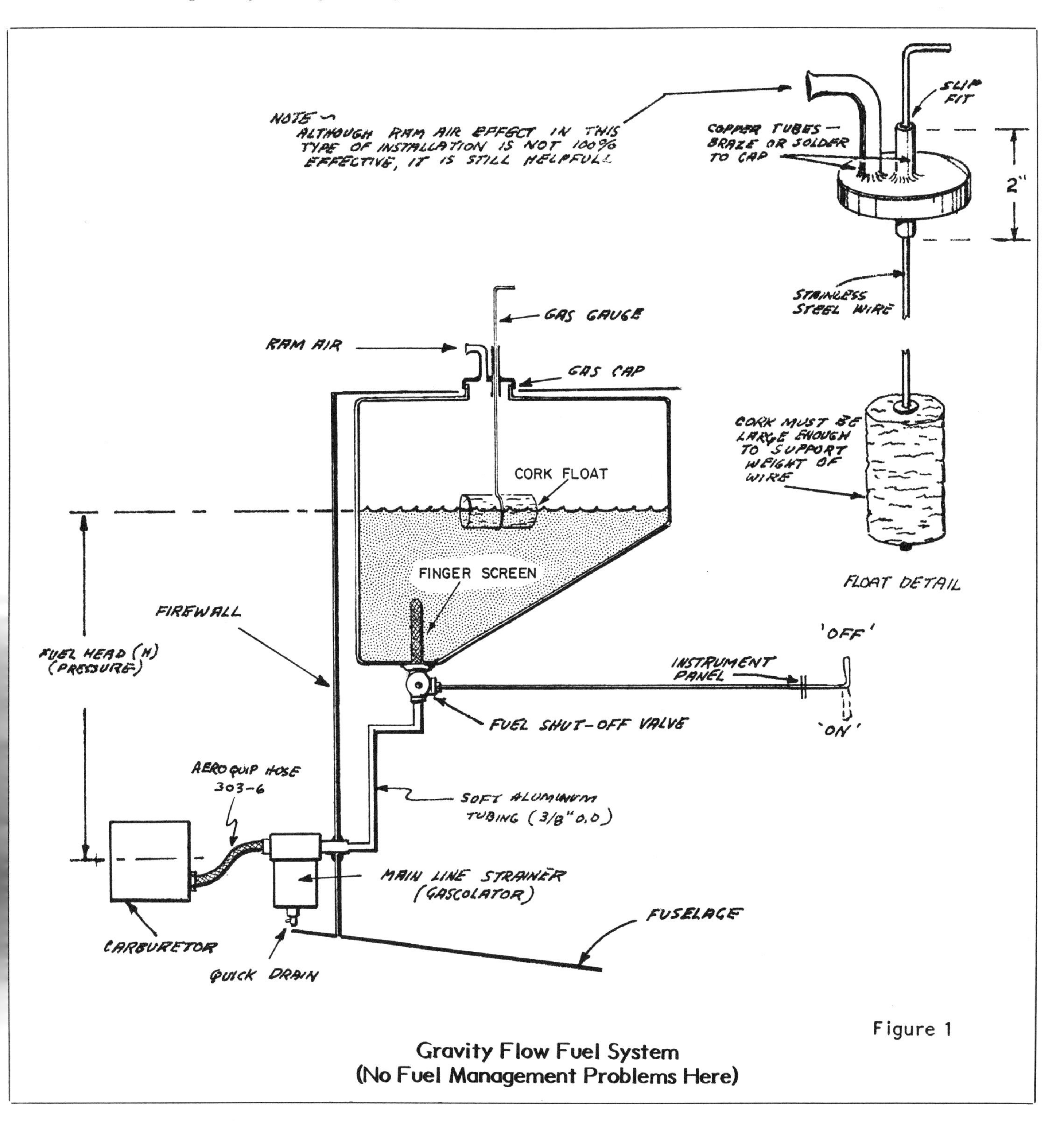

Figure 1

Gravity Flow Fuel System
(No Fuel Management Problems Here)

The only objection to a single BOTH ON arrangement for a two-tank installation is the inability to shut off each tank individually. Ordinarily, there is no need to do so except in the remote event of one of the tanks or its fuel line developing a leak, thereby endangering your fuel supply by draining both connected tanks or creating a fire hazard. If you were able to shut off one tank, the opposite tank would still be available for use.

A two-tank gravity flow fuel installation must have a cross-over vent line interconnecting the two tanks in order for the fuel to flow equally from both tanks. This is essential when only one of the tanks is vented to the outside atmosphere. You can see that when you opt for additional fuel tanks, installation as well as fuel management becomes more demanding and complex.

A two-tank installation in a low-wing aircraft introduces even greater complexity simply because the tanks are often situated below the carburetor level. In order for the fuel to flow up to the carburetor, a fuel pump must be installed. This is usually an engine-driven pump. Since fuel will not flow uphill, and since the engine cannot operate if the engine-driven fuel pump fails, a back-up system for the engine is required. An electrical fuel boost pump is, therefore, installed in most

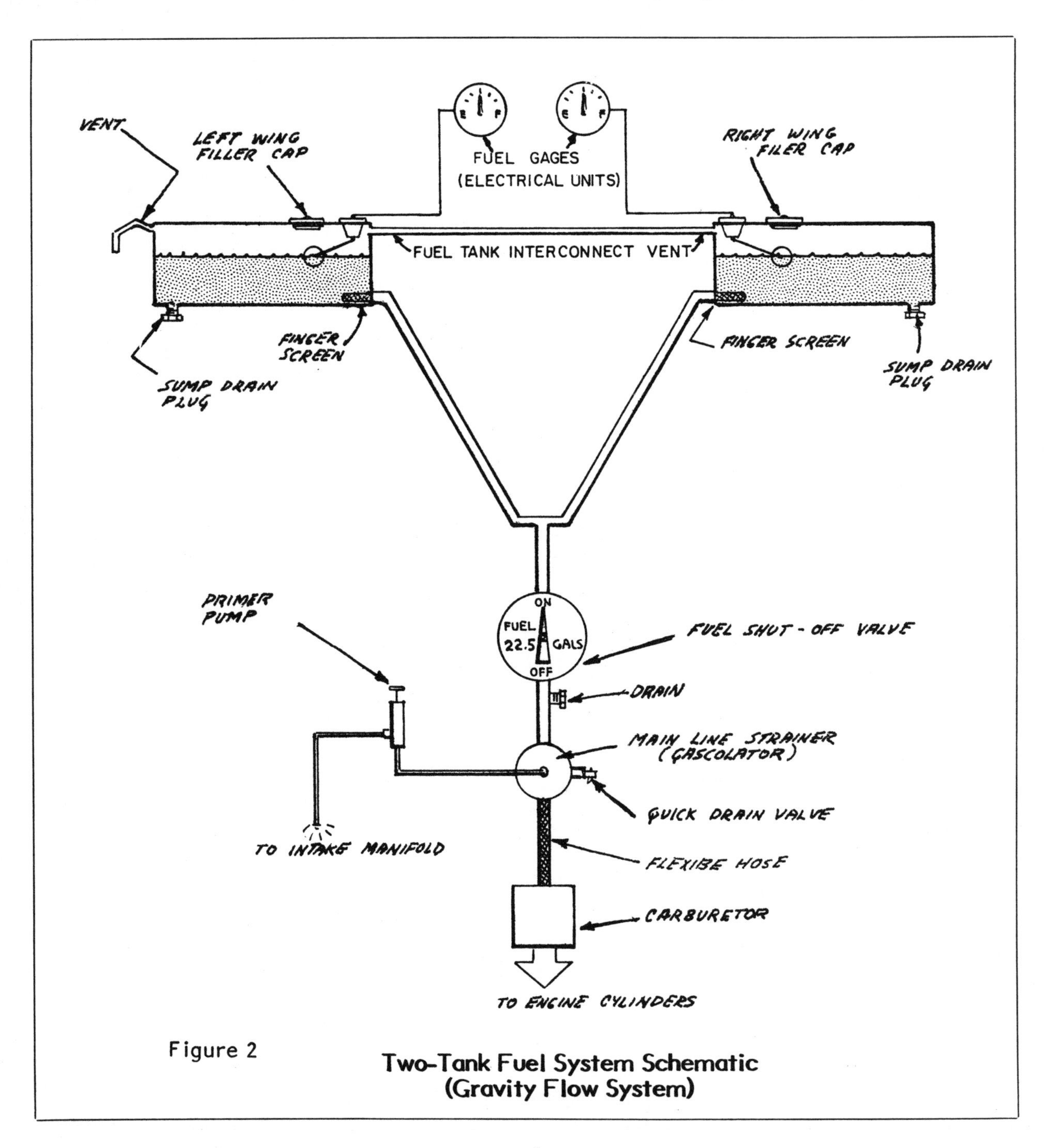

Figure 2

Two-Tank Fuel System Schematic (Gravity Flow System)

aircraft that do not have a gravity flow fuel system. In addition to providing the fuel flow needed to start the engine, the electrical fuel boost pump could keep the engine running in the event the engine driven pump fails. Of course, all this leads to the need for pilot intervention.The complexity grows as the number of tanks increases. In addition, a fuel selector is needed for still more options....and unfortunately, each option adds more ways to goof up.

If, as in the case of quite a few single-seaters and other, larger homebuilts, a second tank is installed in the fuselage as a transfer tank, you will have to install a shut-off valve, a transfer pump, a separate fuel gauge for that tank, the usual plumbing, and perhaps a one-way valve and some sort of warning light or device to let you know when the transfer pump is pumping fuel into the main tank. In such an installation it is possible to forget that the fuel transfer is going on. This would result in precious fuel being pumped overboard if the pump were not switched off in time.

Anyone can install a complex system — all it takes is time and money, but not many can come up with a complex system that is not dependent on the pilot.

Fuel Tanks

The construction of fuel tanks is beyond the scope of this book, but a few essential details may be helpful to you in the selection or construction of your fuel tank and in making your engine installation.

A variety of commercially fabricated tanks are available to homebuilders. For the most part, these commercially produced tanks are made of weldable (soft) 5052H32 aluminum sheet .040-inch to .050-inch thick. Each tank is built according to the plans for a particular design considered popular enough to merit production in some quantity. Almost all of these tanks are heliarc welded. Although welding aluminum is considered to be an "art", don't be fooled by a nice appearing, slim, straight welded bead in a fuel tank. It may have been produced by a newcomer to the field and the appearance of such a weld could imply that little or no welding rod was added in the process. Merely fusing the edges of an aluminum tank results in a weak joint which might split from vibration and fluid loads. It is more reassuring to see a tank with generous weld beads even if they aren't exactly uniform and pretty.

Most homebuilts have relatively small fuel tanks — 12- to 30-gallons is about standard. Fuel tanks larger than five gallons in capacity should have internal baffles (about 12 inches apart, more or less) to keep fuel from surging during banks and turns. The presence of baffles can usually be recognized by the rows of welded-over rivet heads.

Quite a few builders prefer to fabricate their own gas tanks of fiberglass and epoxy resin. The advantage of this type of tank is that the builder can design its shape to fit almost any size area in the aircraft. If he is not careful, however, the tank could turn out to be much heavier than it needs to be.

Before you accept any tank for installation be sure that it has at least a 2-inch or larger filler neck, a vent outlet or connection, a fuel sump fitting large enough to accept a finger screen....and that the tank does not leak. If you want to use an electrical fuel gauge, an opening will have to be provided for it.

Each fuel tank is held in place in the aircraft with two steel straps, lined, preferably, with a nonabsorptive material such as Neoprene. However, cork or felt strips are frequently used by builders to prevent their tanks from chafing. The steel straps ordinarily have an adjustable end fitting which permits the strap to be tightened snugly against the tank.

Remember, you should be able to install and remove the fuel tank and disconnect your

Fuel tank installation in a Sonerai II.

fuel system without disassembling half the airplane.

High-wing aircraft typically have one or two tanks located in each wing and a fuel system that operates through gravity flow. Low-wing planes often can also utilize a gravity flow system if the tank is mounted in the nose of the aircraft just ahead of the instrument panel and high enough above the carburetor to establish the necessary fuel head. Otherwise, a pressure fuel system will be required, complete with an engine-driven fuel pump and the necessary back up electric pump. A gravity flow system must be capable of providing fuel pressure at the carburetor of at least 1/2-psi (you can measure this by disconnecting the fuel line from the carburetor and attaching a low-pressure gauge to the line). To achieve this pressure, your tank may have to be at least 19 inches above the carburetor. Some marginal installations manage to attain adequate fuel pressure with the tank being somewhat lower. However, the fuel head pressure must be augmented by a ram air vent in the top of the tank or filler cap.

FUEL TANK FILLER CAP

The fuel tank filler cap (or the area near it) should be marked with the word FUEL, the minimum usable octane and the tank capacity. Your fuel caps should be red in color; the oil filler cap, yellow.

The area around the filler neck ought to be sealed to prevent raw fuel and fumes from getting into the aircraft compartment areas below. This is particularly important with recessed or flush filler cap installations. Submerged filler cap recesses must be hooked to an overboard overflow line and must drain properly.

Know whether your filler cap is of the vented type (small vent hole through the top of the cap or under the lip) or of the nonvented type. A curved vent tube, soldered, brazed or welded into a filler cap (when used) must always be installed with its open end facing the slipstream to allow ram air to enter. If a line worker inadvertently installs it backwards (facing aft), fuel starvation could result.

Fuel Tank Vents

Each fuel tank must be vented in order for the fuel to flow freely through the system. Most tanks have a fitting either on top of the tank or near one of the tank's upper edges to which a vent line may be connected. This line exits at some point under the aircraft. An obvious outlet location to avoid is one around the exhaust pipe area. The vent fitting used in the tank usually takes either a 1/4-inch or 3/8-inch line. This is generally an aluminum tube, although some builders are installing vent lines of plastic tubing.

The open end of the vent line should be bent to face into the slipstream to induce a positive ram air pressure in the vent. Facing it aft could induce a suction-like effect and result in the loss of fuel.

Fuel vents can become clogged by mud daubers, other insects and debris. This creates the immediate danger of fuel starvation on takeoff or in flight. To help keep your vent line clear of debris and insect mischief, you should fasten a small piece of aluminum screen over the vent outlet. If the vent tube is plastic you may have to clamp or wire the screen on. An aluminum tube vent, on the other hand, can be flared slightly and its screen fitted inside with a little dab of epoxy to secure it. The installation is much neater and, besides, the end of the metal tube can be easily bent to face forward and will maintain that position without any other support.

An alternate vent is good insurance in case one gets clogged. Unless the vent is properly located and installed in the tank, it may allow the loss of fuel when the tank is full or when the aircraft is at a high angle of attack such as during climb out. Vent openings in the tank also should be above the fuel level in normal flight attitudes to allow pressure changes with altitude and to prevent the loss of fuel.

When wing tanks are used that feed simultaneously, the tanks must be interconnected with a vent line to ensure that the fuel feeds equally from both tanks.

A gravity flow system gains greater efficiency from the increase in ram air pressure into the tank when a vent is pointed into the slipstream. This effect is the same as increasing the height of the tank above the carburetor, it increases the fuel head (pressure)....and that is good. (Figure 1)

A fuel tank vent situated in the manner will also provide some ram pressure. However, such an installation should have an aluminum screen over the opening to prevent it from becoming plugged with bugs or debris.

An unusual ram air vent and cork/wire float type fuel gauge housed in the fuel cap. "Line boys" need reminding that such cap openings must face into slipstream.

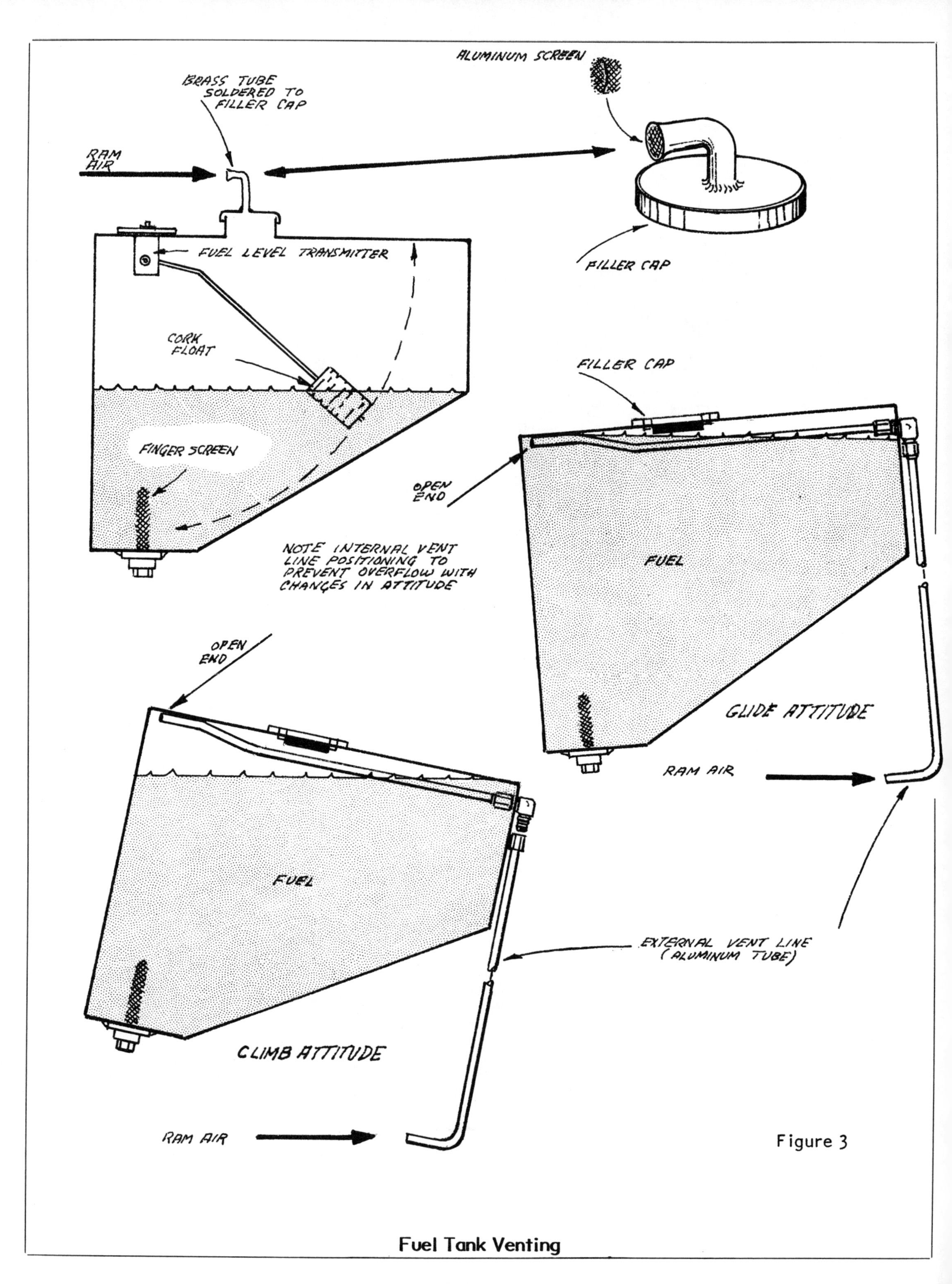

Figure 3

Fuel Tank Venting

A direct reading fuel gauge such as this one requires the tank to be installed so that the gauge is visible to the pilot. This tank fits into the left wing root of a high-wing airplane.

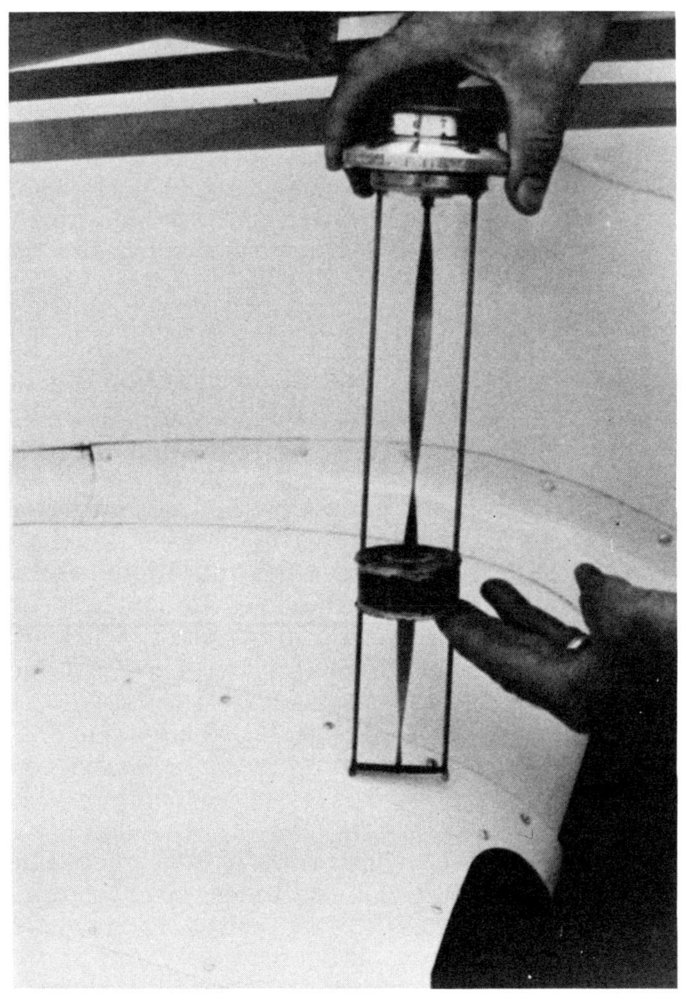

This easy to make fuel gauge consists of a spiral metal strip with a cork sandwiched between the two slotted metal plates which allow the cork to rise and fall turning a dial indicator in the fuel cap.

Fuel Quantity Indicators

Most fuel quantity indicators are of the magnetic type activated by a mechanically connected stiff wire arm on which a cork float is impaled. As the fuel level changes, the cork float rides with the fuel level, operating a variable resistance transmitter attached to the upper end of the float wire. When the float rides to the top, as when the tank is full, the minimum resistance is produced through the tank-mounted transmitter, causing the maximum current flow to the fuel quantity gauge installed in the instrument panel. As the fuel level falls, the resistance in the transmitter increases, producing a lesser current flow to the fuel quantity gauge. The pointer reflects this with a proportionately small deflection.

The sender unit, consisting of the cork float, wire arm and the mechanically attached transmitter, are installed through the top of the fuel tank. The unit is secured to the top of the tank by means of a circular plate. The plate opening is large enough for the transmitter unit to be removed for repair or replacement.

Are you planning to install a used fuel quantity transmitter or reinstall an old unit on an aircraft being rebuilt? Be advised that fuel gauges may give erroneous fuel quantity indications because of broken and/or corroded wires in the fuel quantity transmitter. At any rate, before you install the unit permanently, first determine if the fuel quantity gauge and the transmitter you have are compatible with each other. Then, check the units out by performing a continuity or resistance bench check of the indicating system.

Carefully follow the wiring instructions that are included with a new instrument or it may be damaged.

Shorten (cut) the float arm and adjust the float so that it swings freely through its entire range of movement. You want accurate EMPTY to FULL instrument needle deflections. Once adjusted, do not bend the float arm. Check again that the cork float does not stick or wedge against the tank bottom (EMPTY position).

Calibrate each gauge to read zero in a level flight attitude with the amount of fuel remaining down to the unusable level. Remember, in some tanks, a lot of fuel is unusable. If internal baffling is present, assure yourself that the float will not snag on it and be prevented full, free movement. If EXPLOSAFE material is used to fill the tank cavity, be sure that the float and arm are protected by a corral or chamber that provides at least 1/2- to 1-inch clearance all around the float.

Float and transmitter unit (bolted plate in foreground) requires a ground wire connection from one of the bolts to a common aircraft ground.

Finally, a good, separate ground wire connection is essential, particularly for a fiberglass (non-metal) gas tank. Oh yes, use a new gasket under the transmitter mounting plate when installing the float assembly in the tank.

FUEL GAUGES

The lack of standardization discussed earlier is rampant in fuel quantity indicators. Some aircraft have a separate gauge for each tank. Others share one gauge among several tanks by utilizing a switch to obtain the reading for the level of fuel in each tank in turn. Still other aircraft have fuel tank selector controls with a fixed relation between the selector and the fuel quantity indicator, making it necessary to switch fuel flow to a particular tank to obtain a reading of the amount of fuel remaining. Confused? Who wouldn't be. Your fuel system needn't be that complicated.

Remember those wire-and-float fuel gauges on the old J-3 Cubs? Simple, eh? Reliable too, when properly constructed because they required no switches, no selectors and no mathematics to mess with. They were totally trouble-free. It seems to me that an improvement is no improvement at all if it imposes additional requirements on the pilot.

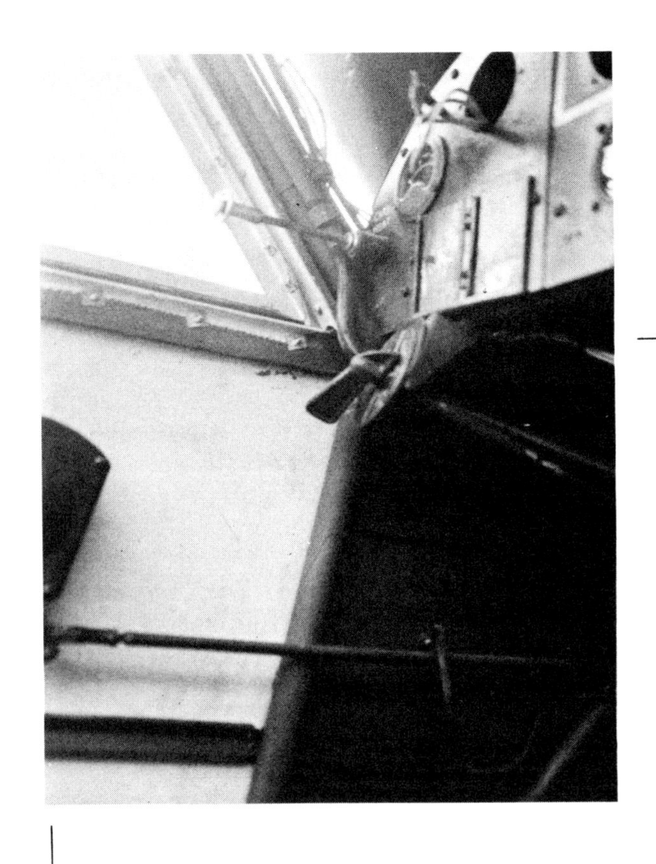

A very convenient, easy to reach location for the fuel shut-off or selector valve.

Protection from engine compartment heat is provided this electric fuel pump and gascolator installation by the enclosed aluminum chamber (Piper).

Fuel Valves

NOTE: Although virtually all fuel valves used in aircraft have an OFF position, some of them are called SHUT-OFF VALVES and some are called SELECTOR VALVES. Technically speaking, all are fuel selector valves because when you turn that handle you are making a selection, even if it is only to turn the fuel OFF. In general, however, we refer to a fuel valve that provides a multiple choice of positions as a SELECTOR VALVE and to one that has only ON/OFF positions as a SHUT-OFF VALVE. Got that?

FUEL SHUT-OFF VALVE

Each fuel tank should have its own fuel shut-off or selector valve with an OFF position. Outside of using the fuel valve to shut off the flow of fuel so that you can work on the fuel system, it really serves little other use during the life of the aircraft. In theory, the shut-off valve might have to be used during an in-flight emergency so the valve must be easily reached in flight. Remember this, however, shut off the fuel and down you go anyway because the engine won't be running. On the other hand, a shut-off valve may be necessary to control the flow of fuel from a transfer tank.

The fuel valves commonly available to homebuilders are basically shut-off valves made of solid brass....and they are HEAVY! It is no wonder that experienced builders caution against screwing a shut-off valve directly into the bottom of the tank. In that location the heavy valve could impose too great a stress on the tank sump due to vibration. Since most of these shut-off valves have an integral mounting bracket, it really is no problem to locate the valve at some convenient, easy-to-reach place. These fuel valves are made in two-way ON/OFF types, in a three-way type, with flow from either side of a two-tank installation, and in a four-way valve for a three-tank installation feeding from any side into a single bottom outlet.

FUEL SELECTOR VALVES

Multiple tank selector valves are not normally available from most homebuilder suppliers. This tends to discourage their use in homebuilts. However, a builder who really needs one usually manages to locate a salvage fuel selector. If you use a second-hand selector it will generally mean that your homebuilt will have a fuel system very much like the one the selector was originally designed for. This may or may not be good, depending upon the reputation of the original system for which it was designed. If the valve is a complex one and was intended for use with several tanks — tanks you don't intend to install — the valve will have a vacant position or two and your installation will become a candidate for inadvertent fuel mismanagement incidents.

If you must use a fuel selector, get one that is no more complex than required for the number of tanks you will install. Here are a few recommendations for choosing a fuel selector valve:

* Look for a valve with a pointer that looks like a pointer and leaves no doubt as to the tank position selected.

* Pick one that has a positive click for each position. If it doesn't have a positive click or feel in each position, it would be easy, when moving the selector handle, without looking at it, to inadvertently stop at a midway position. The immediate result, of course, would be a sudden, eerie silence on the part of the engine. (The NTSB calls that "fuel starvation due to pilot error.")

If you wish to join the trend towards cockpit standardization, consider using the following (natural) fuel selector pointer positions:

* Turning the valve selector to the right should give you the RIGHT tank....to the left, the LEFT tank....forward for ALL TANKS....and rearward for OFF. All other tank positions that may be required must be located between the LEFT and RIGHT tank positions.

* The OFF position should be at least 90 degrees from any tank position selected. Another thing to consider: your fuel selector handle (pointer) should not pass

through the OFF position for obvious reasons. Compare these recommendations with the design of the selector before you buy it. Incidentally, in the event your selector pointer doubles as a shut-OFF handle, its OFF position should be RED....not black, as it may be on some older production line aircraft.

FUEL VALVE INSTALLATION

Locate the selector handle at some point directly in front of you so you can see it and reach it without having to move a pile of junk, like maps, coffee cups, rags, or the seat.

If your aircraft features side-by-side seating, the fuel tank selector should be within reach of both pilots. Such a provision is quite difficult, but not impossible, to make in a dual-control aircraft with tandem seating.

Avoid mounting the selector valve on the cockpit sides if you have wing tanks. It is just that a small human quirk may lead to confusion in time of stress. Why? Well, turning the selector to the right wing in such an installation would point it to the tail or to the nose (depending on the cockpit side used). Don't laugh, strange things do occur, especially during periods of 'tight pucker'. You know Murphy's Law: "If anything can go wrong, it will." You can bet your boots that all mechanical devices, and many humans as well, abide by that edict.

Although designs vary, the handle of a fuel shut-off valve should not, ordinarily, be installed in an UP attitude for the ON position. However remote the possibility, vibration can cause the handle to work downward to the OFF position, shutting off the flow of fuel. It is particularly important to locate the fuel valve at some place where it will not be accidentally activated by your leg or sleeve.

Before permanently installing your fuel valve, it might be a good idea to disassemble it and carefully examine the shaft and bushing for evidence of burrs (yes, even if it is new). Cleaning and polishing the valve should remedy any difficulty in most cases. Coat the parts sparingly with engine oil prior to reassembly. Do not be too quick to condemn any valve as too difficult to operate. Try sloshing some fuel inside it first. Friction is greatly reduced when the valve is worked in its natural environment....wet with fuel.

Valves salvaged from other aircraft may have a placard that can be incorrectly reinstalled by a builder rebuilding or restoring an antique or a classic. Be careful, it is possible to inadvertently reverse the position markings when refinishing the placard. Always check and recheck the correlation of the selector valve pointer, passages, and the placard for their correct relationship to the fuel gauges on final installation.

INSTALLING FUEL LINES AND FITTINGS

Install the fuel lines so that they slope downward from the gas tank to the gascolator (main strainer) without any upward reversals. By the same token, the fuel line should have a constant upslope from the gascolator to the carburetor without any reversals.

The fuel line from the main strainer, mounted on the firewall, needs to be flexible and fire resistant because there is relative movement between the shock-mounted engine and the stationary firewall. Flexing of metal fuel lines will eventually cause them to crack, creating a serious fire hazard.

CAUTION: *When fitting connections, be sure to line up the tubing and its mating nut exactly. If the coupling nut cannot be easily spun on two or three turns with your fingers, STOP! Something is wrong. Don't force it with a wrench. It is misaligned or it may be mismatched. CHECK IT OUT!*

When you make your aluminum lines, cut the tube ends squarely and file them smooth. Always remove burrs and sharp edges before forming the flared ends.

The correct flaring tool to use is one with a 37-degree angle (aviation flare forming tool). Do not use the common 45-degree angle (automotive) flaring tool with aviation type hardware. It is not safe to form flared ends for aircraft fittings with it. You might initially get a connection that will not leak, but the severe flare and the stress imposed by the misfit may contribute to a future fuel system failure.

SUMMING UP

Don't believe for a minute that a selector valve must be used whenever more than a single fuel tank is installed in the fuel system. This is not necessarily true. Most homebuilt fuel systems do not need a fuel selector valve even though two fuel tanks are installed. A simple ON/OFF valve is often sufficient. Such an arrangement is found on most homebuilts and on quite a few production aircraft. Consider the Lake Amphibian and the Cessna 152, for example. The arrangement is an excellent one and virtually eliminates the possibility of mismanagement of the fuel. Both tanks feed automatically all the time....without help from the pilot. Now, isn't that a logical arrangement? Fuel selectors that are poorly designed and confusing in their mode of operation only add to mismanagement problems.

Fuel Strainers

The fuel in the average aircraft fuel system will have to flow through three strainers before it reaches the engine's combustion chamber. These strainers include:

1. The finger screen (inside the gas tanks)
2. The gascolator (usually mounted on the firewall)
3. The carburetor screen (inside the carburetor)

The main function of these strainers is to prevent dirt and other foreign matter from reaching the fuel pump (if one is installed) and getting into the carburetor.

THE FINGER SCREEN

The finger screen is the first strainer that the fuel must pass through. A finger screen must be located in the bottom of each fuel tank you install to screen out the larger foreign particles so they can't enter the fuel system. This is a very important strainer, particularly in a newly built gas tank. Foreign particles are often overlooked no matter how careful a builder might think he is.

Ordinarily, the finger screen is a part of the outlet fitting that screws into the bottom of the fuel tank, even though a few builders, primarily those fabricating fiberglass tanks, sometimes build the finger screen into the tank. Nevertheless, your own finger screen should be made removable for inspection and cleaning if possible.

Make this strainer from a wire screen, preferably galvanized or copper having a comparatively coarse 8- to 16-meshes per inch. Cut the wire mesh to size, roll it into a small cylindrical shape and solder the edge seam. Then, crimp and solder one end. Fit and solder the opposite end into a brass fitting which will screw into the built-in tank sump fitting.

CAUTION: Never use an acid core solder in aircraft work as it has a corrosive effect on material. Use a resin core solder instead.

If you prefer, you can purchase a ready made finger screen from one of the aircraft supply houses catering to the homebuilt trade. These finger screens are made of a 16-mesh brass screen and are mated to the proper brass fitting. These have a 3/8-inch male pipe thread on one end and a 1/4-inch female pipe thread in the other. Be sure that you order the correct size finger strainer and tank sump fitting. Pipe sizes are confusing to anyone not familiar with them. For example, although small aircraft engines use a 3/8-inch fuel line, the pipe thread for the fitting that accommodates it is identified as a 1/4-inch thread.

Small engines and automotive engines such as VW can generally get by with the original diameter fuel lines and the appropriate matching fuel line fittings and sump flange sizes. If your carburetor and fuel pump inlets can only accommodate a 5/16-inch line, there's no use running a larger line up to it and using step-down fittings to make the connection.

GASCOLATOR (MAIN STRAINER)

The second strainer downstream in the system is called the "main strainer", "sediment bowl", or the "gascolator". This is a very important strainer, and the one you will check most frequently. Its screen is finer than that of the finger strainer in the tank (about 60 mesh/inch).

The gascolator must be located in the lowest part of your fuel system, lower than the carburetor inlet and lower than the gas tank(s). It is generally situated on the lower portion of the firewall, but not always.

Pick a location where the gascolator will receive a good flow of relatively cool air. If necessary, make and install a blast tube and/or a cooling shroud to duct cold air to it. This extra care may be necessary in some locations because the possibility always exists that the fuel in the gascolator may vaporize (vapor lock) due the engine heat in the compartment. Some store-bought aircraft have been prone to this problem, expecially at higher altitudes. Usually the problem manifests itself in a partial loss of engine power during cruise conditions.

Where should you place the gascolator? Place it as low as possible on the firewall, but do not situate it directly over (or above) the exhaust pipes or in such a position that it protrudes below the firewall bulkhead contour. This is especially important in retractable gear aircraft. In the event of a gear-up landing, or in the event the conventional gear is wiped off in a ground loop, the incident should not rupture the gascolator, creating a fire hazard. Locating the gascolator nearer one side or the other will make access to it much easier. Often the gascolator can be located so that it is easily reached through the bottom air outlet opening in the cowling.

If you install a quick drain in the bottom of your gascolator, you will be able to frequently and easily check for debris and water in the fuel system. Being in the lowest part of the system, whatever little water may be in the fuel will settle to the gascolator and be trapped there. In addition to permitting you to check for water in the fuel system, this feature will provide you an easy way to drain the entire fuel system. It is a slow method, however, and would take about 30 minutes to drain a 20-gallon fuel tank flowing through a regular 3/8-inch fuel line.

A gascolator with a lazy man's quick drain (that is, one you can work standing up by merely pulling on a spring-loaded handle) has its shortcomings. It is generally a heavier unit than a simple gascolator and is also more expensive. It is not as effective either, because the pilot usually allows the fuel to drain on the ground. This is a slipshod way of checking for water in the system.

If all of the fuel cannot be drained from the tanks through the gascolator, you should also provide a separate drain in each tank. Otherwise water could accumulate in the sump area of the tank and never drain out until so much water builds up that it eventually enters the system. This water build-up is possible from condensation alone, over a number of years.

Most homebuilders use second-hand gascolators. These are usually all right, but they should be checked closely. The one you get may have come off a wrecked airplane or it may be an old one that has not been used for years.

Look closely for cracks and replace the old gaskets, which will undoubtedly have been excessively compressed. Use new gaskets to prevent fuel seepage and resist the temptation to stop a slight seepage through old gaskets by screwing the gascolator bowl extra tight. That could crack or break the bowl, especially if it is glass. You can't always fix a leak by tightening a bolt, screw or nut tighter. A sure fix, however, is a new replacement....why take chances? Who needs a full-blown fuel leak for the initial test flight?

CARBURETOR SCREEN

The third strainer in the system has a very fine mesh (about 200 meshes per inch). It is located inside the carburetor's intake chamber. This small carburetor screen has the capacity of intercepting the smallest of foreign particles which may have gotten through the finger strainer in the gas tank and the main strainer mounted on the firewall.

IMPORTANCE OF STRAINERS

It is easy to see that each of the three strainers is equally important, and, although it would be unusual for one of them to become completely clogged, it is possible. Any one of the strainers could cause an engine malfunction or failure should it become clogged with debris. Individually the three screens don't cost much. However, the absence or clogging of a single one can be costly, indeed. Approximately 11% of all the reported fuel system difficulties were determined to have caused by fuel filters (screens) according to an FAA General Aviation Inspection Aids Summary.

Fuel Flow

If my observations are correct, less than half of the homebuilders completing an aircraft go to the trouble of determining beforehand whether their aircraft has an adequate fuel flow during the maximum angle of climb.

At some time during the preparation and early stages of your test program, you should verify the maximum climb-out angle for your aircraft with a minimum fuel level. This is most essential for a gravity-flow system, as an engine failure at low altitude (if it is the result of insufficient fuel flow from a near-empty tank) may be a fatal event.

Your fuel system should be able to provide a fuel flow rate (gravity systems) of 150% of the take-off fuel consumption for your engine. For an engine-driven pump system the fuel requirement is 125% of the take-off fuel flow of the engine at the maximum approved take-off power.

A fuel system that relies on an engine-driven fuel pump must also have an emergency pump that could supply the engine if the main (engine driven) pump should fail. The emergency pump is usually an electric pump. It is variously known as a booster pump, electric pump, auxiliary pump, and standby pump.

An emergency pump, if used with a float-type carburetor, will put out a relatively low pressure, approximately 3- to 5-psi, while pressure carburetors may require at least a 10- to 16-psi capacity in a pump. Fuel injected engines, on the other hand, may need a fuel pump that can provide approximately 20- to 35-psi.

Your fuel flow can go to pot if excess heat is imposed on fuel system components. That is, considerable fuel vapor can be generated within the fuel lines and gascolator. Under certain conditions this can cause erratic fuel flow and engine-driven fuel pump cavitations.

You can minimize the potential for fuel vapor problems by providing at least one inch of clearance between the fuel line and the exhaust pipes. How the fuel line is routed and

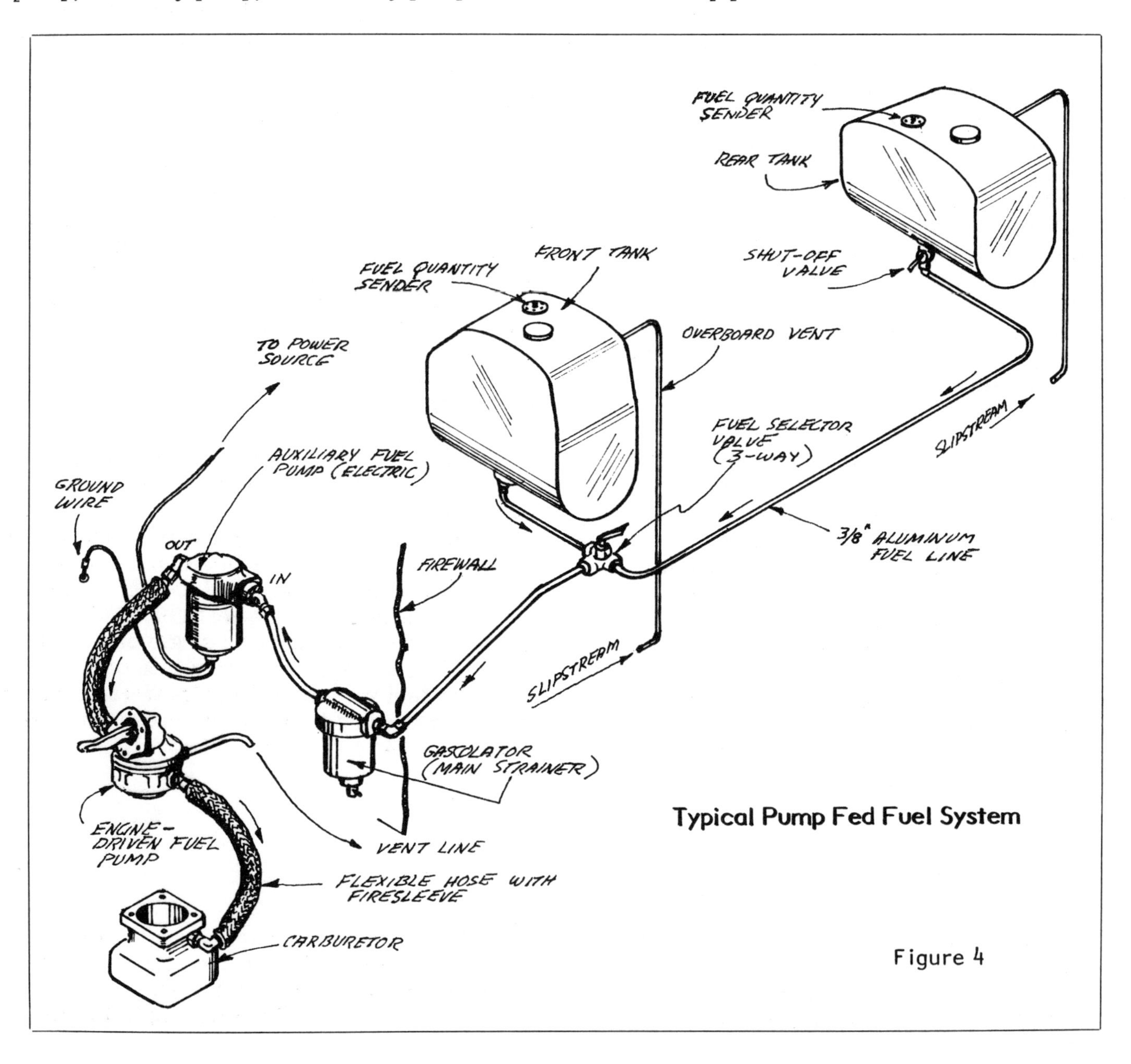

Typical Pump Fed Fuel System

Figure 4

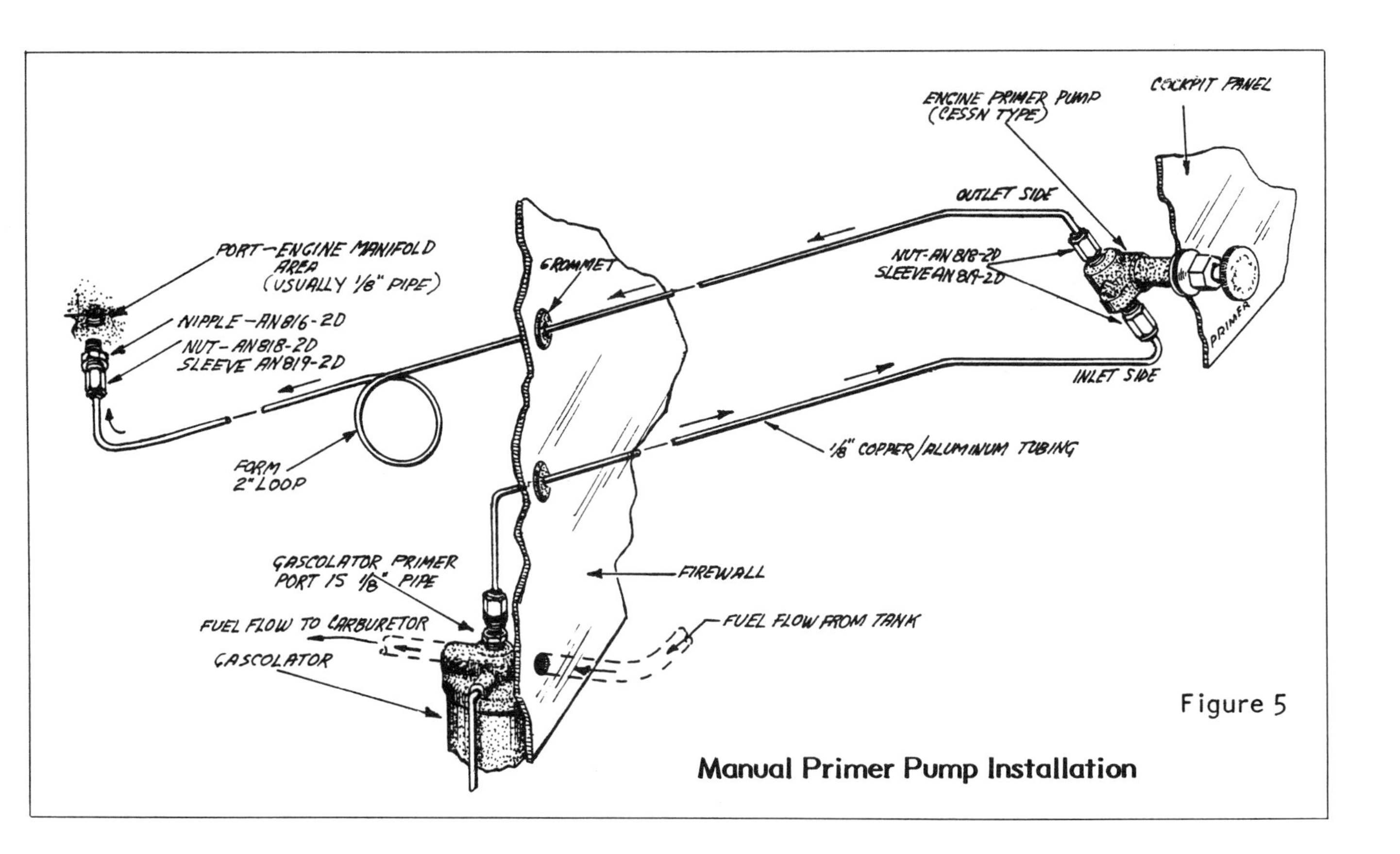

Figure 5

Manual Primer Pump Installation

the restrictions it may have due to severe bends also can affect the purging of vapors. Vapor tends to collect at high points in the fuel line, particularly where the bends are severe and the tube diameters are reduced. One provision sometimes used to help eliminate vapor build-up in the engine fuel system is a separate vapor return line running from the engine-driven pump directly to the main fuel tank. Since the tank is vented, the vapors will be vented harmlessly overboard.

A carburetor that has a vapor elimination connection must likewise be connected to a vent line, which will route vapors back to the aircraft's main tank.

It is not a good practice to feed a pressure fuel system from both tanks simultaneously because air would be sucked up by the pump if one of the tanks ran dry before the other.. You might then have a case of fuel starvation even though plenty of fuel remained in the other tank. This further points up the need for simplicity in your fuel system. To help make a simple system work, you might consider the installation of a header tank when you have two tanks feeding the engine. In this sort of installation the two tanks feed into the header tank first and then the fuel flows from the header tank to the engine. Don't forget to vent the header tank back to the main tank so that it becomes, in essence, a single fuel source.

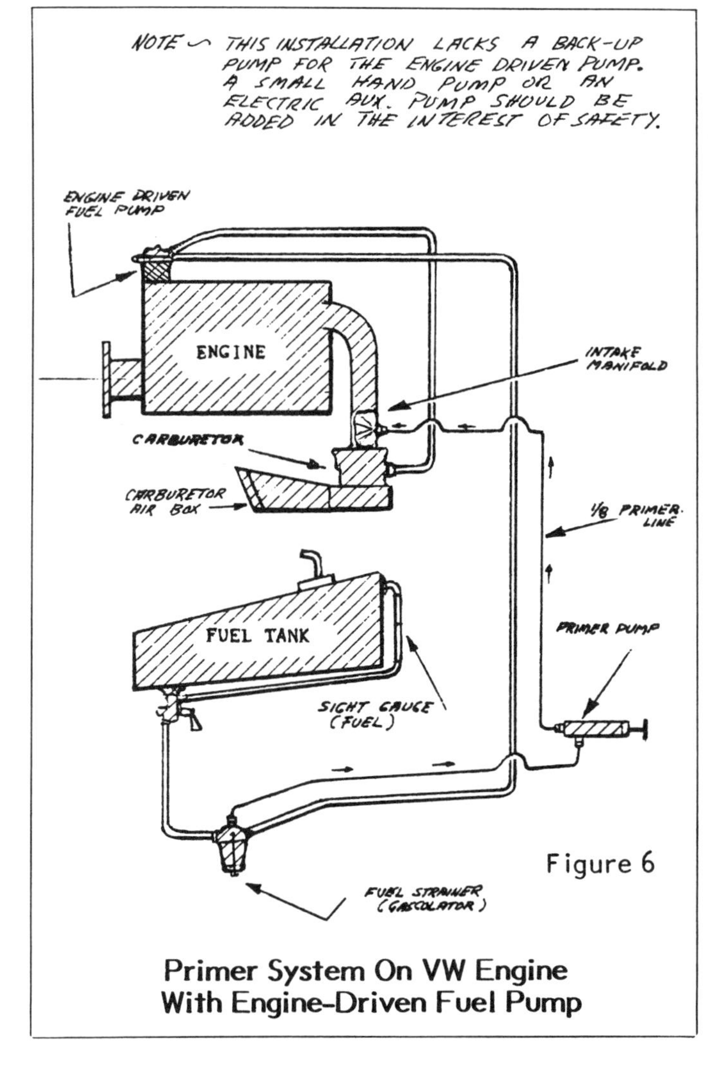

Figure 6

Primer System On VW Engine With Engine-Driven Fuel Pump

A poor gasoclator to carburetor hose installation. That low spot can trap vapors, water or fuel and cause engine problems. Line should run upward without reversal.

A good view of the carburetor and carb heat box installation. Confused? This is on a Breezy pusher so some things are reversed.

Fuel Flow Test

If you've ever drained 20 gallons of fuel from the outlet of a tank fitted with a 3/8-inch opening, you know that it takes a long time— something over 30 minutes. You might conclude from that, that your fuel consumption could be in the neighborhood of 40 gallons per hour (gph). With a trusty 65 hp engine, or even a feisty 180 hp engine, a gravity flow system capable of running that much fuel through the system would provide you with more than enough in the way of fuel flow. Right? Not necessarily.

It seems that odd things happen after you install a tank in an aircraft, connect tubing here and there, install fittings and flexible lines, a gascolator and shut-off valve, etc. The net result sometimes being that you no longer have that anticipated healthy flow of fuel, even though the entire system features standard 3/8-inch (inside diameter) plumbing throughout.

In a gravity flow system, the height of the tank above the carburetor is the single most important element. It has to be high enough so that at least 1/2-psi of fuel pressure can be measured at the fuel line when disconnected at the carburetor. This is not much pressure, but to get it, the tank bottom may have to be at least 19 inches above the carburetor inlet level. Many small homebuilts don't meet that criteria. Oh, some of them operate well enough with the tank full, or even half full, but, sometimes they experience fuel starvation in a climb when the fuel supply is

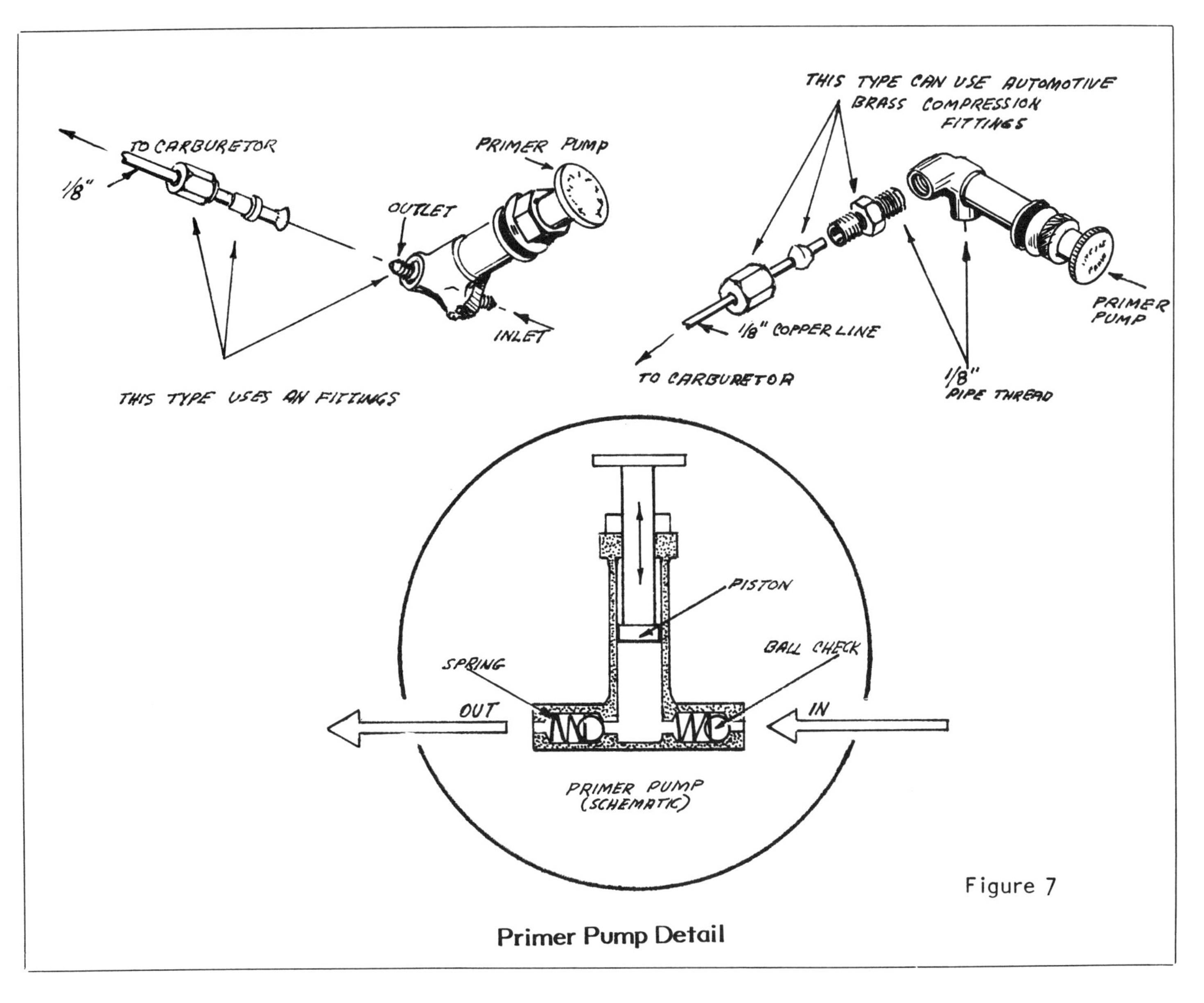

Figure 7

Primer Pump Detail

low or when a full-power takeoff is attempted. Some builders count heavily on a small ram air tube in the filler cap that points boldly into the slipstream to supplement gravity and increase the fuel pressure. This helps, of course, but it is nothing to stake your health on. (Figure 1)

A fuel system that only provides barely sufficient fuel flow when the tank is full or almost full is dangerous. Placarding the panel with a warning not to attempt full-power takeoff when the tank is less than half full is strictly eye wash and cures nothing.

Probably the easiest way to check your fuel flow is by digging a hole for the tail wheel deep enough for the airplane to assume a climb attitude (in the case of a tricycle gear installation, the nose gear will have to be blocked up), then add fuel to the tank. There is a right way and a wrong way to check the flow. The wrong way is by filling the tank and seeing how fast the fuel comes out of the disconnected line at the carburetor. All this will show you is that the fuel flow is sufficient or is not sufficient when the tank is full.

It is much better to start with an empty tank, and add fuel slowly until it begins to flow steadily out of the disconnected line. This demonstrates how much of the fuel in your tank is unusable. At this point, start your timing and add one gallon of fuel to the tank. How many minutes will it take to drain completely?

NOTE: FAA's FAR Part 23, Airworthiness Standards: Normal, Utility, and Acrobatic, 23.955 covers fuel flow requirements. These are equally applicable to homebuilt aircraft and provide excellent guidance. Basically, the requirement is this:

"Gravity systems — The fuel flow rate for gravity systems (main and reserve supply) must be 150% of the takeoff fuel consumption of the engine.

"Pump systems — The fuel flow rate for each pump system (main and reserve supply) for each reciprocating engine, must be 125% of the takeoff fuel flow of the engine at the maximum power approved for takeoff ...."

Here's a typical example: A gravity flow system with an engine that burns 8-gph at full throttle has to have a minimum flow of 12-gph (150%) through the disconnected line at the carburetor to ensure adequate fuel flow to the engine. This must be at a climb attitude.

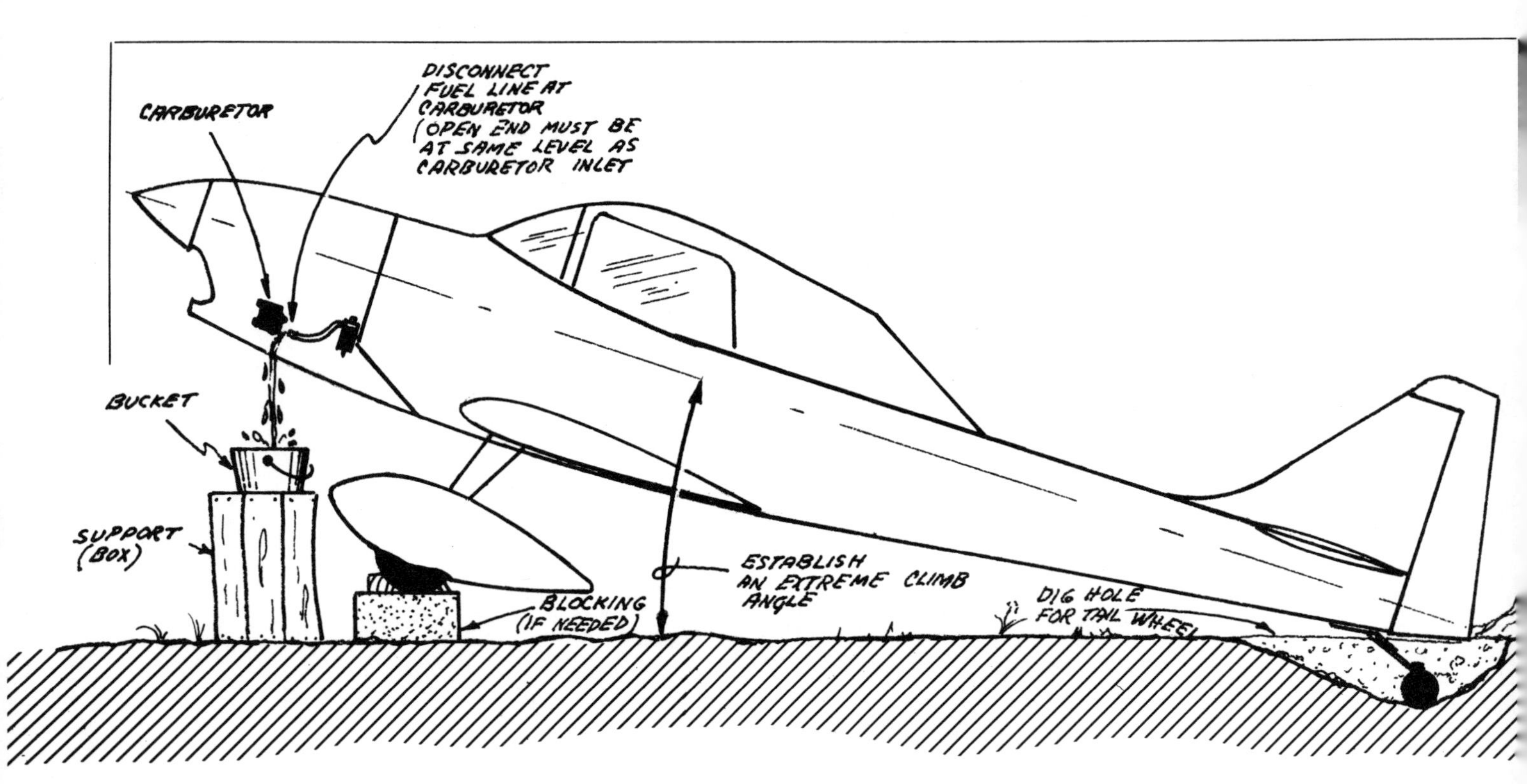

Figure 8

Fuel Flow Test Set-Up
(Gravity Flow Installation)

That means the gallon you poured into the tank must take no longer than five minutes in order for the fuel flow rate to be sufficient to drain.

A fuel flow test of a pump-equipped fuel system may be performed in much the same manner. In this case, however, the auxiliary fuel pump must be turned ON. It has to provide 125% of the required full-throttle needs of the engine rather than 150%.

One more thing you should take into account is the possibility that the acceleration during takeoff with a low fuel level may cause the fuel outlet to become momentarily uncovered, resulting in interrupted fuel flow. Although this is more likely to be a problem with flat, shallow fuel tanks, it is something to consider.

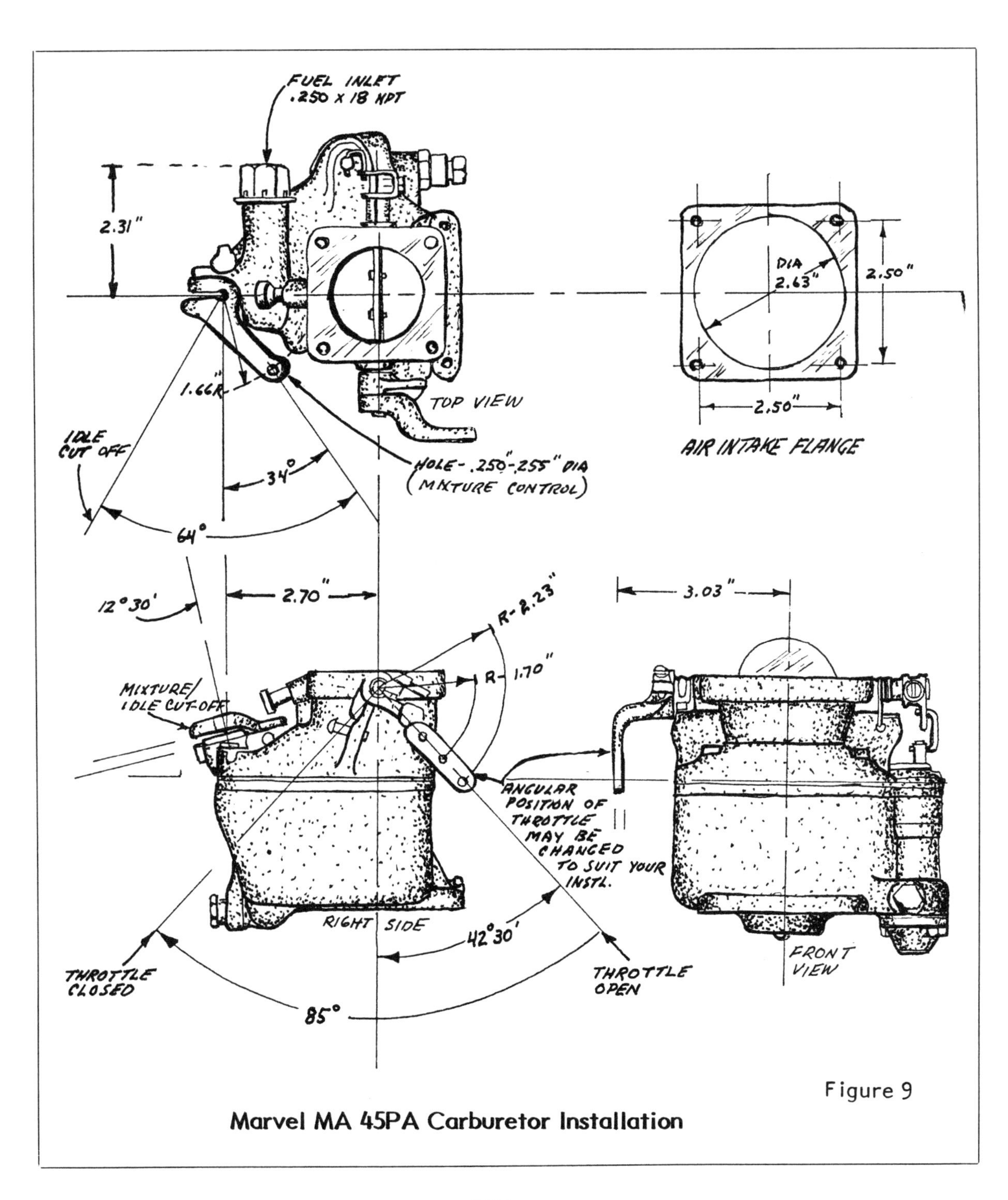

Figure 9

Marvel MA 45PA Carburetor Installation

ignition & electrical

9

A Battery....Will You Need One?

In a conventional aircraft, the battery serves as an energy storehouse for starting the engine and for operating all electrical units and accessories on board. While in flight, of course, all electrical needs are met by the generator (or alternator). Should either of these fail, however, or become ineffective during low engine speeds, the battery will supply the necessary electrical energy .... for awhile at any rate.

Even if your aircraft will not have a complete electrical system (generator or alternator, voltage regulator and starter), you may still want a battery to operate one or more electrical devices and a radio.

Fortunately, advances in solid state electronics have made aircraft radios and many instruments very energy efficient. In fact, the operational current drain in the latest of these devices is so small that it is now feasible to power a radio and all essential engine instruments for fairly long periods on battery power alone. This is quite a bonus feature for owners of small sport aircraft in which a battery is sometimes the sole electrical source on board. Of course, after a day of flying, the battery will probably have to be recharged before the next flying session. For most of us, this is an acceptable inconvenience, particularly when weighed against the alternative choices of installing a complete electrical system, which imposes severe weight and balance limitations on a small low-powered light aircraft, or not having any electrical equipment at all.

AIRCRAFT BATTERIES

Batteries specifically manufactured for aircraft use are, for the most part, lead-acid types. They are capable of being recharged by the aircraft generator, or by a battery charger plugged into the nearest wall plug, be it in the hangar, garage or bedroom. The essential difference between an auto battery and one made for the aircraft trade is the non-spillable cap feature and light weight.

Ordinarily, you can assume that the smaller the battery, the smaller the amount of energy stored in it. Therefore, select a battery primarily for its capacity. It should be large enough to supply the necessary electrical energy to handle the engine starting loads and the in-flight energy requirements imposed on it.

Almost all electrical systems installed by homebuilders are of the standard 12-volt variety (12-volt components are more plentiful). There is, however, a trend among aircraft manufacturers towards using 24-volt systems in order to meet customer demands for more and more electrical accessories, radios and gadgets with less drop in voltage. Undoubtedly, as 24-volt system components and accessories become more plentiful, they will be used by homebuilders in greater numbers.

Regardless of which voltage system you use, there is little or no difference in the installation of the various units or in their operation. Your use of either system, therefore, should not result in unexpected complications.

Typical aircraft battery sizes and weights for a few representative types are tabulated for your convenience in Table 1.

The dimensions given in Table 1 can be useful in determining the battery box size you will need. This may also be the time to consider whether to install a larger battery box. One that could accommodate either the smaller 25 ampere-hour batteries or the larger 35 ampere-hour batteries. (You may desire more power later.) However, the smaller 12-volt, 25 ampere-hour lead-acid aircraft battery is still pretty much the standard among homebuilders. It has proven to be large enough to crank the average four-cylinder aircraft engine in a sprightly manner.

INITIAL SERVICING OF THE BATTERY

Your lead-acid aircraft battery, should you elect to use one, will probably have to be ordered from an aircraft parts supplier, who will undoubtedly ship it via United Parcel Service in a dry charge state. This means that you will have to obtain an electrolyte acid kit locally (auto parts shops) and service the battery yourself. Unfortunately, most auto people seem to package battery acid in large

ESSENTIAL BATTERY DATA AND DIMENSIONS (Representative Types)

Type/Source	Capacity (ampere-hr)	Dimensions (inches)	Full Weight (pounds)
ACE	27	7 3/4 x 5 1/8 x 6 3/8	17 (dry)
* Gill 6-HU-20	17	7 1/2 x 4 1/2 x 6 3/8	21
* Gill PS6-9	25	7 3/4 x 5 1/4 x 7 1/2	21
Willard W-25	25	7 3/4 x 5 1/8 x 7 3/8	22
Rebat S-25	25	7 3/4 x 5 1/8 x 7 5/16	21.5
Rebat R-35	35	9 3/4 x 5 1/8 x 7 5/16	27
* Gill PS6-11	35	9 3/4 x 5 1/4 x 7 1/2	27
Willard W-35	35	9 3/4 x 5 1/8 x 7 3/4	29
MANIFOLD VENTED BATTERIES (No case required)			
Willard W-54M	25	7 5/8 x 5 1/8 x 7 5/16	21.5
Willard W-78M	35	9 11/16 x 5 1/8 x 7 1/8	29
GELL CELL BATTERIES (No case required, fully aerobatic)			
Globe-Union	25	7 3/4 x 5 1/8 x 7	20
Globe	28	7 5/8 x 5 1/8 x 6 1/4	Unk
Power-Sonic	6	6 x 2 1/2 x 4	5.29
Wag-Aero	28	7 3/4 x 5 3/16 x 7 5/16	Unk

*** Do not use automotive electrolyte kit. Requires different specific gravity.**

TABLE 1

1 1/2-gallon plastic containers. All you really need is about two quarts for that little aircraft battery. What a waste!

Enclosed with the new battery, or affixed to it, will be instructions for activating it. Similar instructions are also printed on the electrolytic container. HEED THEM! Especially observe the recommended safety precautions and disposal instructions.

Although some builders do and some don't, experienced maintenance people recommend that you put a charge to that new dry-charge battery after the electrolyte has been added and before putting it into use. Do it in accordance with the manufacturer's instructions to get the best service life for that battery.

MANIFOLD VENTED BATTERIES

What about manifold vented batteries? Well, for one thing, they are well liked by builders because they permit the elimination of the weight and expense of a battery box. Some even claim that they produce more power per pound and are quite inexpensive to purchase as well. The smaller 25 ampere-hour size weighs about 22 pounds and has dimensions similar to those for the standard small-case aircraft batteries. Although no weight penalty is incurred, a few builders may find it difficult to devise a means of securing the battery to the structure without a battery box.

NI-CADS (NICKLE CADMIUM BATTERIES)

Ni-Cad batteries are good tough batteries but they are expensive and they require a special charging system that measures temperatures, voltage and charge. The system must have a means for cutting off the charging action when any cell in the battery reaches 115 degrees F. An over-charge current could result in thermal runaway, with battery temperatures reaching 1,200 degrees F! Best recommendation for the present is to play it safe and use an economical, time-tested lead-acid battery.

Mounting a battery box in the aft fuselage to alleviate a weight and balance problem is one thing. Providing easy access to it is another. Don't overlook the necessity to protect control cables from battery acid and fumes. Ensure box is vented.

How would you remove this battery? That's right, you would have to remove the cowling. Also looks like the engine mount tubing, ducting to cabin heat, and the inaccessible rear hold-down bolt would make removal a very time-consuming chore. This may be all right for store-bought, but you should provide better access.

GELL CELLS

What about Gell Cell batteries? The idea is great and the battery has proved to be very attractive to many homebuilders, particularly builders and owners of aerobatic aircraft. An acceptable charge rate is obtained when the aircraft's regulator is set at 14.5 volts. However, there is a general agreement among maintenance folks, as of this writing, that Gell Cell batteries have not yet reached the level of efficiency equal to that of the conventional lead-acid type. For example, a 20-amp battery charger must be used when the battery is flat dead. Your aircraft generator simply will not charge a Gell Cell battery that is flat dead.

SOLAR CELLS

Solar cells are not really batteries, but rather, a means for generating electrical energy. You will be hearing more about these energy producers in the future. At present, however, they are extremly expensive and produce very little electrical energy.

DETERMINING BATTERY LOCATION

Your first thought about the battery will probably be: "Where will I put it?" Battery location is always a problem. The battery is heavy and quite bulky. You don't simply slide it under something where it will be out of your way. Anyway, its location is primarily a weight and balance consideration. After all, it does weigh between 21 and 35 pounds. An airplane with a 160 to 180 hp engine, complete with a constant speed metal propeller, possibly a propeller spacer and a few other goodies up front in the engine compartment just about has to have its battery located somewhere aft of the cockpit area. On the other hand, one equipped with a wood propeller in place of a heavy metal one would have very different center of gravity considerations that might make installation of the battery in the engine compartment necessary.

Also of importance is the need for the battery to be located with due regard to its primary electrical load. The cable length between the starter and battery must be as short as practical to keep the voltage drop to a minimum (say, to less than a half a volt).

If you must put the battery at a considerable distance from its largest current guzzler, the starter, you will be forced to use a higher capacity battery, larger cables or both, to obtain enough electrical gusto for the starter action.

Paradoxically, some locations in the engine compartment are poorly suited for batteries and yet they are the most convenient for the aircraft owner. For example, take an engine installation utilizing the efficient Cessna 150 twin exhaust mufflers. When the battery is located on the firewall, nestled in between the engine mount tubes, it will be positioned very close to one muffler, whose large end-diameter will act like a high heat radiating element. Extreme heat from the nearby muffler is detrimental because battery separators cannot tolerate high tempertures. They deteriorate rapidly when exposed to continuous temperatures much over 100 degrees F.

In any high temperature location, the battery must be shielded from the direct heat and the area must be well ventilated, otherwise the battery may have to be replaced frequently. After engine shut-down, there is a temperature rise inside the cowling and this, too, is not good for the battery. In addition, being the rather bulky object that it is, it could interfere with the normal outflow of engine cooling air. It seems there are more reasons for not locating a battery on the firewall than for doing so.

One final point regarding battery location. It should permit easy installation and removal. This consideration, however, ought to be secondary to ease of inspection and servicing. Without easy accessibility, the battery will suffer in the future from neglect. Believe me, it is most unlikely that you will check the electrolyte level of the battery as frequently as you should if you have to remove cowlings, seat, inspection plates or fairings to get at it. Nor, should it be necessary for you to have to use a flashlight, creeper and mirror to aid in viewing the electrolyte levels.

When located in the baggage compartment or cockpit area, the battery must be protected from inadvertent damage by either the pilot or passenger (batteries are not made to stand on). In other locations, it must be protected from vibration and from careless handling of tools, such as wrenches, screwdrivers, etc.

BATTERY BOXES

Ordinarily, a lead-acid battery has to be enclosed in a battery box made of fiberglass or aluminum and be attached to the structure using aluminum brackets and rivets or bolts. You may either purchase a battery box or make your own. In any case, you will have to install it yourself. If it is to be located in the cockpit area, it should be encased and vented so occupants will be protected from any fumes or electrolyte spillage resulting from battery overcharging, inverted flight or minor accident.

The immediate vicinity of a battery and the areas adjacent to a battery drain (which may become contaminated with battery acid) should be protected with an application of acid-proof paint.

The paint traditionally designated for this purpose is an asphalt- or rubber-based paint. These are sometimes hard to find. You can, however, almost always find a two-part epoxy paint, which will provide similar protection and will look better besides.

Install the battery box (or the hold-downs if no battery is used) so that the battery is secured in place without exerting excessive concentrated pressures on it that might distort or crack the battery case. If the battery box fit is too loose, use wood blocks or rubber spacers to prevent the battery from shifting and possibly shorting the cables.

CAUTION: Do not use wood in Ni-Cad battery boxes as it could eventually become contaminated and conductive, causing a current flow from the battery to ground.

WHAT ABOUT G LOADS?

Your battery, together with its case will be rather heavy — at least 22 pounds. That concentrated mass requires a mounting support rigid enough to withstand any load you would expect to encounter in flight. That is, it must have the capacity to support the battery even with considerable G loading.

The loading requirements imposed by the FAA on manufacturers of light aircraft stipulate that the battery installation must retain its integrity under vibration and loads to something like 6.6 G downward, 3.0 G upward, 9.0 G forward and 1.5 G sideways. As an example, our 22 pound battery under a 7 G load would weigh 154 pounds (22 x 7). It shouldn't be too difficult to mount a battery box capable of taking a load like that. Of course, if you are going to be building an aerobatic aircraft, even higher G loads must be anticipated and the battery support requirements will become slightly higher (9.0 G downward, 4.5 G upward, 9.0 G forward and 1.5 G sideways).

VENTS AND DRAINS

A battery gives off highly corrosive and explosive fumes in the form of hydrogen gas, so some means must be made to carry off these gases rapidly. Fumes and overflow from batteries located in the aft fuselage, for example, have been known to literally eat away rudder and elevator cables. Sometimes, the condition is detected in time, sometimes not!

Vent the battery compartment, or at least induce a sufficient airflow around the battery to prevent the accumulation of explosive gases. A flow of air of

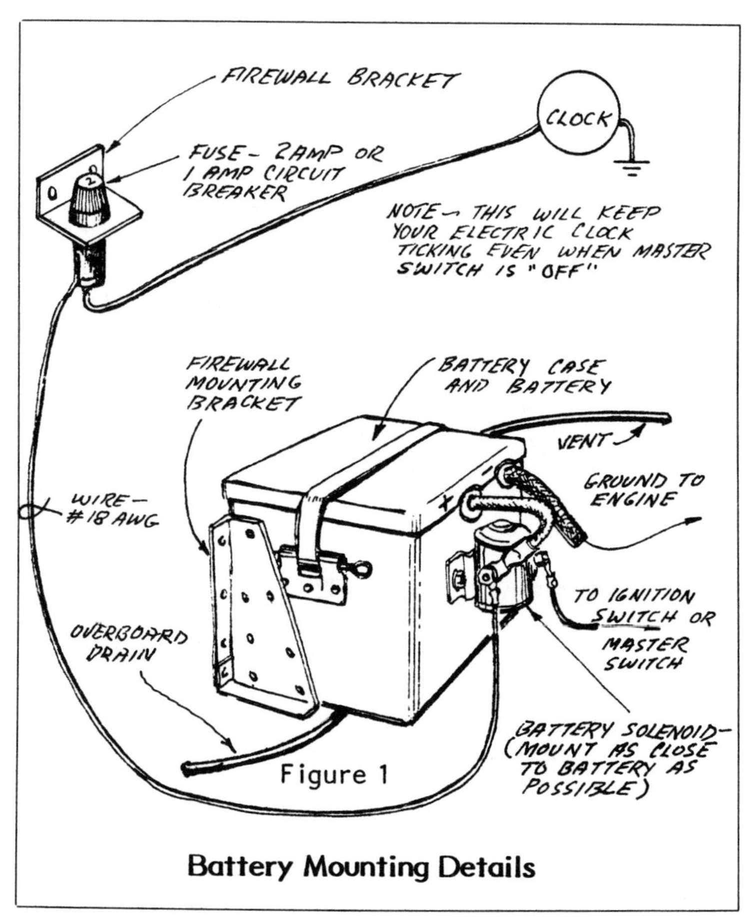

Figure 1

Battery Mounting Details

approximately five feet per minute (and that's not much) around the battery or through the compartment is considered adequate to continuously purge the gases.

When a battery box is used, it should have a vent in its lid and a rather large drain opening in the bottom of the box to carry off any electrolyte that might have spilled over.

The drain, of course, has to be in a low point of the box and not be blocked by the bottom of the battery. Use small wood spacers if you have to, to raise the battery slightly so that the drain remains uncovered.

Connect a plastic tube to the drain nipple and route it to some point overboard where vapors and fluid can be safely dumped into the slipstream without any of it getting on the aircraft's structure.

The plastic tube used should be at least 1/2-inch in diameter to minimize possible clogging. No spillage must be allowed to come into contact with the aircraft at any time, because it will discolor, corrode, and otherwise damage the structure or external surface. In this regard, the aft side of a landing gear leg might be a good location for terminating the end of the vent tube. Some builders run the plastic drain tube all the way out to the aft end of the fuselage just to be safe.

Battery boxes with louvers or holes built into their lids do not need a separate venting arrangement provided the fumes are not given off in a confined compartment or are brought into the cabin somehow.

Although a manifold-vent battery requires no battery box, it does require ventilation. Connect one of its vent nipples to a plastic tube and route it to a positive pressure area of the aircraft and the other downstream to some location where its contents will be discharged overboard. (Pick a relatively low pressure or negative pressure area for this one.)

BATTERY CABLES AND CONNECTIONS

Use the size battery cables and the battery location that is designated in your plans. If neither is specified, you are on your own. In that event, study a few installations in uncowled production aircraft for ideas.

Try to keep the cable lengths as short as possible but long enough to avoid putting a strain on the battery terminals, which have been known to break off.

If your battery requires installation in some aft area, you will be faced with the prospec of routing a long, heavy and expensive power cable (and possibly a grounding cable as well) all the way back to the battery. The happy thought probably has already occurred to you that aluminum battery cables would be much lighter than the copper cables normally used. Aluminum cables are indeed lighter and, for that reason are used in some homebuilts and even a few manufactured aircraft. There is, however, a problem unique to the use of aluminu cables. Under certain conditions they oxidize heavily at the crimped lugs. This results in high resistance connections which may, at the very least, cause starting problems, particularly in cold weather. Worse than that, however, is the possibility that the high resistance connections could overheat and cause serious damage to the cable insulation, creating a scary cockpit environment with the smoke and fumes that are generated.

To minimize the oxidation problem, always install and crimp the lugs immediately after you strip an aluminum cable to the core in preparation for its installation. And, be sure the battery cables, the terminals, and the connecting hardware are clean and tightly secured when connected. Loose cable connections add resistance to the electrical system and can cause a hard-start or no-start condition, as well as prevent the charging system from supplying enough current to fully charge the battery. Additionally, loose battery cables may contribute to radio frequency interference.

MAKING THE BATTERY CONNECTIONS

1. Connect the battery cable from the positive battery terminal (through the battery and starter solenoids directly to the terminal of the starter. (Figure 6)

2. Connect the battery ground cable directly to some point on the engine. This simple hook-up will establish an effective circuit and guarantee the starter's vitality. Of course, the engine must have a good ground to the firewall and/or fuselage.

NOTE: *Always connect the positive cable to the battery first, and then the ground cable. That reduces the risk of shorts and spectacular arcing due to accidental contact from screwdrivers or other tools. Never connect or disconnect the battery while the circuit is loaded. Switches must be off!*

After the connections have been made, the lugs and terminal areas should be coated with an inhibitor to minimize the formation of oxide. Petroleum jelly or Dow-Corning 24 Silicone lubricant is often used. At the very least, lubricate the terminal post areas with plain old heavy grease.

Your battery installation is not complete until the cables, terminals, relays and bus bars have been given some sort of protection from possible shorting to the battery case (if metal) or to the hold-down brackets. This is an important precaution since these units are not ordinarily protected by circuit breakers or fuses. An electrical short in this part of the circuit may create a fire hazard. Installing protective rubber caps over each battery terminal to protect it from falling wrenches and screwdrivers may be one possible solution. As for the other units, they can best be protected by locating them out of harm's way.

Make your cable installation so that it will be impossible for anyone to inadvertantly reverse the cables on the terminal posts. Making one of your cables shorter than the other solves this problem handily. Also saves weight! Another solution is to color code the cables for recognition (red for hot, black for ground).

At any point where the cable passes through a compartment wall or opening of any sort, protect the cable against chaffing with a rubber grommet.

BATTERY CONNECTOR (RELAY)

A battery cut-off relay (battery connector/battery solenoid) is installed to provide a means for isolating the electrical circuit. Mount this relay as close to the battery as practical, certainly no farther away than a couple of feet. Mount it on the battery box if you can. Such close proximity lessens the risk of a short and a possible fire within this section of cable.

Be sure the battery connector you obtain is a continuous duty relay and not a starter relay. A starter relay is designed to work intermittently and it will get hot and burn out when required to do continuous duty. Be sure your relays are designated for the correct voltage, be it 12 volts or 24 volts.

BATTERY-REGULATOR COMPATIBILITY

There is always the chance in a new installation that your original voltage regulator setting may not be compatible with the battery's requirement and over-charging may result. Be sure that the regulator is set to produce no more than 14.6 volts. Actually, 13.2 to 14.6 volts at maximum output is sufficient for a 12-volt system. If the voltage is greater than that, you may find that you are "cooking off" your battery water. However, don't jump to conclusions because a poor ground connection to the engine could induce a similar symptom. Consider both possiblities if a charging problem develops.

Ignition Switches

The ignition switch is unique. It is a kind of switch not duplicated anywhere else in your sport plane.

Using a design adaptation of the rotary switch principle, it permits you to open one circuit and close another as you turn the ignition key (or lever) through its different positions. Because this single complex unit can switch several circuits at once, it assures a simplified installation. For example, a single standard ignition switch, as might be installed in any homebuilt with a full electrical system, controls two magnetos and the battery circuit, individually, and in all essential combinations.

Although ignition switch designs vary in complexity, all versions provide control of the ignition circuit with at least an ON and an OFF electrical function for each magneto. Additionally, some switches provide control over the starter circuit and even over the engine priming circuit.

As builders, we need to be reminded now and then, that the aircraft ignition switch functions in a manner opposite to that of a conventional battery circuit switch. That is, to make a battery ignition system operative, its switch must be closed. On the other hand, when an ignition switch controlling a magneto is "closed", its circuit, in contrast, is inoperative and the magneto is really in the off position. Clear as mud, right?

The ignition switches most readily available to builders are the Bendix and Gerdes switches. However, antique rebuilders can still obtain the old AC Type A-9 switch and other lever operated models similar to those used in a number of classic and old aeroplanes. Check with homebuilder suppliers.

IGNITION SWITCH FUNCTIONS

You can obtain ignition switches that are operated by a lever instead of a key if you prefer that type. Otherwise, most ignition switches are very much alike and the greatest difference among them is in the number of functions they can accommodate.

A builder with a single-ignition engine often uses a single toggle switch to control the magneto because ignition switch positions other than ON and OFF are not necessary. No need to "check the mags" when the engine is already operating on the only one you have!

I've noticed that a number of small light aircraft, having dual ignition engines but no electrical systems, utilize individual toggle switches to control each of the magnetos. There is no law that says you must use a standard ignition switch when two magnetos are installed. Still, there is something nice about having a real store-bought ignition switch, even if the cost makes your eyes water a bit.

SWITCH SELECTION

When selecting a switch for your aircraft there is no need to get one with an excess number of functions. If yours is to be a rather simple system, a four-position switch would be adequate and, in fact, should be suitable for most any kind of engine except single-ignition versions.

Builders of aircraft with extensive electrical systems, and those requiring an electric

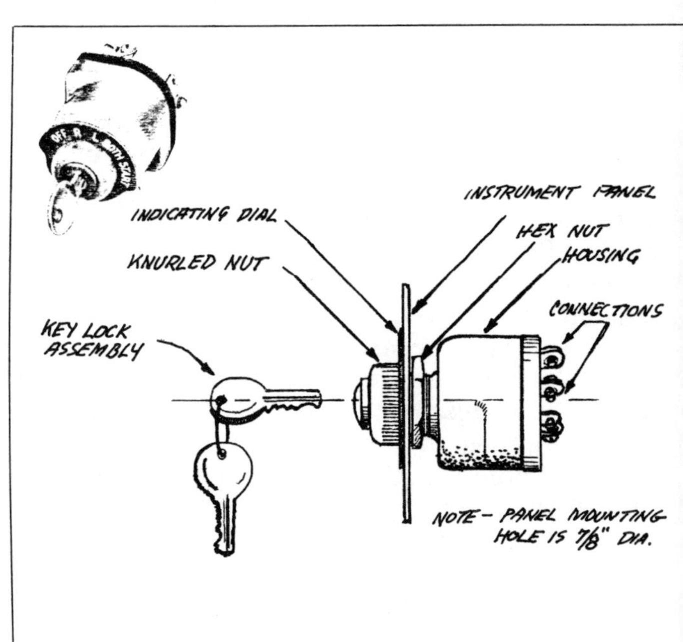

Ignition Switch Installation Detail (Bendix Type)

Figure 2

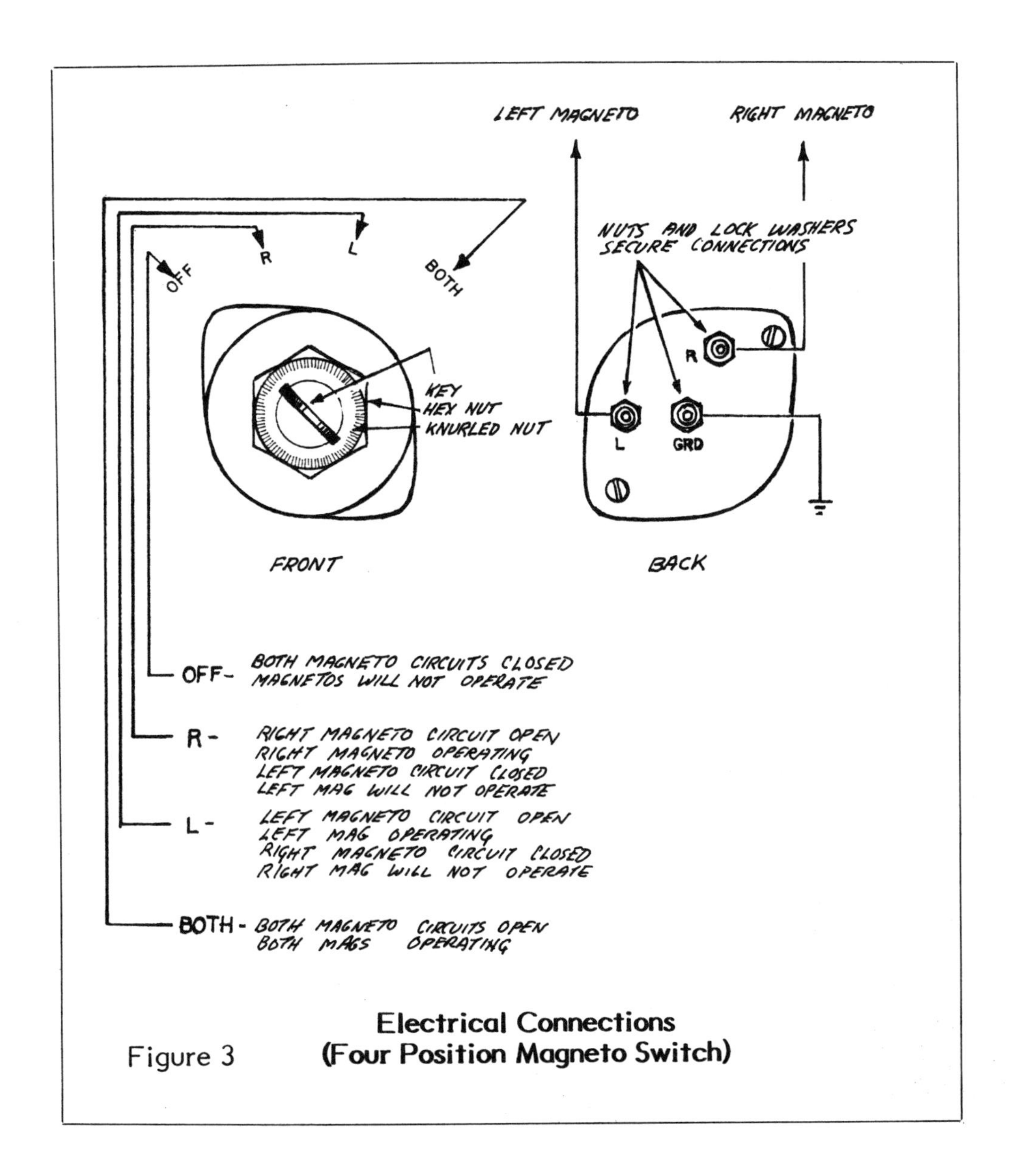

Figure 3 **Electrical Connections (Four Position Magneto Switch)**

start capability would probably want to install a five-position switch with a START position. All five-position ignition switches have similar functions in that the switch controls each essential phase of ignition and the starting of the engine. Some of these switches have a "PUSH-to-start" feature while others have "TWIST-to-start" action. (Figures 2 – 3)

Magnetos without an impulse coupling need some other means to generate a better spark for starting the engine. Some years back the Bendix people introduced the retard breaker magneto which effectively eliminated the need for an impulse coupling. They built a magneto incorporating two breakers instead of one, both activited by the same cam. This arrangement permits the left, or retard breaker to be positioned so that its contacts open a predetermined number of degrees after the main breaker contacts open. As part of the system, a separate battery-operated starting-vibrator must be installed with this magneto to provide the necessary retarded ignition for starting.

The starting-vibrator accomplishes this by providing interrupted battery current to the magneto's primary coil. Transformer action then steps up this pulsating direct current and produces a "shower of sparks" at the plug thereby assuring a positive start even under adverse conditions.

A standard ignition and starter switch is used whenever a starting-vibrator incorporating a relay is used. However, the Bendix combination ignition and starter switch must be used with vibrators that do not incorporate a relay. This switch automatically grounds the right magneto during the starting sequence.

Vibrators incorporating a relay may be recognized by their four-connection terminal blocks and those without, by their two-terminal connection. The wiring requirement for incorporating a vibrator is shown in Figure 4.

INSTALLING THE IGNITION SWITCH

Before you install any type of ignition switch, check its operation and make sure it operates easily and with little or no drag between stops. Incidentally, you should be able to feel a positive stop for each position.

An ignition switch having a START function should spring back when released, but not beyond the BOTH position. If the switch has a "PUSH-to-start" feature, it too, should work freely and with the proper return-spring action when released.

Just about all of the key-type ignition switches require a 7/8-inch hole in the instrument panel for mounting. Of course, the old time lever switches ordinarily require a standard 3 1/8-inch instrument panel cut-out.

Locate your switch in an easily accessible, but out of the way spot on your panel. Remember, a protruding key can make quite a crease in one's skull in the event of an accident involving a sudden stop. Here again, I believe builders should strive for a degree of standardization in cockpit layout. Locate the switch, if you can, in what may be considered to be a traditional location for your aircraft (lower left hand side of the panel or subpanel).

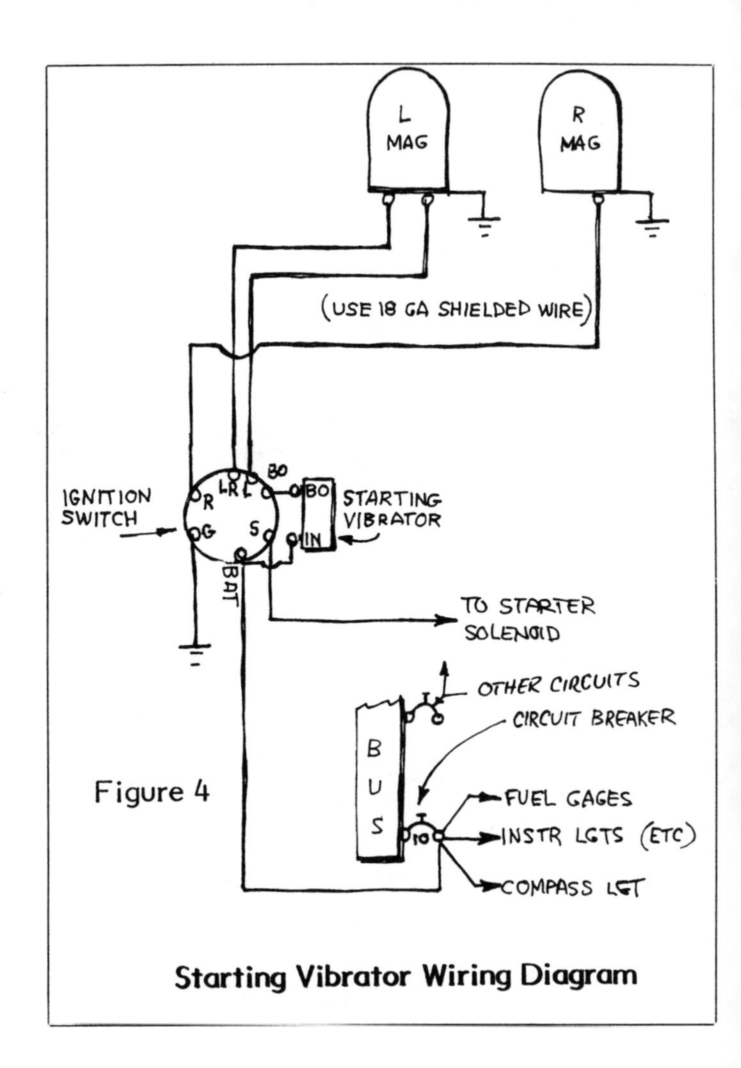

Figure 4

Starting Vibrator Wiring Diagram

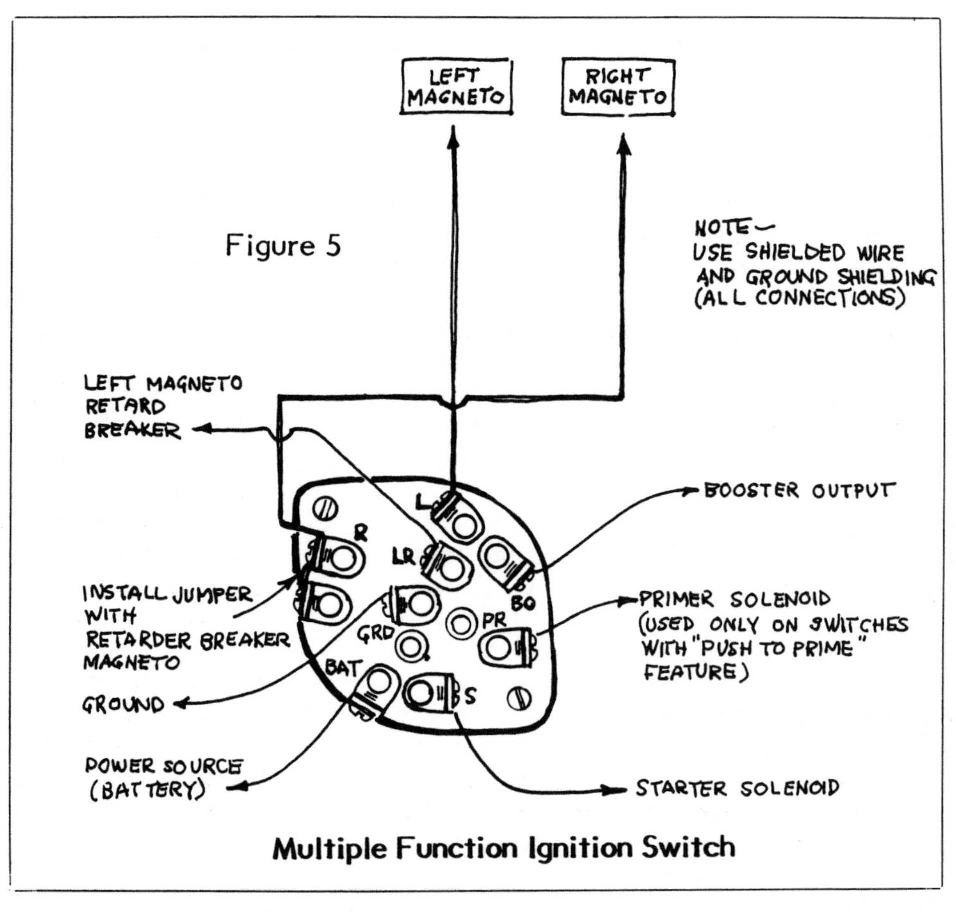

Figure 5

Multiple Function Ignition Switch

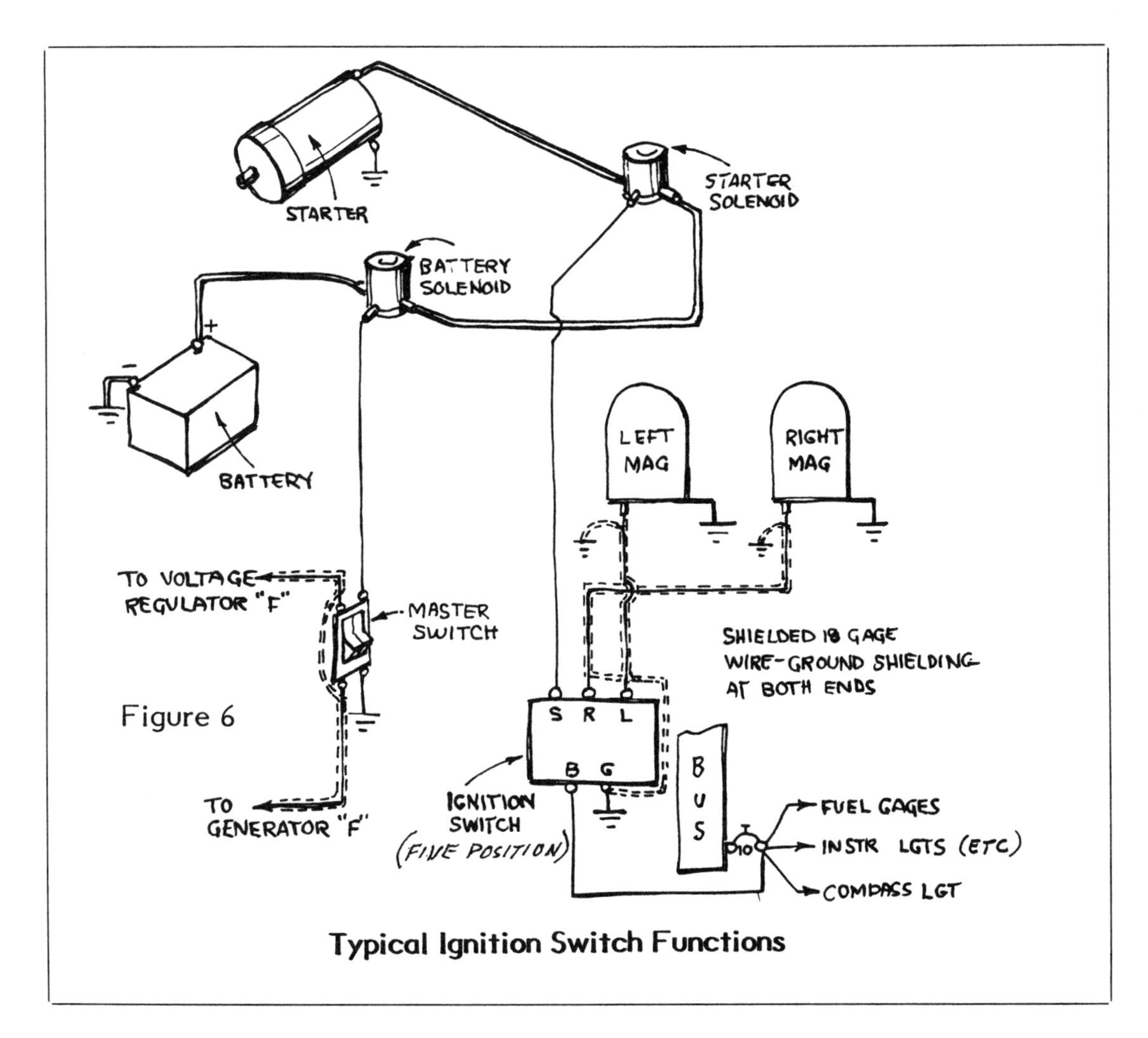

Figure 6

Typical Ignition Switch Functions

The hook-up of the ignition switch is usually much easier to accomplish when you connect (and label) the wires to the proper terminals before mounting the switch in the panel. It is very difficult to work from behind a panel without seeing what you are doing. Of course, if you have a hinged panel or easy access from behind, you've got no problem.

Mounting a standard ignition switch is simple enough. After inserting the switch in the panel hole, slip the labeled (position indicating) dial on the threaded portion protruding from the panel and align it with the lug on the threaded shaft of the housing. Add the knurled (round) nut over the threaded portion and secure the switch by tightening the hex nut against the rear of the panel. If you do the tightening by turning the round knurled nut against the front of the panel you might ruin your pretty paint job.

Switch terminals are normally provided with screws and lock washers for securing the connections. Make sure that the wire at each terminal is secure and does not touch an adjacent connection.

The switch and wiring requirement for a magneto installation utilizing an impulse coupling is simpler than that for a magneto requiring a starting vibrator system.

Only a single wire from the ignition switch to the magneto (primary coil) is necessary for the operation of the ignition system. Referred to as the "P" lead, this wire should always be shielded if a radio is to be used in the aircraft. Shielding suppresses ignition noises which could interfere with the radio's operation.

It is nice to know that the ignition system is independent of the electrical system. Should an electrical problem arise, it will not affect the continued operation of the ignition system or the engine.

ONE FOR THE ROAD

Remember! An installed magneto, even though not connected, is assumed to be HOT! Be careful, amigo.

Electrical Systems

You will need several different gauges of standard plastic insulated stranded copper wire. A solid core wire is unacceptible in aircraft as it is not as long lived under vibration induced flexings which are often severe. Save that stuff for rewiring your doghouse or birdhouse.

The gauge (size) wire you select is important. Always use the smallest size wire having the capacity to serve the unit to which it is connected at its rated current requirement without overheating and without exceeding the maximum allowable voltage drop.

In general, the allowable voltage drop from the bus bar (aircraft's power source) to the unit is 0.5 volts in a 12-volt system and one volt in a 24-volt system. This allowable voltage drop limitation is for continuous operation of an electrical unit. For an intermittent current load, however, a slightly higher voltage drop of one volt is acceptable in a 12-volt system (two volts in a 24-volt system). (See Table 5)

Sometimes the cable size selected can carry the load as far as the heating effect is concerned, but, because of its length (battery cables, in particular) the voltage drop exceeds the allowable minimum. It must be replaced by the next larger size cable.

Electrical cable is quite heavy, particularly in the larger gauges. Some units don't require much current and even a fine wire would adequately serve the unit. However, some old timers believe that aircraft cables smaller than 20-gauge copper or 18-gauge aluminum are too flimsy to withstand much handling and generally lack the tensile strength necessary. Still, aircraft manufacturers are currently installing 22-gauge wire for light loads in an attempt to keep weight down, so, why not the homebuilder, too?

ALUMINUM CABLES

Aluminum cables are not generally used in homebuilt aircraft except in the larger sizes — mostly for battery cables.

In addition to being poorer conductors of electricity (about 40 percent less efficient than copper wire), aluminum cables are notoriously oxidation prone in their teminal connections. This tendency often turns them into high resistance connections which have, on occasion, produced smoke and fumes in the cockpit from overheating.

If you do intend to install aluminum cables, use only aluminum lugs for the terminal connections. And, when you get ready to crimp the aluminum cables, don't leave their stripped ends exposed for any length of time before completing the connections.

Protect cable terminals with an approved inhibitor. Aluminum lug barrels are normally filled with petrolatum-zinc dust to remove the oxide film and to later inhibit oxidation of the crimped cable terminal. This minimizes the risk of a gradual build up of high resistance in the joint.

SHIELDED WIRES

Any cable carrying a current that might radiate electromagnetic waves (alternating current) could mess up your radio reception.

As a rule, you should use a shielded wire in any circuit which could cause radio interference. Now, which circuits are we talking about? The worst generators of radio noise are the spark plug leads. They must be shielded Here are some other circuits that commonly use shielded cables, and the minimum wire gauges normally used:

Main bus — Generator circuit breaker to generator or alternator (batt terminal), 8-gauge

Alternator field — (F terminal) to F terminal on voltage regulator, 18-gauge

Alternator field — (Batt terminal) to A+ terminal on voltage regulator, 18-gauge

Master Switch — To S terminal on the voltage regulator, 18-gauge

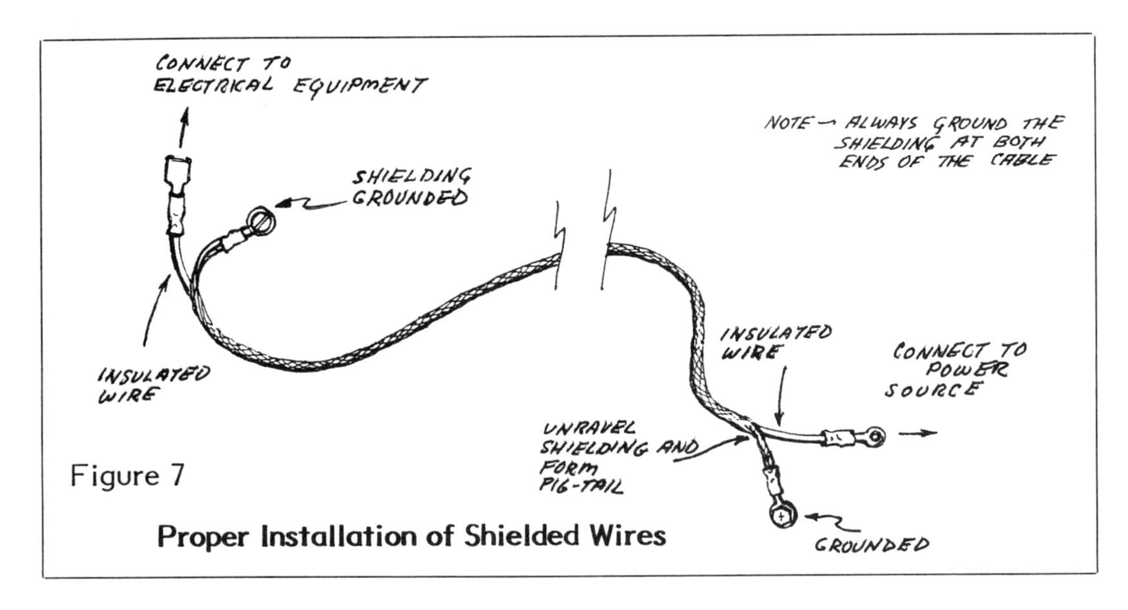

Figure 7

Proper Installation of Shielded Wires

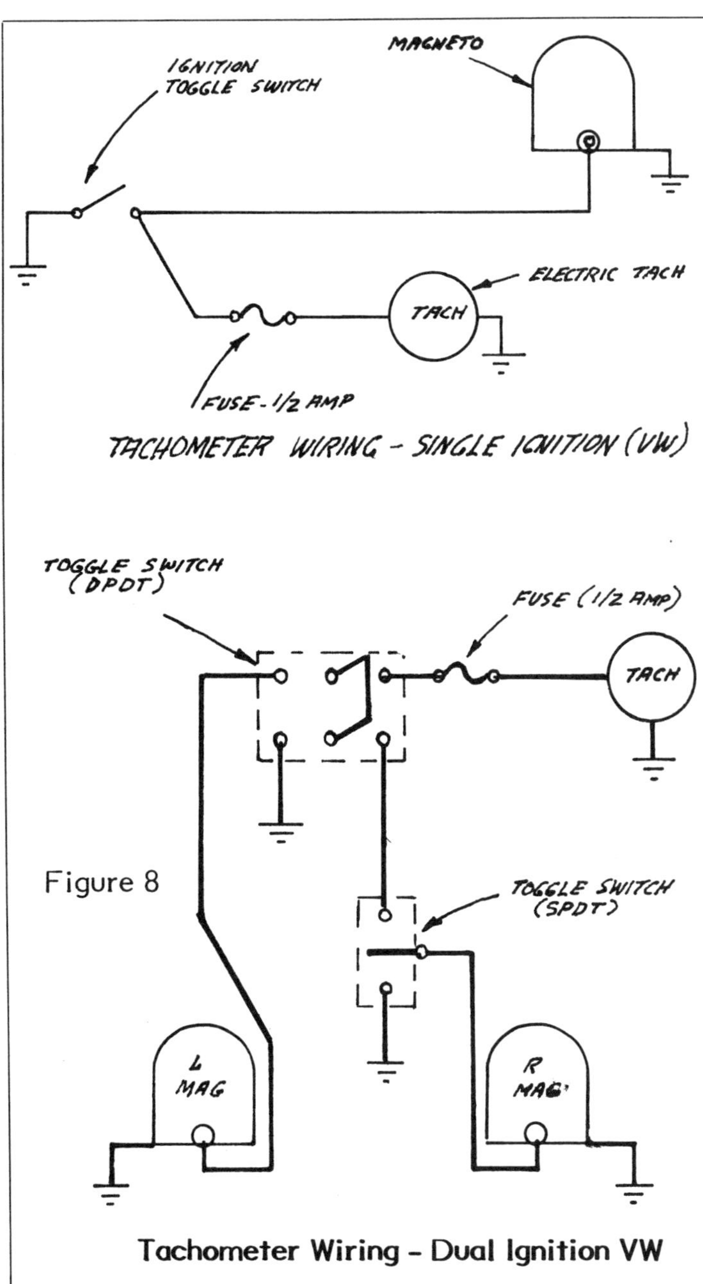

Figure 8

Tachometer Wiring - Dual Ignition VW

Ignition switch — (P-leads) to magnetos, 18-gauge

Strobe light circuit

It should be noted that many homebuilders tend to use larger gauges, with 14-gauge cable being quite prevalent, even though cable lengths are shorter than in production aircraft.

Shielded cable is identical to the regular stranded variety except that it comes already encased in a tinned copper braid which provides a path for the electromagnetic waves to follow rather than allowing them to radiate in all directions. The radiation travels along the shielding and is harmlessly grounded at both ends. To ground the shielding properly, a bit of shielding braid is unraveled at each end and twisted into a pigtail which can be fastened to the nearest common ground. (Figure 7)

If you can't get shielded wire, you could try to obtain some braided hose-like shielding to slip over the regular wire. But then, where do you get that stuff? Anyway, if you experience radio noise and have ordinary insulated wires installed, you can replace them one at a time with shielded wires until you find the culprit. Don't be surprised if several of them need the shielding cure.

CIRCUIT PROTECTION

Each circuit in your aircraft that is essential to safe flight must be protected by a separate circuit breaker or fuse. Less important circuits can be doubled up to share one or more circuit protectors.

The purpose of these protective devices is to minimize the hazard associated with an electrical fault (malfunction, short, or whatever) and to isolate the trouble from the remainder of the electrical system. Circuit breakers and fuses will not prevent an electrical malfunction, nor can they magically restore the function they interrupt when a fault causes one to pop.

Circuit breakers and fuses are normally installed between the bus bar and the circuit load (unit being served). It is customary to install fuses by soldering one terminal of each fuse to a common bus bar.

Circuit breakers, on the other hand, are installed with short machine screws rather than being soldered. Any number of fuses or circuit breakers can be attached to a single bus bar. This allows them to be conveniently grouped together and centrally located where they are easily accessible to the pilot.

Which type of circuit protector is better? Well, fuses are cheaper and lighter, but you'll hear the argument that circuit breakers are better because if one pops all you have to do is push it back in and the circuit is back in business again. In a fuse installation, however, once the fuse melts, that circuit will forever remain open and inactive until the fuse is replaced. The argument is really academic because, if there really is a fault in the circuit, resetting a popped circuit breaker or replacing a fuse will not cure anything. Always use actual amp rating for unit installed.

What individual circuits merit a separate circuit breaker or fuse? These will vary with the complexity of the aircraft and your own preference. Here are some of the more common circuits often protected separately:

* Alternator/Generator — 30-65 amps, depends on rated capacity
* Auto Pilot — 10 amps
* Auxiliary Fuel Pump — 10 amps
* Cabin Lights — 10 amps
* Clock — (usually) 1 amp fuse
* Flaps — 15 amps
* Fuel Quantity Indicator — 10 amps
* Ignition Starter Switch — 10 amps
* Instrument Lights — 10 amps
* Landing Gear — 10 amps
* Landing Lights — 10-20 amps
* Master Switch — Same as alternator field
* Navigation Lights — 10 amps
* Pitot Heat — 10 amps
* Radio (each separately) — 5-10 amps
* Rotating Beacon — 10 amps
* Stall Warning — 2 amps
* Strobes — 5 amps
* Taxi Lights — 20 amps
* Turn Coordinator — 5 amps

Various map lights, door lights, dome lights and compass and instrument lights often share protective devices. So, too, may some electrical gauges and warning lights. These are usually about 10 amps capacity each.

It is not uncommon, however, to add some minor circuit to one of the so-called essential individual circuit protectors.

When in doubt regarding the size circuit breaker or fuse capacity to use, select one based on the gauge of the wire used in the circuit.

One last thought regarding fuses and circuit breakers. You should identify the circuit that each protective device serves. Small instrument panel transfer decals can be used for this purpose with nice results. Or, if that's too much trouble, at least label a small piece of masking tape and stick it on next to the circuit breaker or fuse until you get around to something better.

SWITCHES IN GENERAL

There are thousands of different kinds of switches and I daresay a hundred or more varieties have found their way into homebuilts. An aircraft switch will normally have its nominal current rating stamped on it somewhere. That is, its continuous current rating in its closed position. Any switch you select should have a snap-action design to ensure a speedy opening and closing of the contacts. This minimizes contact arcing.

It is important that the switch have the electrical capacity to handle the electrical unit being operated without causing the wire to overheat and pop the circuit breaker. Other than that, you can use any style switch you prefer.

Locate your switches between the circuit breaker (or fuse) and the unit to be operated by that circuit.

MASTER SWITCH

Install a master switch so that you can activate or shut off the entire electrical system. This does not, of course, affect the ignition system as it is an independent system obtaining its electrical energy from the magnetos.

Since the master switch controls two circuits, we normally install a double pole, single throw (dpst) switch. The standard aircraft switch for this purpose is a RED rocker type and is usually embossed with the label MASTER.

TYPICAL ELECTRICAL LOAD ANALYSIS
(12-volt Systems)

ELECTRICAL UNIT	*RANGE OF AMPERES REQUIRED (APPROX.)*		
Auxiliary Fuel Pump	*3.0*	-	*7.0*
Battery Connector (solenoid)	*.6*	-	*.7*
ADF	*1.2*	-	*2.0*
Auto Pilot	*2.0*	-	*3.0*
Cabin Lights	*2.5*	-	*3.3*
Cigar Lighter	*10.0*	-	*20.0*
Clock	*Insignificant*		
Cylinder Head Temperature Unit	*.2*		
Compass Light	*.1*		
DME	*3.0*	-	*5.0*
Flap Motor	*10.0*	-	*15.0*
Fuel Quantity Indicator	*.4*		
Glide Slope	*.2*	-	*.3*
Instrument Lights	*1.3*	-	*2.0*
Landing Light	*8.4*	-	*15.6*
Landing Gear Motor	*10.0*		
Marker Beacon	*.2*		
Navigation Lights	*1.7*	-	*5.6*
Nav/Comm Radio (older sets)	*4.5*	-	*7.0*
Nav/Comm Radio (transisterized)	*1.3*	-	*1.4*
Post Lights	*2.0*		
Pitot Heat	*6.5*	-	*10.0*
Rotating (flashing) Beacon	*7.0*		
Stall Warning	*.2*		
Strobes	*2.0*	-	*4.0*
Turn Coordinator	*.2*	-	*.8*
Transponder	*1.0*	-	*3.0*

TABLE 2

Split-type master switches are also available. One of these will permit you to activate both the battery (BATT) and alternator (ALT) simultaneously or singularly. This type of switch gives you greater flexibility in controlling the electrical system.

For example, the right half of the split switch labeled BATT, controls all the electrical power throughout the aircraft. If you want, you can use the BATT side of the switch to check or activate equipment while on the ground. The engine need not be in operation. The left side (ALT) controls the alternator. When you turn it OFF, the alternator is cut out of the electrical circuit.

Install your master switch so that it must be flipped up for the ON position and down for the OFF position.

MAIN BUS

In most aircraft, the battery and generator (alternator) output cables connect to the main bus bar forming the central power distribution point in the airplane.

The bus bar is merely a strip of copper (1/16-inch to 1/8-inch x 3/8-inch) to which one terminal of each circuit breaker or fuse is connected. Some store-bought aircraft use a thin strip of aluminum. I guess this is all right. however, an incident traced to a bus bar that could only handle a total 105 amp load proves that it may not always be acceptable. Not when the actual inflight load of all the equipment in use totaled 120 amps. The predictable result was smoke in the cockpit. It could have been worse.

LOAD ANALYSIS

There is a scientific way of determining load but it is too complex and gets into temperatures, continued and intermittent operation and stuff like that. Still, it would be a good thing to know how much current would be needed if you were to turn on all systems and perhaps even use the cigar lighter. (See Table 2)

For all practical purposes, all you really want to know is if your generator or alternator can adequately handle all of the connected equipment under the most adverse flight conditions without drawing current from the battery.

Base your figures on the most unfavorable flight conditions you are likely to encounter. For example, you can assume such a condition will occur on a night flight. You will be using the navigation lights, strobe lights, instrument panel lights and radios continuously. Intermittently, you will be using the landing and/or taxiing lights. How much current does all that require? Some representative current requirements for different equipment are shown in Table 2. You should always use the current rating for the type of units you have installed if you know what they are.

How much capacity do you need in your generator (or alternator)? An IFR-rated airplane should have at least a 60-amp generator or alternator for obvious reasons. Most of us can get by with a far lower capacity, say 25 to 30 amps. Still, when you add all of the current draw you expect to use, you'll see for yourself what your particular situation will be.

Don't forget that after a prolonged or heavy cranking of the engine the battery's charge will be drastically reduced. Obviously, it would not last very long inflight if the generator (alternator) were to give up the ghost. Of course, if all this were to transpire during the cold, cold days of winter, you could count even less on any help from the battery.

WIRING TIPS

Wiring an aircraft is not an awesome, overhwelming task. Simply wire one circuit at a time, check its operation, and go on with the next circuit. Study the wiring diagrams and the simplicity of the undertaking will surprise you. (See Figures 15 - 22)

The environment in which your airplane will be operating in the future should be kept in mind as you procede with your wiring installation. The airplane will be subjected to varying degrees of vibration during its operational life. It will be exposed to sun, rain, high and low temperatures, possibly even snow. All of this can be conducive to the development of corrosion. Electrolytic action between dissimilar metals is another potential source of corrosion. This type of corrosion can happen any time moisture is introduced. Wiring, therefore, should be protected not only from rub-

COPPER CABLE CURRENT CARRYING CAPACITY/CIRCUIT PROTECTION

CABLE SIZE A/N GAUGE	*SINGLE CABLE MAX. AMPS.*	*MAX. RESISTANCE OHMS/1,000*	*INSTALLED WT. LBS./100*	*CIRCUIT BREAKER (AMP)*	*FUSE (AMP)*
22	06	N/A	N/A	5.0	5.0
20	11	1.025	.56	7.5	5.0
18	16	.644	.84	10.0	10.0
16	22	.476	1.080	15.0	10.0
14	32	.299	1.710	30.0	15.0
12	41	.188	2.500	25 - 30	20.0
10	55	.110	4.270	35 - 40	30.0
08	73	.070	6.920	50.0	70.0
06	101	.043	10.270	80.0	70.0
04	135	.027	16.25	100.0	70.0
02	181	.017	24.76	125.0	100.0

SOURCE: CAM 18/U.S. Dept. of Commerce

NOTE: *For aluminum cables, select a wire size two gauges larger than for copper. For example, to replace a 10-gauge copper wire, you should use a number 8 aluminum wire.*

TABLE 3

bing against any metal objects, but also against the penetration of moisture. Silicone rubber is excellent for this purpose and may be purchased almost anywhere.

Keep your cable (wires) routing as simple and direct as possible and your bend radii large.

Equipment having exposed arcing contacts or brushes ought not to be located anywhere where flammable fluids could collect if a leak were to develop.

Both battery terminal posts should have protective electrical terminal nipples slipped over them.

Solderless crimp-type terminals are the preferred type to use. Soldered terminal connections are more likely to break from vibration or inadvertent bending because the soldered portion becomes stiff with solder, making them more brittle.

Mark your wires with an indelible ink. The marking may be on the wire or on an attached tab. Both ends of the wire should be marked. it is helpful also to have a few additional markings along the cable's entire length. (Figure 9)

An all-wood or composite aircraft requires the addition of a ground return wire from all installed electrical units. Don't make problems for yourself. Determine from the start that you will use a different color wire for all the ground leads, preferably black. Then, when you have to make or break connections, your wires will not get inadvertently crossed.

Any place you have a couple of wires spliced with connectors, make one shorter than the other so both splice connections are not side by side where they can accidentally come into contact with each other. Better still, insulate both connections to eliminate future problems.

BONDING

Wood and composite aircraft have a greater need for bonding than do metal aircraft that are blessed with a common electrical ground already. Bonding strips or jumpers (braided copper) are used to electrically connect isolated metal parts. They join all the metal parts with what amounts to a common ground. This bonding may be necessary in reducing radio noise, but it could also lessen the probability of localized damage in the event of a lightning strike. (Lightning? What are you doing out there, man??) Parts most commonly connected by bonding jumpers are control hinges and other larger metal assemblies not connected to the aircraft's ground. (Figure 10)

To keep from having a lot of ground wires running here and there, it might be useful to install a short terminal block in some convenient location to serve as a "ground bus". If a lot of ground wires run parallel to each other, tie them together into a neat bundle. They should be supported every foot or so.

Wires passing through any bulkhead or structure must be protected by a rubber grommet or equivalent means. Do not form dissimilar objects into a single bundle, for example, metal tubing, flexible metal cables and electrical wire. Also, don't run them through the same opening.

It is always a problem to satisfactorily route wires into the cockpit area without having them exposed and running hither, thither and yon. In low-wing aircraft the spar is the complicating problem. Drilling holes through the spar to run wires through is a risky proposition unless provisions have been provided for such an installation by the designer.

GENERATOR VS. ALTERNATOR

Claiming to have an electrical system implies that you have a generator or an alternator on board to supply the electrical current needs inflight and to replenish the battery's energy needs when necessary.

A alternator is considered to be superior

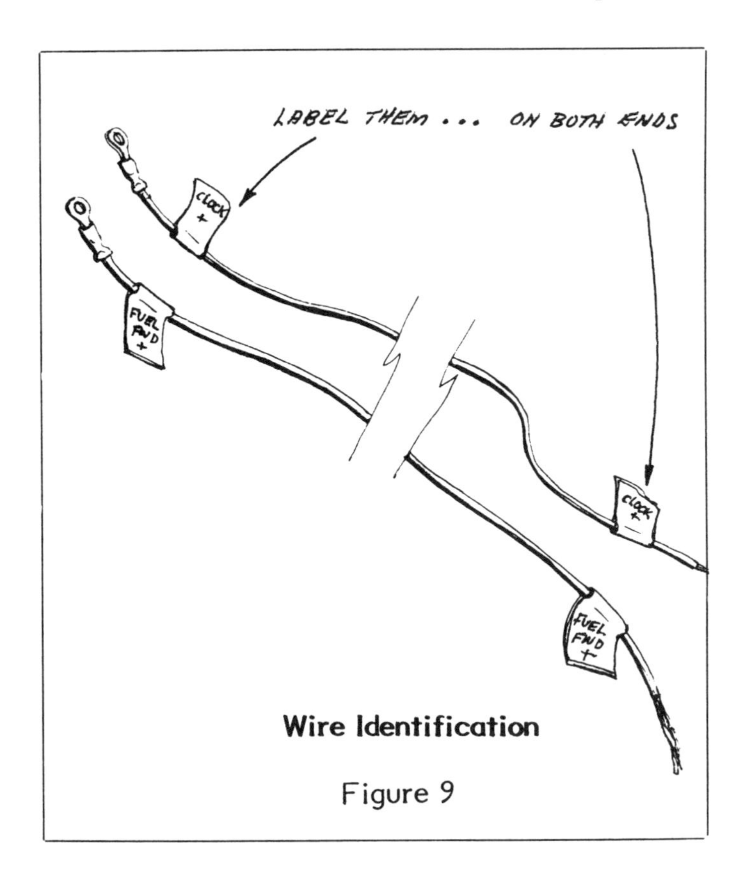

Wire Identification

Figure 9

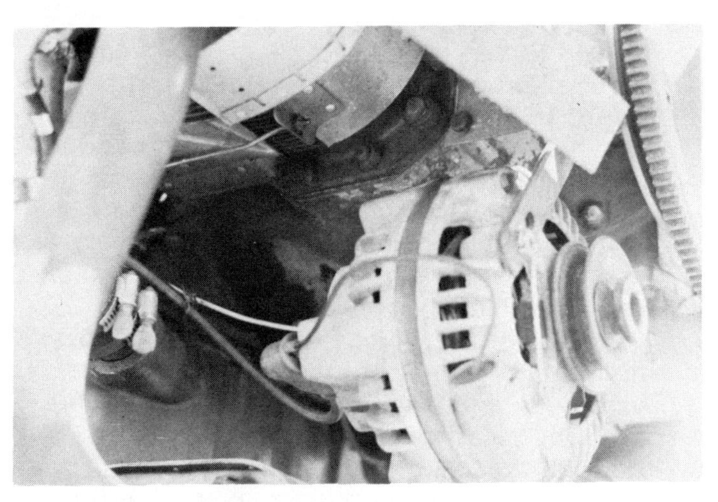

An up-front belt-driven alternator is one of the distinguishing features of a Lycoming engine. It is readily accessible but requires careful alignment of the pulleys and the proper adjustment of belt tension.

This gear-driven alternator plugs into the rear accessory section. Installation is typical of the Continental engines. This one is on an 0-200 model.

An automotive alternator is sometimes installed by a builder because it is more economical.

to the DC generators that have been in use since the early days of aviation. For one thing, alternators are lighter and less expensive. Furthermore, an alternator starts producing electrical energy at approximately 460 rpm, while a generator does not cut in until the engine is revved to about 1,200 rpm. You can see, then, that while taxiing out for takeoff, a generator will be loafing while any electrical units in use (navigation lights, radios, strobe lights, etc.) are busily draining your battery. An alternator, by contrast can be expected to provide you with at least 20 amps at the same low taxing engine rpm.

In general, the generators or alternators used in small Continental engines are of the gear-driven variety, which are plugged into the accessory case of the engine. Lycoming engines, on the other hand, have generators or alternators up front mounted on brackets. These are belt driven. They demand more attention as the pulleys must be accurately aligned and the belt tension maintained. Brackets do crack and break, so if you have to make your own, make them out of 4130 steel at least 1/8-inch (.125-inch) thick.

More and more builders are using automotive or motorcycle alternator units because they are considerably less expensive. However, the automotive and motorcycle units are not designed to operate at high altitudes, nor are they generally as durable, at least that's what the maintenance facility personnel tell us.

SPARK PLUG INSTALLATION

There is more to plug installation than most of us realize. Let's look at a few installation rules:

* Use a good six-point spark plug wrench and the proper tools to install and remove plugs.
* Before installing plugs, inspect each visually (yes, even new ones) for thread damage, cracked insulator tips, barrels, and obvious damage.
* Never install a spark plug that you have dropped. Even at $10.00 each they are cheaper than a new airplane.
* Check or set the gap before installing the plugs.
* Use a thread lubricant (anti-seize compound) sparingly in order to ensure easy future removal (but don't use the lubricant on the first thread)
* Use a good, preferably new, outside copper gasket. Ensures proper reach, sealing and heat transfer.

* Use the recommend torque on the installed plugs. According to the Champion Aviation Spark Plug manual, their plugs are torqued to 25-30 ft. lbs. for Continental engines and 30-35 ft. lbs for Lycomings. The 14mm plugs require 20-25 ft. lbs. If you don't have a torque wrench use a short handled wrench to keep from overtightening.
* Use shielded spark plugs if you are radio equipped. If not, unshielded spark plugs are cheaper and just as efficient.

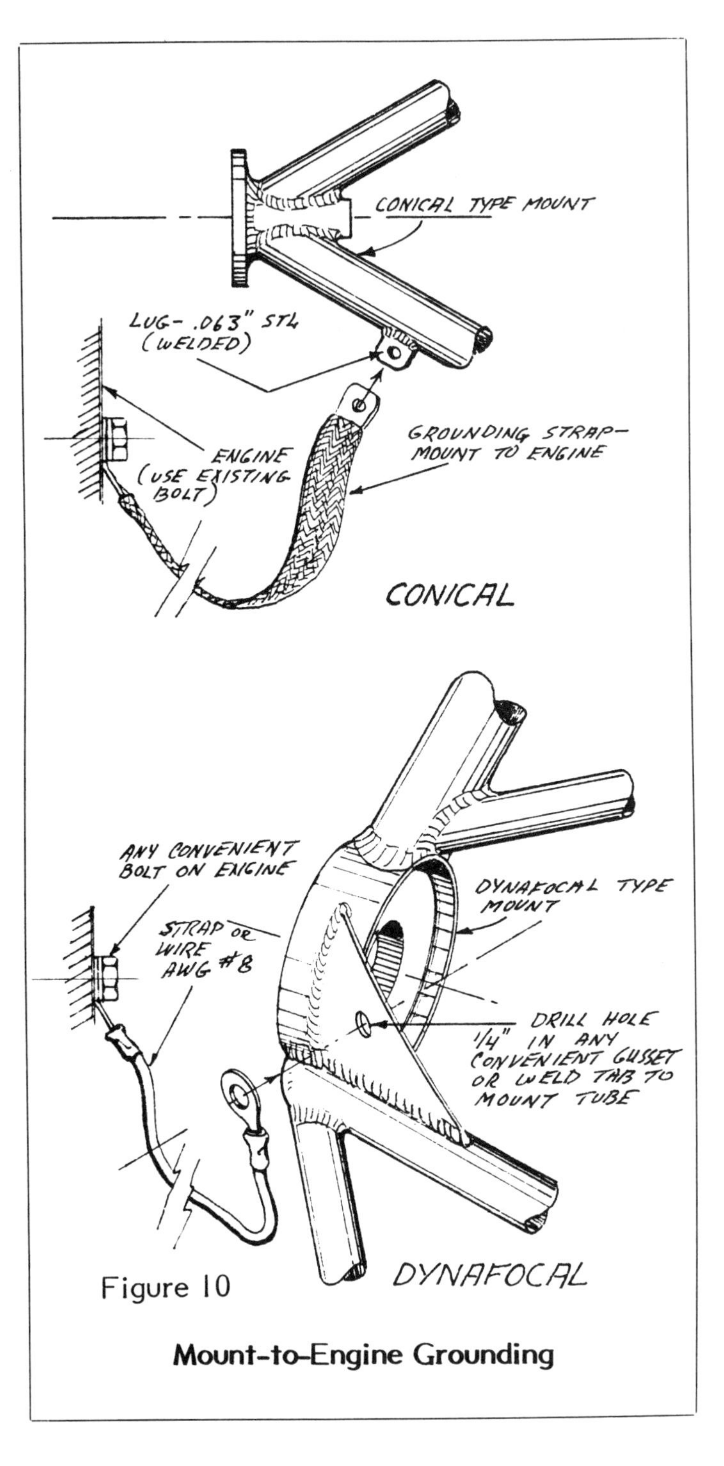

Figure 10

Mount-to-Engine Grounding

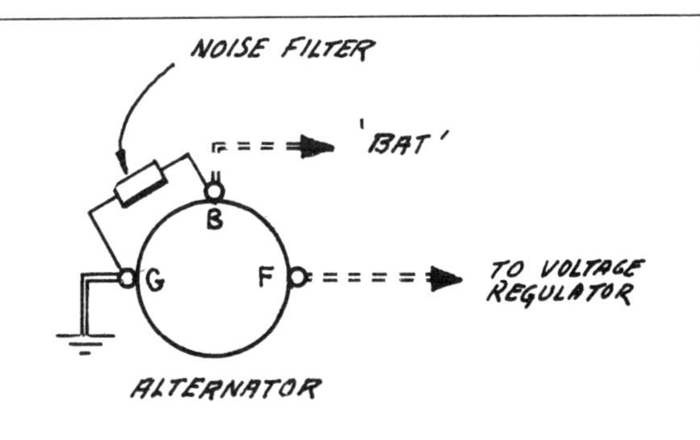

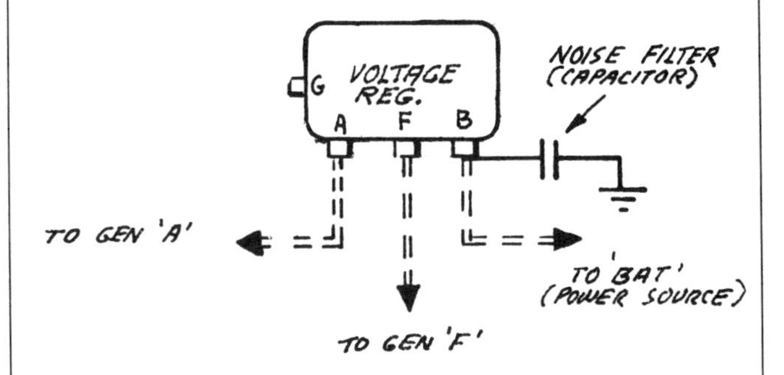

NOTE–

RADIO INTERFERENCE (NOISE) SOURCES . . . . .

1. CHECK ENGINE-TO-FUSELAGE BONDING (⏚).
2. CHECK FOR SOURCES OF METAL-TO-METAL (MECHANICAL) FRICTION AND ELIMINATE.
3. INSTALL A 1 MF CAPACITOR (CONDENSER) AT GENERATOR OR VOLTAGE REGULATOR AS ILLUSTRATED ABOVE.
4. AVOID RUNNING OTHER WIRES CLOSE TO IGNITION WIRES ('P' LEADS).
5. INSTALL SHIELDED WIRES WHERE REQUIRED. SEE WIRING DIAGRAMS ELSEWHERE. BOTH ENDS OF SHIELDING MUST BE GROUNDED.
6. CHECK INSTALLATION AND CONDITION OF IGNITION HARNESS. CHECK KNURLED NUTS FOR TIGHTNESS.

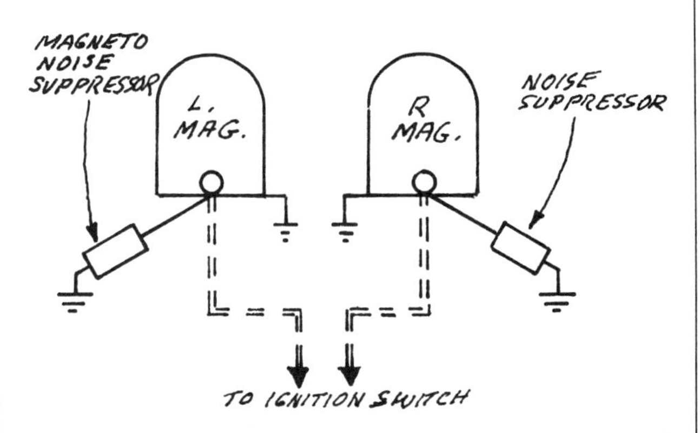

Figure 11

Noise Filter Installations

IGNITION HARNESS INSTALLATION

If a serviceable ignition harness did not come with your engine you will have to install one yourself. You will most likely install one made up of high voltage shielded wire and shielded terminals (current flow is low but the voltage is high in ignition leads).

Connecting the leads to the proper cylinders is not difficult but it does take a little thought. As you may know, a magneto "fires" each lead in sequence as the engine/magneto rotates. It is, therefore, imperative that each individual lead runs to the correct cylinder to obtain the proper firing order. For example, the No. 1 lead from the magneto's distributor goes to the No. 1 cylinder to fire. The No. 2 lead doesn't go to the No. 2 cylinder, however, but to No. 3, which is the second to fire, and so on.

FAA APPROVED SPARK PLUGS (SHORT REACH) FOR CONTINENTAL AND LYCOMING AIRCRAFT ENGINES

ENGINE MODEL	CHAMPION																					
	62-S	C-27	C-27S	D41N	EM41E	EM41N	EM42E	HM41E	M41E	M41N	M42E	REM37BY	REM38E	REM38P	REM38W	REM39N	REM40E	RHM38E	RHM38P	RHM38W	RHM39N	RHM40E
CONTINENTAL																						
A65, A75	s	u	s	u	s	s	s	s	u	u	u			s			s		s	s		s
C75, C85	s	u	s	u	s	s	s	s	u	u	u			s	s		s		s	s		s
C90		u	s	u	s	s	s	s	u	u	u			s	s		s		s	s		s
O-200			s		s	s	s	s	u	u	u			s	s		s		s	s		s
C145		u	s	u	s	s	s	s	u	u	u		s	s	s	s	s		s	s	s	s
O-300-A-B-C-D		u	s	u	s	s	s	s	u	u	u		s	s	s	s	s		s	s	s	s
LYCOMING																						
O-235-C-E-H					s			s	u					s	s		s		s	s		s
O-235-F-G-J																						
O-235-K-L-M												s	s	s	s	s	s	s	s	s		s
O-290 Series					s			s	u					s	s		s		s	s		s
O-320-A-C-E					s			s	u			s		s	s		s		s	s		s
0320-B-D-H												s	s	s	s		s	s	s	s		s
IO-320-B-F												s	s	s	s		s	s	s	s		s
AIO-320												s	s	s	s		s	s	s	s		s
LIO-320-B												s	s	s	s		s	s	s	s		s
IO-320-A-D-E					s			s				s		s	s		s		s	s		s
AEIO-320					s			s				s		s	s		s		s	s		s
O-340-A Series													s	s	s		s	s	s	s		
O-340-B Series					s			s						s	s		s		s	s		
HIO-360-B												s	s	s	s		s	s	s	s		s
HO-360												s	s	s	s		s	s	s	s		s
O-360-A-C-E-F												s	s	s	s		s	s	s	s		s
IO-360-B-E-F												s	s	s	s		s	s	s	s		s
AEIO-360-B-H												s	s	s	s		s	s	s	s		s
LO-360-A-E												s	s	s	s		s	s	s	s		s
O-360-B-D					s			s				s		s	s		s		s	s		s
IVO-360												s	s	s	s		s	s	s	s		s
VO-360-A-B												s	s	s	s		s	s	s	s		s
IO-360-A-C-D													s	s	s			s	s	s		
LIO-360-C1E6D													s	s	s			s	s	s		
HIO-360-A-C-D-F													s	s	s			s	s	s		
AIO-360													s	s	s			s	s	s		
AEIO-360-A													s	s	s			s	s	s		
LHIO-360-C-F													s	s	s			s	s	s		

ENGINE MODEL	AC SPARK PLUG CO.																
	A-88	HS88	HSR-83P	HSR831R	HSR-86	HSR-87	HSR-88	HSR-93	S88	S88-S88D	SR-83P	SR831R	SR-86	SR-87	SR-88	SR88-SR88D	SR-93
CONTINENTAL																	
A65, A75	u	s					s			s						s	
C75, C85	u	s	s	s			s	s		s	s	s				s	s
C90	u	s	s	s			s	s		s	s	s				s	s
O-200	u	s	s	s			s	s		s	s	s				s	s
C145	u	s	s	s			s	s		s	s	s				s	s
O-300-A-B-C-D	u	s	s	s			s	s		s	s	s				s	s
LYCOMING																	
O-235-C-E-H			s			s	s	s	s		s			s	s		s
O-235-F-G-J			s			s		s			s			s			s
O-235-K-L-M			s		s	s		s			s		s	s			s
O-290 Series			s			s	s	s			s			s	s		s
O-320-A-C-E			s			s	s	s	s		s			s	s		s
0320-B-D-H			s			s	s	s			s			s	s		s
IO-320-B-F			s			s	s	s			s			s	s		s
AIO-320			s			s	s	s			s			s	s		s
LIO-320-B			s			s	s	s			s			s	s		s
IO-320-A-D-E			s			s	s	s	s		s			s	s		s
AEIO-320			s			s	s	s	s		s			s	s		s
O-340-A Series			s			s	s	s			s			s	s		s
O-340-B Series			s			s	s	s	s		s			s	s		s
HIO-360-B			s			s	s	s	s		s			s	s		s
HO-360			s			s	s	s	s		s			s	s		s
O-360-A-C-E-F			s			s	s	s	s		s			s	s		s
IO-360-B-E-F			s			s	s	s	s		s			s	s		s
AEIO-360-B-H			s			s	s	s	s		s			s	s		s
LO-360-A-E			s			s	s	s	s		s			s	s		s
O-360-B-D			s			s	s	s	s		s			s	s		s
IVO-360			s			s	s	s			s			s	s		s
VO-360-A-B			s			s	s	s			s			s	s		s
IO-360-A-C-D			s		s			s			s		s				s
LIO-360-C1E6D			s		s			s			s		s				s
HIO-360-A-C-D-F			s		s			s			s		s				s
AIO-360			s		s			s			s		s				s
AEIO-360-A			s		s			s			s		s				s
LHIO-360-C-F			s		s			s			s		s				s

ENGINE MODEL	BENDIX AUTO-LITE CORP.																
	18-1A	B4	B4S	BH-4S	BR-4	BR4SB	H15	PH26	PH260	SH2M	SH15	SH15R	SH20A	SH26	SH150	SH200A	SH260
CONTINENTAL																	
A65, A75	u	u	s	s			u			s	s	s	s		s	s	
C75, C85	u	u	s	s			u				s	s	s		s	s	
C90	u				u	s	u				s	s	s		s	s	
O-200							u				s	s	s		s	s	
C145					u	s	u				s	s	s		s	s	
O-300-A-B-C-D					u	s	u				s	s	s		s	s	
LYCOMING																	
O-235-C-E-H								s	s		s	s	s			s	
O-235-F-G-J																	
O-235-K-L-M								s	s					s			s
O-290 Series								s	s		s	s	s			s	
O-320-A-C-E								s	s		s	s	s			s	
0320-B-D-H								s	s				s	s		s	s
IO-320-B-F								s	s				s	s		s	s
AIO-320								s	s				s	s		s	s
LIO-320-B								s	s				s	s		s	s
IO-320-A-D-E								s	s		s	s	s			s	
AEIO-320								s	s		s	s	s			s	
O-340-A Series								s	s				s	s		s	s
O-340-B Series								s	s		s	s	s			s	
HIO-360-B								s	s				s	s		s	s
HO-360								s	s				s	s		s	s
O-360-A-C-E-F								s	s				s	s		s	s
IO-360-B-E-F								s	s				s	s		s	s
AEIO-360-B-H								s	s				s	s		s	s
LO-360-A-E								s	s				s	s		s	s
O-360-B-D								s	s		s	s	s			s	
IVO-360								s	s				s	s		s	s
VO-360-A-B								s	s				s	s		s	s
IO-360-A-C-D								s	s					s			s
LIO-360-C1E6D								s	s					s			s
HIO-360-A-C-D-F								s	s					s			s
AIO-360								s	s					s			s
AEIO-360-A								s	s					s			s
LHIO-360-C-F								s	s					s			s

USEFUL INFORMATION -

- SPARK PLUG SIZE = 18MM.
- NEVER INSTALL A SPARK PLUG THAT HAS BEEN DROPPED.
- SPARK PLUG GAP =

  LYCOMING ENGINES-SET AT .017 TO .021.

  CONTINENTAL ENGINES - GAPS VARY WITH PLUG TYPE (MOST .015 TO .018, SOME .019 TO .022, SOME LARGER). USE SPARK PLUG MANUFACTURER'S RECOMMENDATION CHART.
- APPLY ANT-SEIZE COMPOUND SPARINGLY TO THE FIRING END THREADS BUT NEVER TO THE FIRST THREAD NOR TO THE SHIELDING BARREL TERMINAL THREADS.
- USE A SIX-POINT SOCKET (7/8") TO AVOID DAMAGING YOUR SPARK PLUGS.
- ENGINES REQUIRING LONG REACH PLUGS ARE IDENTIFIED BY YELLOW PAINT ON CYLINDER FINS BETWEEN SPARK PLUG HOLE AND ROCKER BOX COVER.
- TORQUE FOR ALL SPARK PLUGS =

  LYCOMING, 360 TO 420 IN. LBS.

  CONTINENTAL, 300 TO 360 IN. LBS.
- ALWAYS USE NEW GASKETS WHEN REINSTALLING SPARK PLUGS.

CODE -

s SHIELDED
u UNSHIELDED

TABLE 4

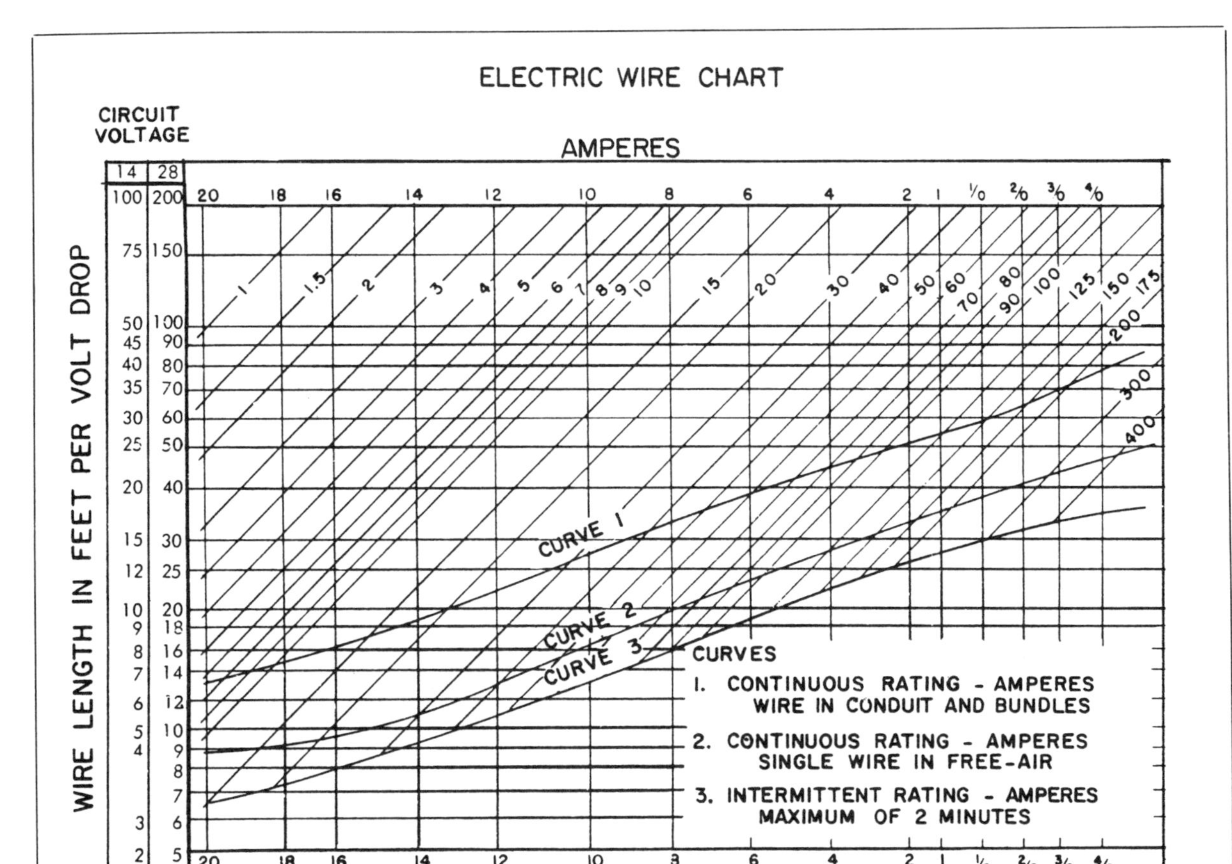

CONDUTOR CHART (Applicable to copper wire conductors)

SOURCE - U.S. DEPARTMENT OF COMMERCE (CAA)

HOW TO USE THE WIRE SELECTION CHART

A. TO DETERMINE A CABLE'S CURRENT CARRYING CAPACITY -

1. SELECT AMPERE LOAD THE WIRE MUST CARRY (DIAGONAL LINES).
2. FOLLOW THAT DIAGONAL DOWN UNTIL IT INTERSECTS THE CURVE (#1, #2 OR # 3).
3. DROP STRAIGHT DOWN FROM POINT OF INTERSECTION TO READ REQUIRED WIRE SIZE.

EXAMPLE -
IF A SINGLE WIRE HAS TO CARRY 20 AMPS - SELECT THE 20 AMP DIAGONAL AND FOLLOW IT DOWN UNTIL IT INTERSECTS CURVE #2 (SINGLE WIRE). DROP STRAIGHT DOWN AND YOU WILL NOTE THAT A 16 AWG WIRE SHOULD BE USED.

B. TO DETERMINE THAT A CABLE SIZE IS ADEQUATE (NO MORE THAN A 1 VOLT DROP).

1. SELECT HORIZONTAL LINE FOR LENGTH OF CABLE YOU MUST INSTALL.
2. DETERMINE THE AMP LOAD FOR THE CIRCUIT AND FOLLOW THE APPROPRIATE DIAGONAL DOWN UNTIL IT INTERSECTS THE SELECTED HORIZONAL LENGTH LINE.
3. READ STRAIGHT DOWN FOR CABLE SIZE REQUIRED (VOLTAGE DROP 1 VOLT OR LESS).

EXAMPLE -
YOUR BATTERY CABLE IS 22' LONG AND THE STARTER WILL DRAW 150 AMPERES - FIND 22 FOOT HORIZONAL LINE (BETWEEN 20 AND 25). SELECT 150 AMPERES LINE AND FOLLOW IT UNTIL IT INTERSECTS WITH 22' HORIZONTAL LINE. READING STRAIGHT DOWN INDICATES CABLE SIZE REQUIRED IS #4AWG TO KEEP VOLTAGE DROP AT OR UNDER 1 VOLT.

TABLE 5

Messing around with Magnetos

The magneto ignition system used with small aircraft engines (VW's too) is a spartan affair which ordinarily performs as expected. However, as we homebuilders ruefully discover during the early stages of construction, the term "ordinarily", describes a condition or state applicable to someone else's project, not our own. Where one's own project is concerned, it is just as well to assume that Murphy's Law (Anything That Can Go Wrong, Will) prevails.

When operating properly, the magneto generates a high voltage impulse which forces a spark to jump across the spark plug gap at the precise moment the piston is on the proper stroke (compression), and at a specified number of crankshaft degrees before top dead center (BTDC piston position).

Well now, if that is all there is to the magneto, why are so many of us perplexed by this little black box? Attribute it to our lack of familiarity with these shocking devices. After all, not many of us work on engines and it is only when our project approaches the engine installation stage that we realize how unsure we are about timing procedures.

Mechanically simple, magnetos are reliable, efficient and .... POTENTIALLY DEADLY! The installed magneto might be likened to a rattlesnake ready to strike. It won't hurt you if you are alert and retain control of the situation, but if you relax your guard, it can kill you. If there are latent fuel vapors in the engine cylinders and a faulty ground or switch connection, LOOK OUT! Don't move that prop even the slightest amount! The engine might fire up briefly .... just long enough to get you. Why? Blame it on the impulse coupling of that innocent looking magneto bolted to the engine.

THE IMPULSE COUPLING

Contrary to what many people believe, in order to start a magneto-equipped engine, it isn't necessary to grasp the propeller with both hands and snap it through smartly with one leg gyrating through the air 'a la Hollywood'. Not at all. Not with an impulse-coupled magneto.

How can you know if the engine has an impulse coupling? You can bet on it. Without an impulse coupling (or booster mag, or vibrator) the starter won't crank the engine fast enough to get the magneto up to sparking speed. A good hot spark is not produced until the magneto is turned above a certain number of revolutions, let's say, more than 100 rpm. This is known as the "coming in speed" of the magneto. Because of this characteristic, it is a standard industry practice to utilize an impulse coupling to generate a better spark for starting. It does this by causing the magneto to spin faster than the engine cranking speed and, at the same time, automatically retarding the spark for starting.

If you tried to start the engine with its normal advanced (early) spark timing, it would kick backwards like a mule and maybe break something .... like your starter or your arm. Therefore, the retarding or late spark capability provided by the impulse coupling is essential.

After the engine starts, the impulse coupling performs as a drive coupling for the magneto, allowing it to function in the full advance spark position.

VW engine buffs frequently encounter a unique problem affecting the impulse coupling's action. If the homemade adapter for attaching the magneto to the engine does not provide the proper spacing between the engine and the bolted-on magneto, the impulse coupling binds and cannot function.

You can recognize the presence of this condition when you turn the propeller and do not hear that familiar "clack" noise of the coupling. In addition, while turning the prop, the crankshaft will feel unusually "tight" and display a noticeable absence of end play. The remedy? Shim the magneto away from its mounting as necessary to free the coupling.

IS IT OFF?

The scary thing about magnetos is you can never be sure they are "OFF" when the switch is in the OFF position. Very few light aircraft engines have an IDLE-CUT OFF feature with their mixture control. So, most of these engines are normally stopped by turning the magneto switch to the OFF position. That's not a bad setup because if the engine stops,

Seldom seen now is the Vertex magneto which builders simply plugged into the engine distributor port. VW engine developments have left that sort of easy conversion behind.

you know the switch is functioning, and the magneto, at least at that moment, is safe (electrically grounded).

An ignition switch for a magneto system functions in a manner opposite that of ordinary switches. In the OFF position, the switch is closed, causing the breaker points to be electrically grounded. The magneto, therefore, cannot operate. If, on the other hand, the engine does not cease firing in the OFF position, the magneto ground lead (also referred to as the "P" lead) is open. This means trouble and must be corrected.

The external magneto wiring hook-up is simple. The magneto's primary wire, or "P" lead (18-gauge, shielded wire is all right), is connected to the ignition switch. The only other wire that may be installed would be a short wire running from the magneto case to some point on the engine that serves as a ground.

A shielded magneto such as the Scintilla has a built-in device that automatically grounds the magneto whenever the switch wire is disconnected. Unfortunately, this requires a rather fussy connector. You can't just twist the switch wire around a terminal and snub it down as in some unshielded installations.

It is necessary to obtain a Terminal Assembly Kit for the magneto if one is not already installed and dangling from the magneto (most unlikely). The kit consists of a contact washer, insulators, ferrules, and a union or connector nut. The switch wire is assembled into a unit of the exact length required to make proper contact, inside the magneto, with the breaker grounding spring. (Figure 12)

Also required is some 18-gauge shielded switch wire. In addition, a radio noise suppressor is usually installed in the magneto switch wire circuit to eliminate electrical interference picked up from the switch wire.

Don't assume for a minute that if the primary ground lead ("P" lead) is disconnected the magneto is rendered harmless. IT MAY NOT BE! Many magnetos, such as the Scintillas, do have a special breaker grounding spring in the magneto to short circuit the primary .... when the switch wire is <u>not</u> installed. However, other magnetos are NOT provided with automatic grounding springs at the switch wire terminals. After months or years of flying, it is easy to forget whether or not your magneto has the built-in safety feature of a grounding spring, so always be careful when messing around with the magneto, propeller or engine.

Play it safe. Whenever you do any magneto maintenance requiring the removal of the "P" lead from the magneto, remove the top spark plugs or, at least disconnect all spark plug cables from the installed spark plugs. Often recommended but not as safe, of course, is to make sure the magneto switch is off and that there is a completed connection from the switch to the engine and from the magneto to the switch.

When installing the magneto "P" lead connection, insert and secure the switch wire terminal assembly before replacing the breaker cover plate so you can visually ascertain that the end of the switch wire terminal is making positive contact with the breaker contact spring. If not, reposition the breaker grounding spring as necessary. Otherwise, the magneto might remain grounded even when the switch is moved to the ON position. If you can't get a sparking in your plugs from the magneto, that is probably your problem. It is most likely to occur after an engine or magneto overhaul.

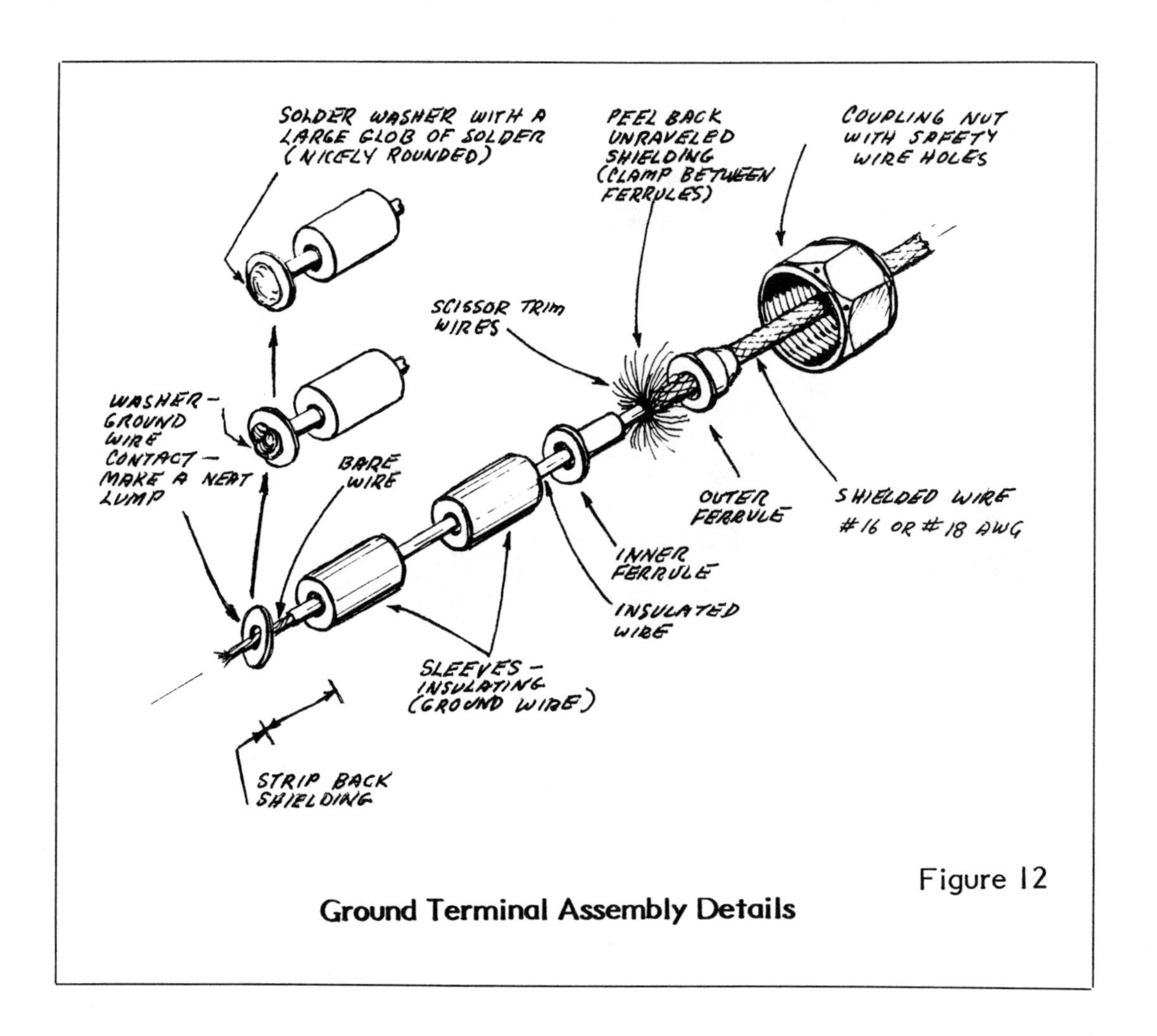

Figure 12

Ground Terminal Assembly Details

TIMING THE MAGNETO TO THE ENGINE

Homebuilders use many types of magnetos but most of them are timed to the engine using the same general procedures. Realize, of course, that there may be many minor variations in the magneto-to-engine timing ritual. Most of the methods in use are good to excellent and any one may be preferred by the individual mechanic (builder) simply because he happens to have the particular gadget, feeler gauge, buzzer, piece of cellophane, light or other tool required. Use whatever you have. If you don't have anything, make or borrow something. You will be able to do a more precise job of timing.

The best guidance for timing an aircraft engine will be found in the maintenance manual for that engine. Its instructions should be followed faithfully. Your only alternative data may be the information obtained from the aircraft engine data plate. VW engines do not have standard aircraft-use manuals nor data plates for guidance and most builders grope around on their own for the best results.

Why such a fuss about timing? A bad timing job can cause all sorts of problems, failure of all cylinders to fire, irregular operation of all cylinders, low horsepower output, loud exhaust, overheating of the engine, detonation, backfiring and difficult starting. As if that's not enough, improper ignition timing, especially advanced timing can lead to preignition, resulting in piston burning, stuck and broken rings and scored cylinders. And that can seriously damage an engine in a very short time.

ONE MAN'S TIMING PROCEDURE

1. Remove all upper spark plugs for safety!
2. Determine the number of degrees BTDC. As previously mentioned, refer to your engine data plate or engine manual. Most engines are timed to fire from 25 degrees to 30 degrees BTDC. For example the C85-12 is usually timed at 28 degrees BTDC for the left mag and at 30 degrees BTDC for the right mag. The last VW aircraft conversion I worked with was

timed at 27 1/2 degrees BTDC.

3. Locate top dead center (TDC) for the No. 1 cylinder. This means, of course, that you will have the No. 1 piston on the compression stroke and both valves closed. To find the compression stroke, you can press your finger or thumb over the spark plug hole while turning the crankshaft in its normal direction of rotation. The compressed air in the cylinder will try to escape past your finger with a "pfuph" sound. Be sure that you have one plug installed in the other side of the cylinder (dual ignition engines), or else you'll be turning that prop all day waiting for the "pfuph". At this point you might ask how TDC is determined. Most aircraft engines have engraved marks on the propeller hub which are aligned with the centerline split of the crankcase. Other engines and converted auto engines might have no such references to use and TDC will have to be established from scratch. A 'screwdriver and hammer' mechanic would probably poke a stick or a wire through the spark plug hole and feel for the piston's rise in the cylinder. When the rising sensaton ceases, the piston is somewhere around top center, not TDC maybe, but close to it. With the piston at or near TDC, there is very little motion of the piston in relation to the crankshaft rotation and, as a result, it is not easy to establish the exact TDC of the piston. You might be off as much as five degrees on successive tries.

   An exact TDC determination using a Piston Position Indicator is possible. There are store-bought types, but you can make your own piston position or top dead center indicator from a piece of wire and an old shielded spark plug with its guts punched out Mechanics have been using such devices for decades. The hollow spark plug shell screws into the cylinder and serves as a very convenient support for the inserted indicator wire. Minute movements of the wire indicator are observed more accurately from this steady base. (Figure 14)

4. Set the crankshaft at the proper position for firing No. 1 cylinder. After you have found TDC, back off on the crankshaft 30 to 40 degrees and then begin to come back slowly in the direction of rotation until you reach the prescribed degrees ahead of top dead center (Figure 13) (i.e. BTDC) as specified by engine manufacturer. When this position is reached, you are ready to mate the magneto to the engine.

5. Rotate the magneto shaft to the position where the No. 1 distributor block lead is ready to spark. Different magnetos have different features built in to assist you in establishing the point where the magneto breaker points should be just beginning to open (ready to fire). On some magnetos look for a marked tooth on the distributor gear which shows up in a small window in the cover of the magneto at the drive end ...

   or

   There may be an opening in the distributor cap through which the position of the distributor finger may be seen ...

   or

   In some cases there may be matching lines on the distributor gear and magneto housing ...

   or

   The magneto may have a step cut on the timing collar secured to the cam ...

   or

   If all else fails, install a spark plug wire in the No. 1 distributor block opening and cause the magneto shaft to be rotated until that lead sparks when held against a metal ground. (This could be a shocking experience if you are not careful.)

6. Now, with both the crankshaft and magneto positioned to fire No. 1, install the magneto on the engine and draw the bolts up snug but not tight. Rotate the magneto in both directions as far as the elongated mounted holes will allow. During this rotation, check the breaker points to see that they open and close. If not, the magneto will have to be removed and its drive shaft turned slightly before reinstalling for another check.

7. As a final check, rotate the magneto in the direction opposite to the normal rotation until the breaker points are just opening. The exact moment of opening may be determined by using a feeler gauge (.0015-inch) or a piece of cellophane. Insert the cellophane between the breaker points and rotate the magneto case slowly by tapping it while maintaining a slight pull on the feeler gauge or cellophane strip. it will begin to slip as the points are just beginning to open. At this point, stop and permanently tighten the mounting bolts. Your timing is complete. The other magneto should be

timed in the same manner but using different firing position if applicable. Next, hook up all the other spark plug leads. The No. 2 distributor lead of the magneto goes to the second cylinder to fire (not to No. 2 cylinder), the 3rd distributor lead goes to the third cylinder to fire, etc.

If both magnetos are timed to fire at the same crankshaft position, you can check their synchronization with the cellophane strip as described above, by inserting a strip in each magneto. Rotate the engine crankshaft backwards about 45 degrees and bring it back in the normal direction slowly. Both sets of breaker points should open simultaneously.

Don't be surprised if the engine starts easily and runs well .... you probably expected otherwise, but Murphy's Law doesn't always apply.

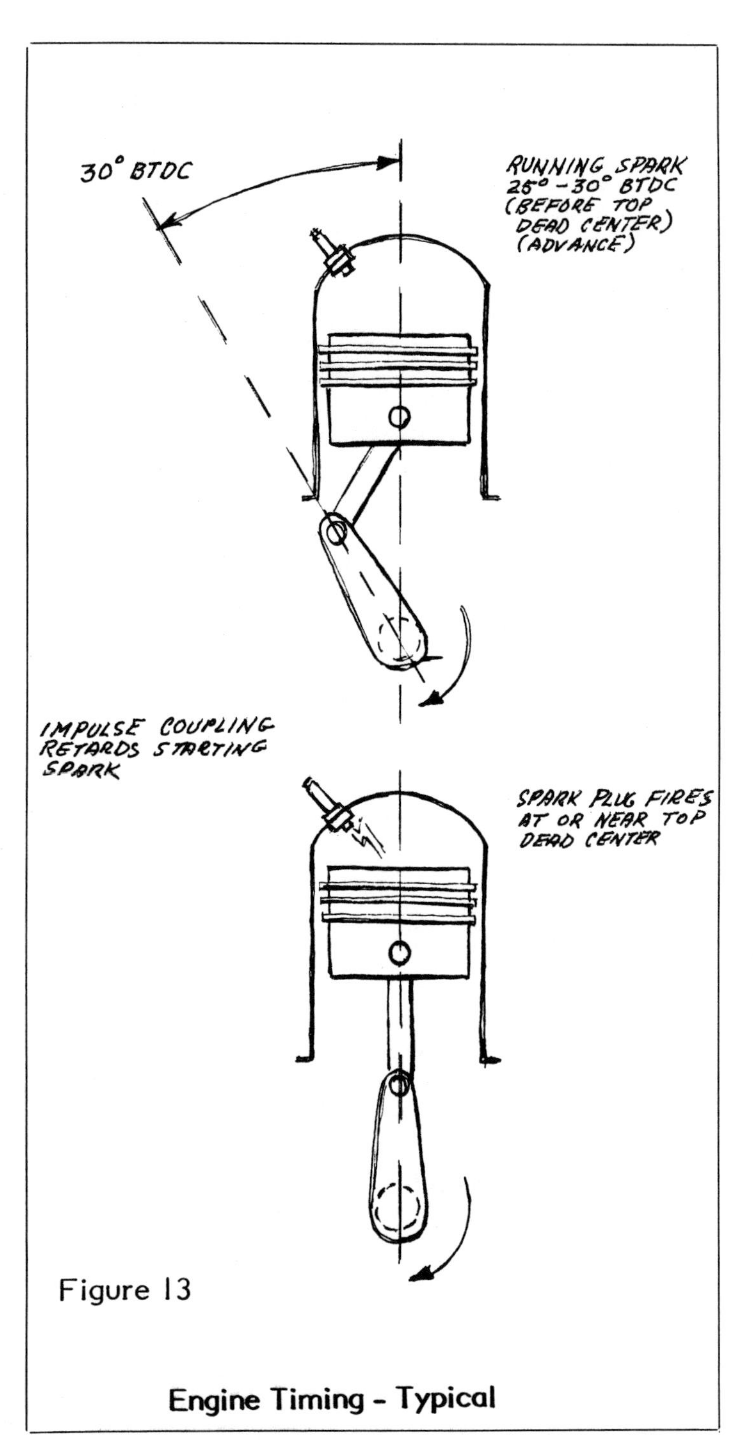

Figure 13

Engine Timing - Typical

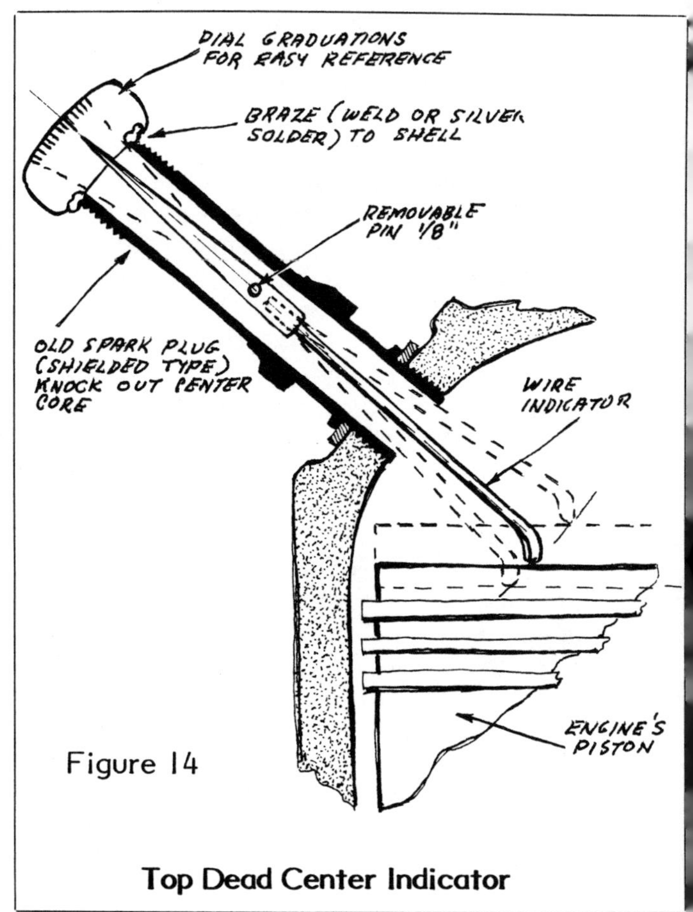

Figure 14

Top Dead Center Indicator

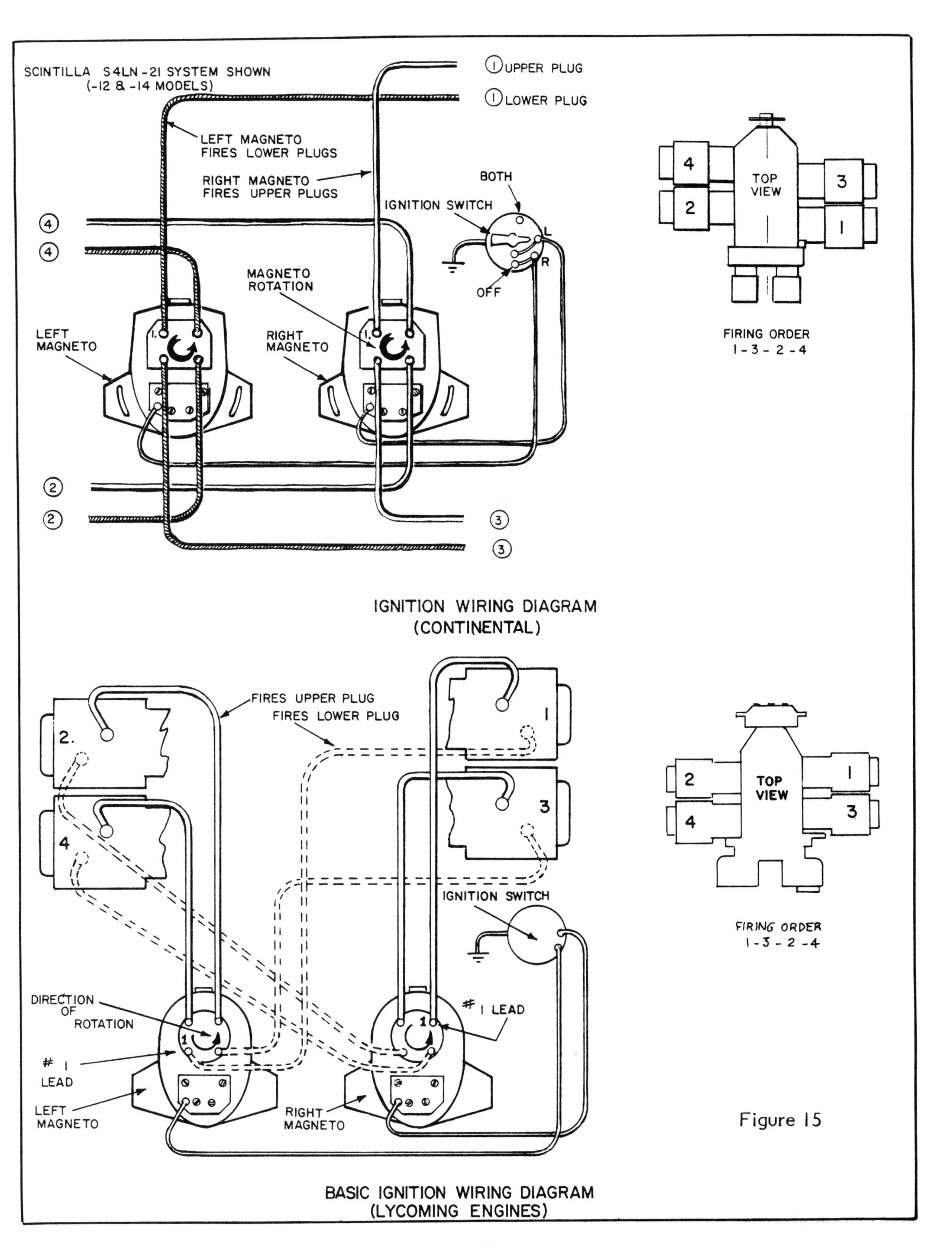
SCINTILLA S4LN-21 SYSTEM SHOWN
(-12 & -14 MODELS)
1 UPPER PLUG
1 LOWER PLUG
LEFT MAGNETO
FIRES LOWER PLUGS
RIGHT MAGNETO
FIRES UPPER PLUGS
BOTH
IGNITION SWITCH
L
R
OFF
MAGNETO
ROTATION
LEFT
MAGNETO
RIGHT
MAGNETO
4
4
2
2
3
3
TOP
VIEW
FIRING ORDER
1-3-2-4
IGNITION WIRING DIAGRAM
(CONTINENTAL)
FIRES UPPER PLUG
FIRES LOWER PLUG
IGNITION SWITCH
DIRECTION
OF
ROTATION
1
LEAD
1 LEAD
LEFT
MAGNETO
RIGHT
MAGNETO
TOP
VIEW
FIRING ORDER
1-3-2-4
BASIC IGNITION WIRING DIAGRAM
(LYCOMING ENGINES)

Figure 15

STANDARD VW FIRING ORDER →
1-4-3-2
(SEE NOTE BELOW)

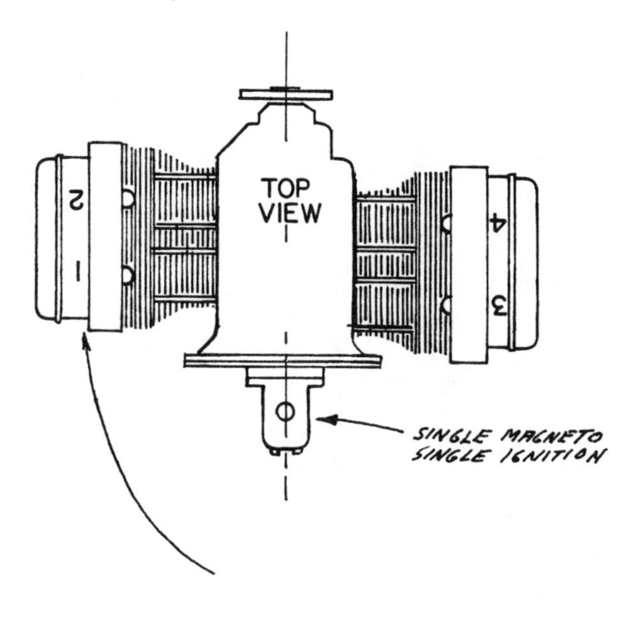

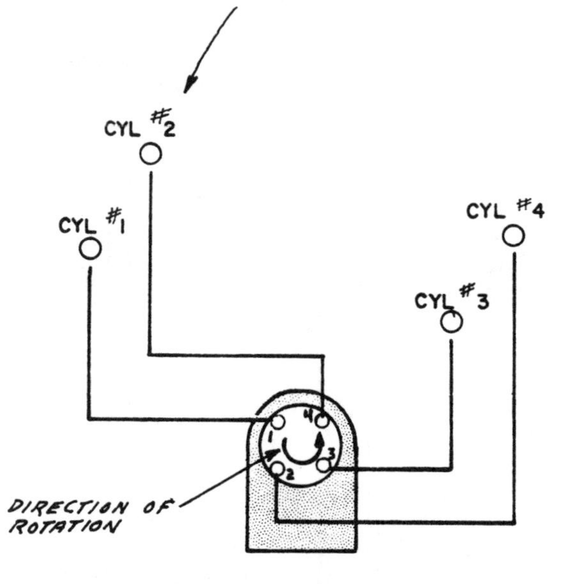

IMPORTANT NOTE → SOME VW AIRCRAFT ENGINE CYLINDERS ARE RENUMBERED TO PRODUCE AN EASY-TO-REMEMBER 1-2-3-4 FIRING ORDER. INTERNAL FIRING SEQUENCE REMAINS THE SAME

STANDARD VW FIRING ORDER	1-4-3-2
MODIFIED 'FIRING ORDER'	1-2-3-4

(#4 CYLINDER BECOMES #1, #3 BECOMES #2 ETC)

Figure 16 IGNITION WIRING DIAGRAM (VW ENGINE)

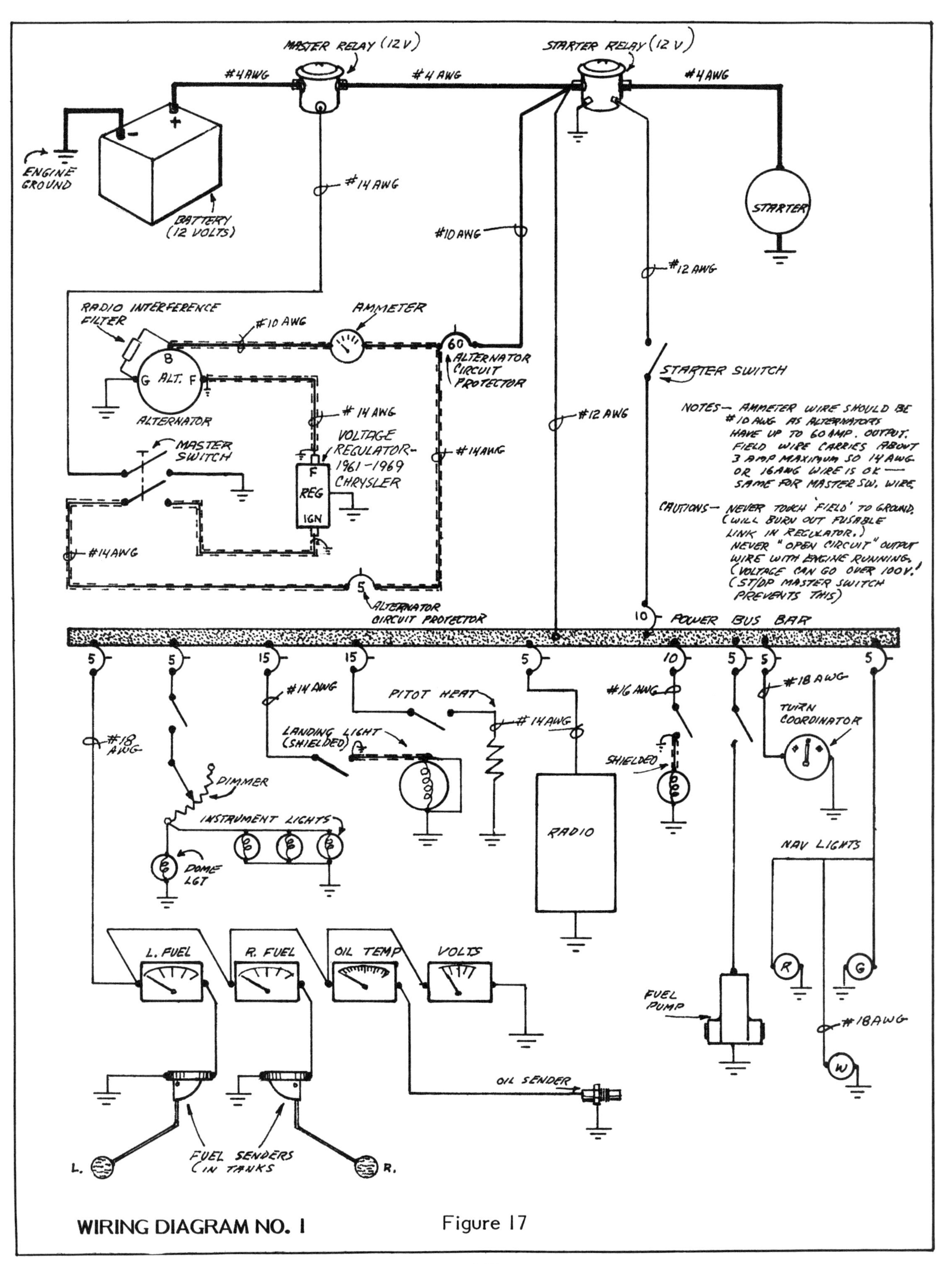

WIRING DIAGRAM NO. 1

Figure 17

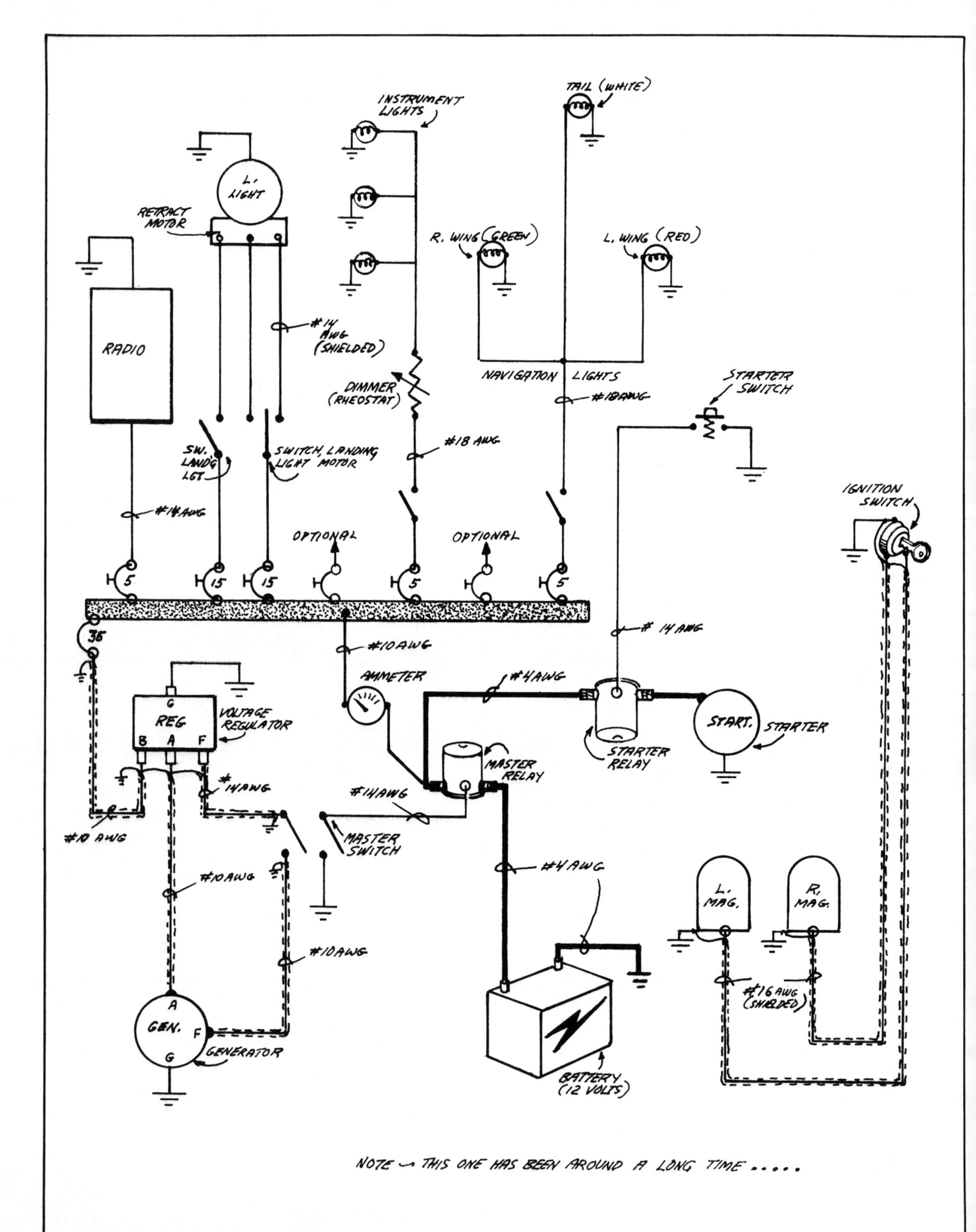

WIRING DIAGRAM NO. 2 Figure 18

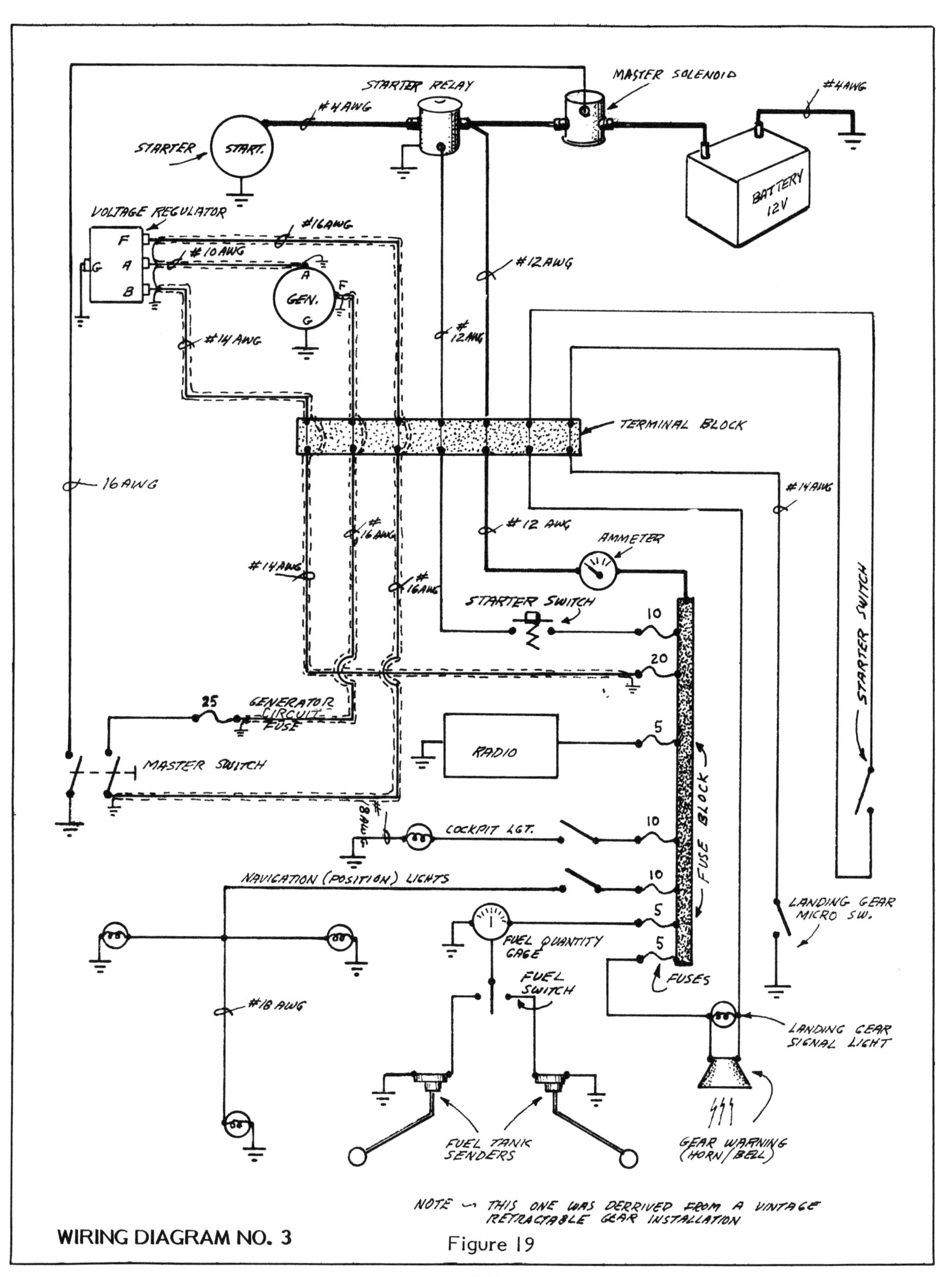

WIRING DIAGRAM NO. 3

Figure 19

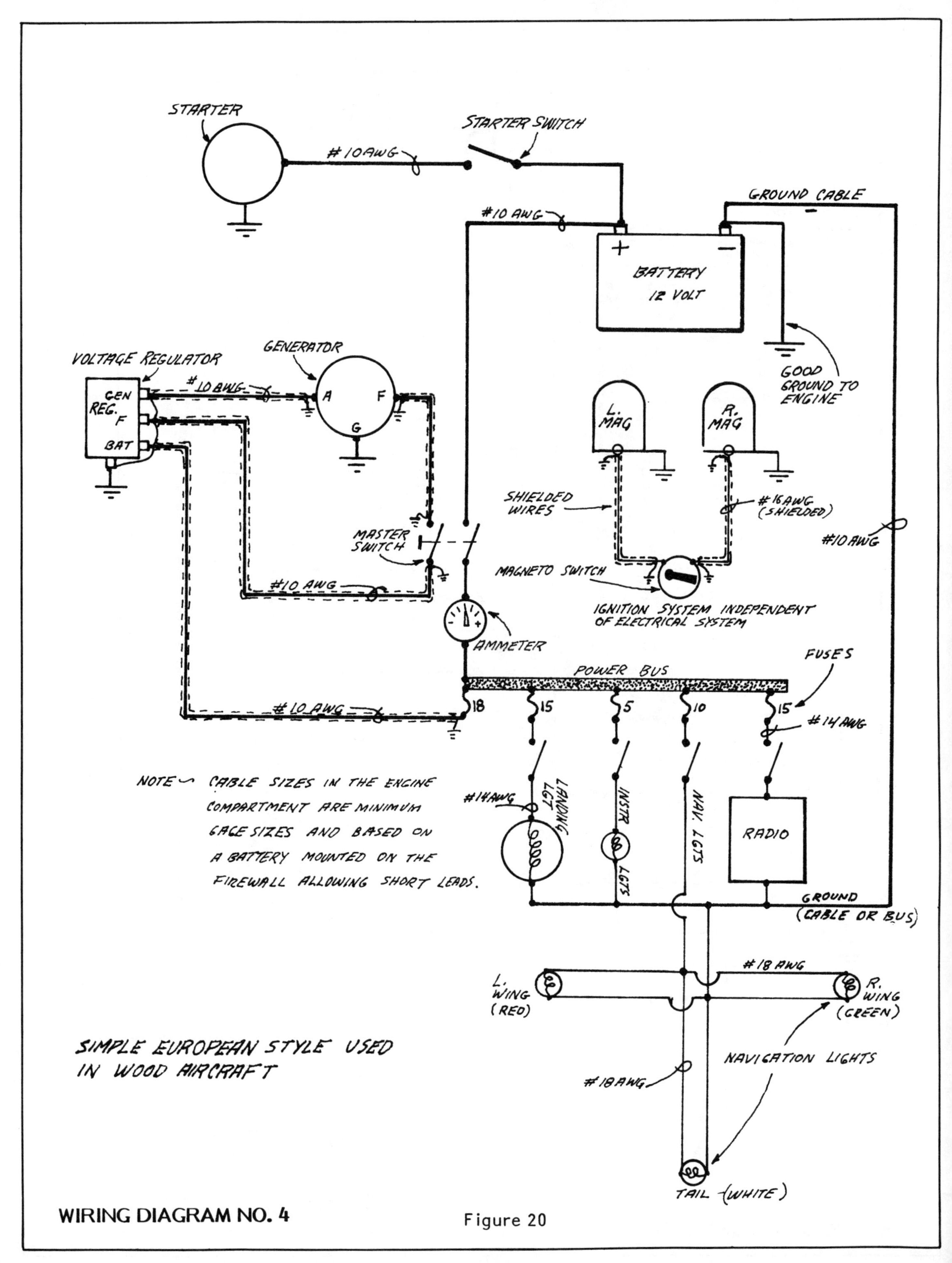

WIRING DIAGRAM NO. 4

Figure 20

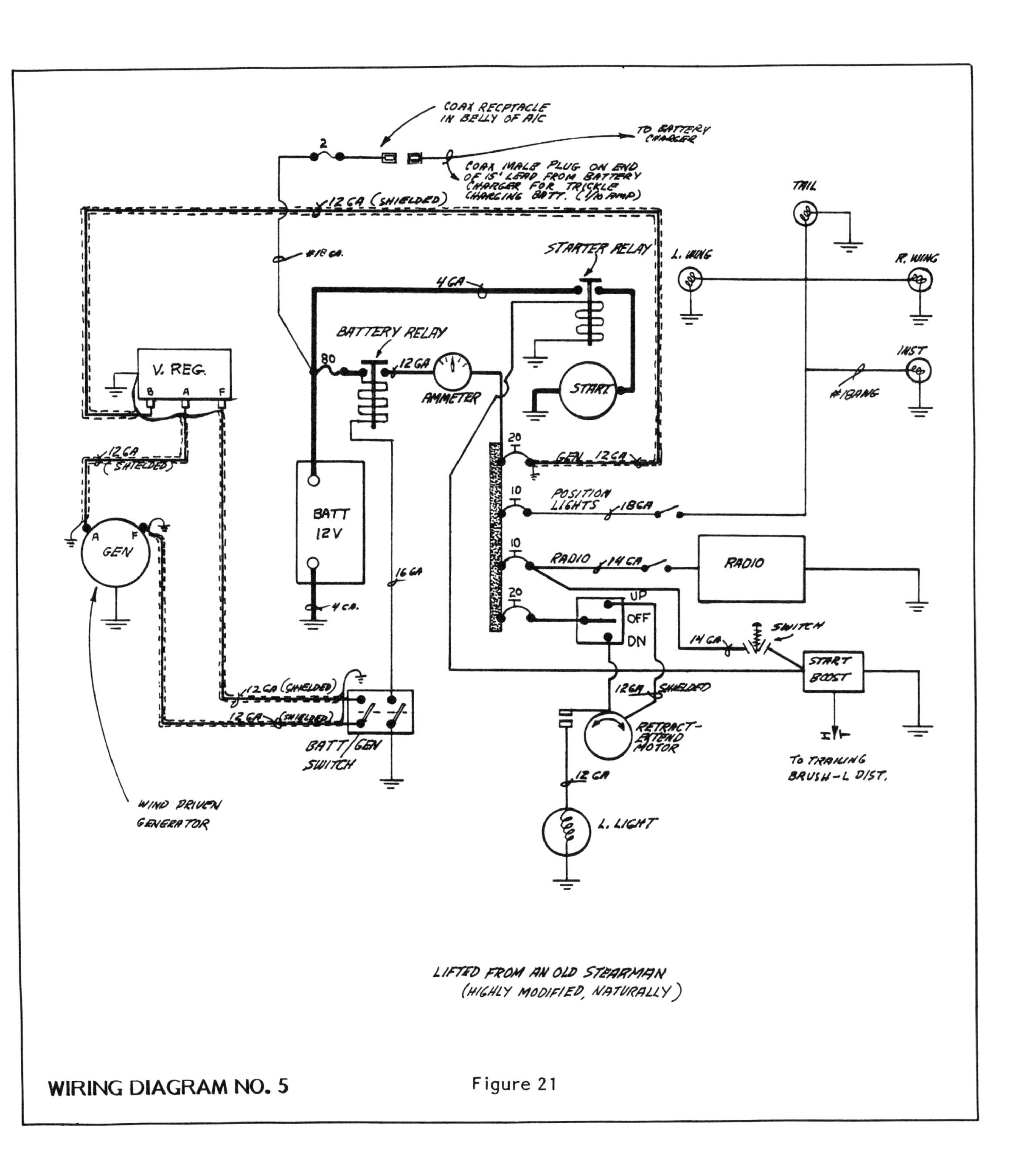

WIRING DIAGRAM NO. 5

Figure 21

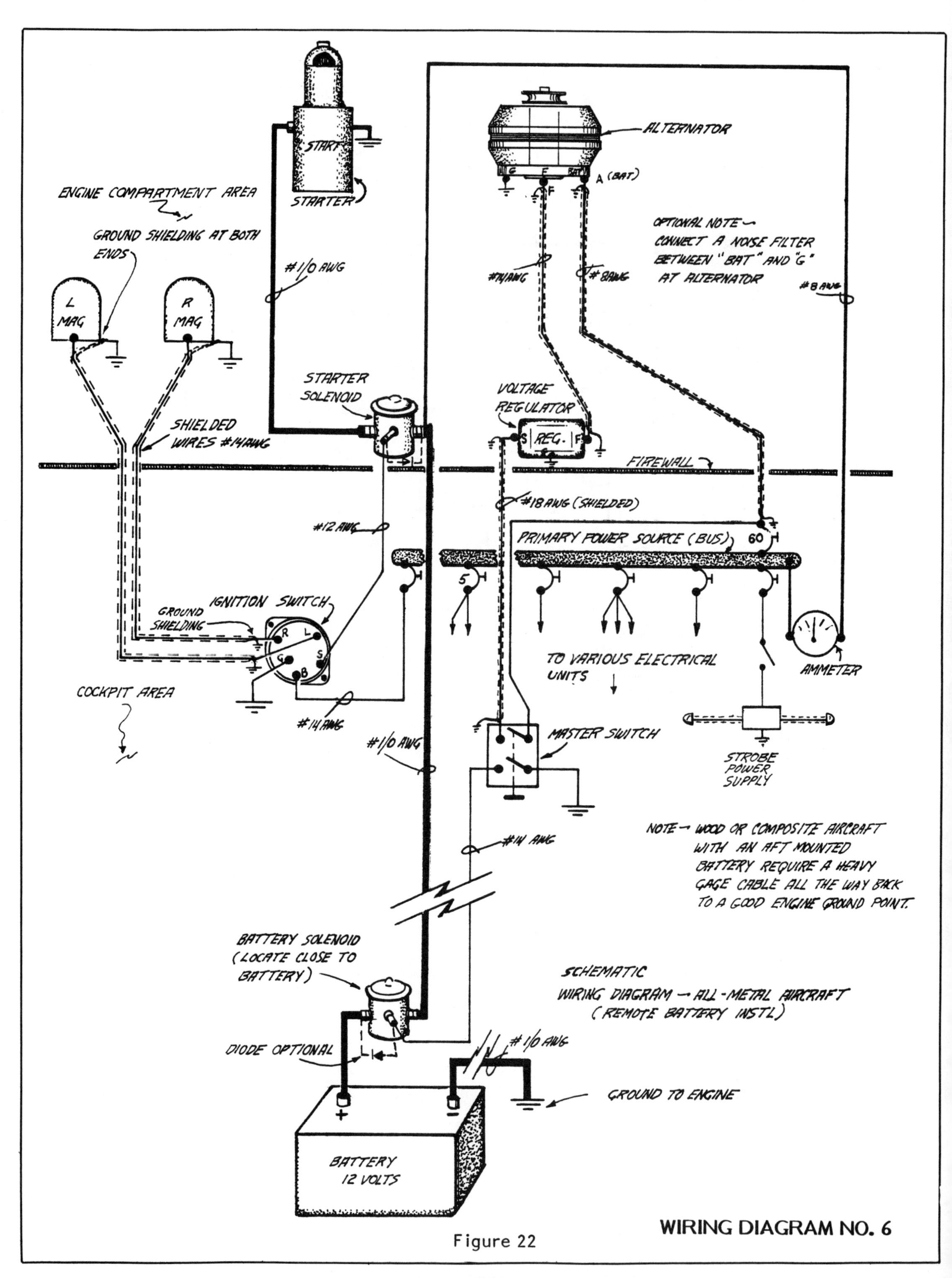
ALTERNATOR
G
F
BAT
A (BAT)
START
STARTER
ENGINE COMPARTMENT AREA
GROUND SHIELDING AT BOTH ENDS
L MAG
R MAG
#1/0 AWG
#14AWG
#8AWG
OPTIONAL NOTE — CONNECT A NOISE FILTER BETWEEN "BAT" AND "G" AT ALTERNATOR
#8 AWG
STARTER SOLENOID
VOLTAGE REGULATOR
S
REG.
SHIELDED WIRES #14AWG
FIREWALL
#18 AWG (SHIELDED)
#12 AWG
PRIMARY POWER SOURCE (BUS)
60
5
IGNITION SWITCH
GROUND SHIELDING
R
L
G
S
B
TO VARIOUS ELECTRICAL UNITS
AMMETER
COCKPIT AREA
#14 AWG
#1/0 AWG
MASTER SWITCH
STROBE POWER SUPPLY
NOTE — WOOD OR COMPOSITE AIRCRAFT WITH AN AFT MOUNTED BATTERY REQUIRE A HEAVY GAGE CABLE ALL THE WAY BACK TO A GOOD ENGINE GROUND POINT.
#14 AWG
BATTERY SOLENOID (LOCATE CLOSE TO BATTERY)
SCHEMATIC WIRING DIAGRAM — ALL-METAL AIRCRAFT (REMOTE BATTERY INSTL)
DIODE OPTIONAL
#1/0 AWG
GROUND TO ENGINE
+
−
BATTERY 12 VOLTS
WIRING DIAGRAM NO. 6

Figure 22

Be sure your switches operate in the correct direction and that they are properly marked.

Ignition harnesses soon get frayed and damaged unless adequately supported.

Dragonfly

powerplant instruments

10

Introduction to Instruments

What engine instruments do you really have to have? Well, if your airplane is to be certificated in the amateur-built category you will need a tachometer, an oil pressure gauge, an oil temperature gauge and a fuel quantity indicator — this assuming, of course, that your airplane engine is air-cooled. You would, of course, also install a coolant temperature gauge, possibly in lieu of an oil temperature gauge when operating a liquid-cooled powerplant.

Incidentally, that fuel quantity indicator need not be a standard needle-type gauge. The old Cubs with their cork and wire indicators were in compliance with FAA requirements.

If your engine uses a fuel pump, you should have a fuel pressure indicator. Likewise, if you have an oil tank that is separate from the engine, it should be provided with an oil quantity indicator.

It is obvious that the basic instrumentation requirements are rudimentary indeed when compared with some of the lavish displays many a homebuilder provides for his pride and joy. The options are yours provided, of course, you at least install the minimum instruments and indicators mentioned.

FAA Airworthiness Standards: Normal, Utility and Aerobatic (FAR Part 23) provides the necessary guidance regarding powerplant instrumentation. Although this regulation does not necessarily apply to amateur-built aircraft, its provisions are normally adhered to by most builders. Part 23.1305, Powerplant Instruments, states that the following are required powerplant instruments:

A. A fuel quantity indicator for each fuel tank
B. An oil pressure indicator
C. An oil temperature indicator
D. A tachometer
E. A cylinder head temperature gauge for each air cooled engine with cowl flaps
F. A fuel pressure indicator for pump-fed engines
G. A manifold pressure indicator for each altitude (turbo-charged, or super-charged) engine

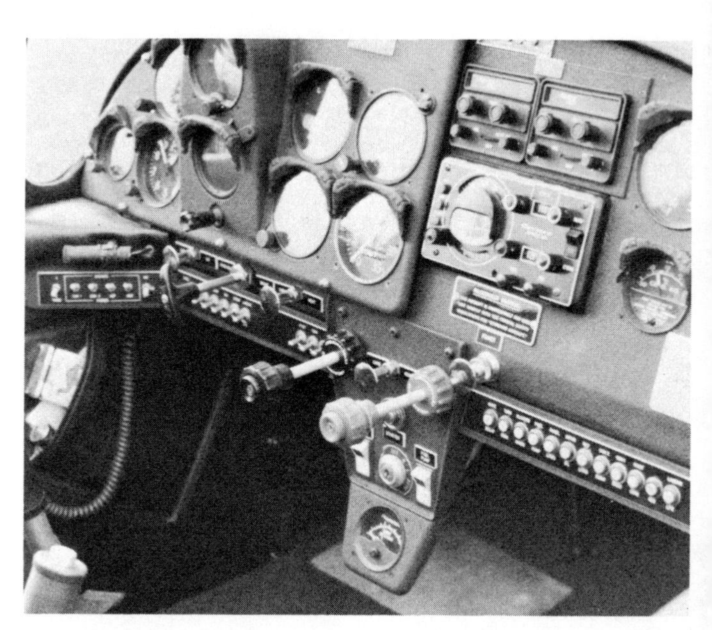

This switch and circuit breaker layout is neat and the grouping makes wiring behind the panel easier to accomplish.

H. An oil quantity indicator for each oil tank

NOTE: The regulation lists additional instrumentation requirements for turbine engines that would be of little or no interest to the average homebuilder.

ABOUT BUYING YOUR ENGINE GAUGES

You will find quite a few different types and kinds of engine instruments to choose from for your aircraft. This can be confusing. It need not be if you give serious consideration to obtaining the BEST INSTRUMENTS YOU CAN AFFORD. Be wary of the inexpensive instruments. Often they are unreliable and could provide you with inaccurate readings or they may fail completely before they have given sufficient service. Your initial investment added to the replacement cost would then exceed what you might have paid for a better instrument in the first place. Besides, think of the frustration, operational risks and extra work that might result.

It is always a risky business to buy used instruments although a used serviceable aircraft instrument may be better than a new "cheapie" one. At Fly-Ins it is hard to resist buying what may appear to be the bargain of the century. However, unless you know a lot about the delicate mechanism you purchase, you are taking a chance that the instrument may be better suited for the scrap heap than for your airplane.Of course, an instrument rebuilt by an authorized instrument repair facility should be as good as a new one. Look for a yellow Serviceable Tag attached.

Before you buy any instrument be sure that it will be suited to your type of installation. We can classify engine instruments by the type and power source required for their operation. It would be a shame for you to inadvertently purchase an instrument requiring a 12-volt electrical system if your aircraft doesn't have an electrical system.

Furthermore, just about every type of engine gauge or engine related instrument is manufactured and sold for use in aircraft equipped with a 12-volt or 24-volt electrical system. Be sure the instrument you acquire is suited to your aircraft's system. Some instruments, however, are made for use without an electrical system, and this can be very important to many builders.

INSTRUMENTS AND THEIR POWER SOURCE REQUIREMENTS

Although the following list is not exhaustive, it will give you a good idea of the options you have. The first consideration in making your instrument selection is to assure yourself that it is compatible with your power source.

ELECTRICAL TYPE GAUGES -- The following gauges are typically available for use in aircraft equipped with a 12-volt or 24-volt electrical system:

- Oil pressure gauge
- Oil temperature gauge
- Tachometer/tachourmeter
- Cylinder head temperature gauge
- Ammeter
- Voltmeter
- Fuel pressure gauge
- Fuel level gauge
- Carburetor air temperature gauge
- Hourmeter

THERMOCOUPLE (ELECTRICAL) TYPE GAUGES — No electrical system is required. These instruments get their power from electricity produced by a thermocouple probe or gasket:

- Cylinder head temperature gauge
- Carburetor air temperature gauge
- Exhaust gas temperature gauge

MAGNETO/DISTRIBUTOR/ALTERNATOR (ELECTRICAL) TYPE — No electrical system is required. These receive their power from one of the above sources.

- Tachometer
- Carburetor air temperature gauge

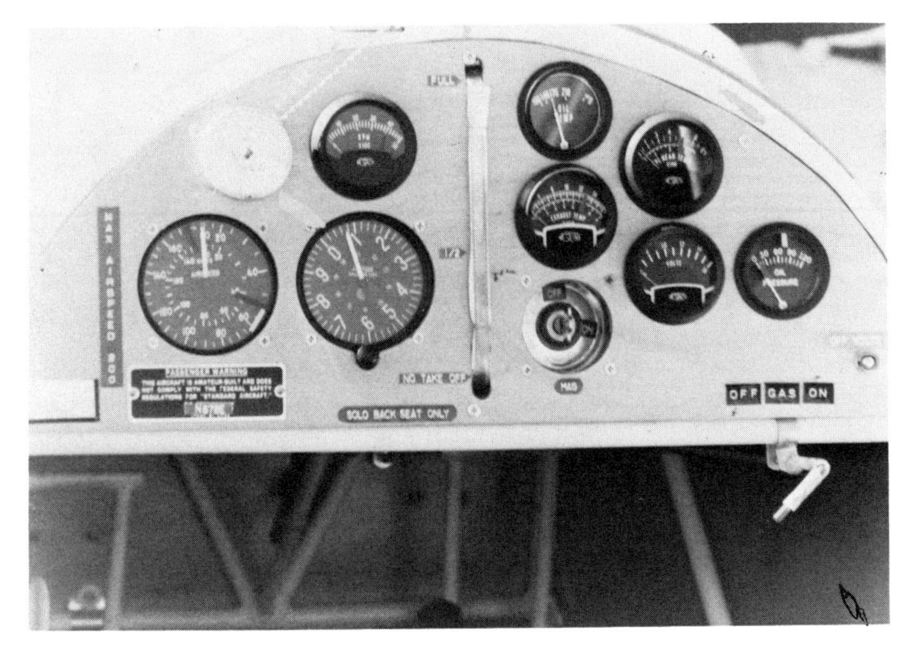

Here's a well-placarded instrument panel with a clear plastic tube-type fuel gauge. A better location for the fuel shut-off valve handle would be an improvement.

MECHANICAL TYPE — No electrical system is required. The instruments are connected mechanically to the engine or operate by some other method than electrically:

Tachometer/tachourmeter (cable operated)
Fuel quantity gauge (direct reading)

MECHANICAL (ELECTRICAL) TYPE — No electrical system is needed. A tiny generator screws into the tachometer drive of an aircraft engine:

Tachometer

DIRECT READING TYPE — No electrical system is required. This type of instrument is directly connected to the engine by tubing or other means:

Oil pressure gauge (tubing)
Fuel pressure gauge (tubing)
Oil temperature gauge (capillary type)
Carburetor air temperature gauge (B-5 probe)

A FEW POINTERS

Lighted instruments generally do require a 12-volt or 24-volt aircraft electrical system to power them. (Most instruments have the lighted option although the costs will be higher.)

It is easier to select "the best instruments you can afford" when you have a better understa[illegible] ing of their function and means of operation. Unfortunately, many builders learn too late that some instruments purchased in good faith are far from being reliable or long lived.

A newly installed gauge should always be considered suspect until it has proven itself. Do not be too quick to assume personal blame for a problem indicated by an abnormal gauge reading, or to assume that the engine is malfunc- tioning. The gauge could be wrong, you know, even if it is a new one.

Noteworthy features. Engine instruments conveniently clustered for easy reading, power quadrant control knobs, and cut-out for future radios.

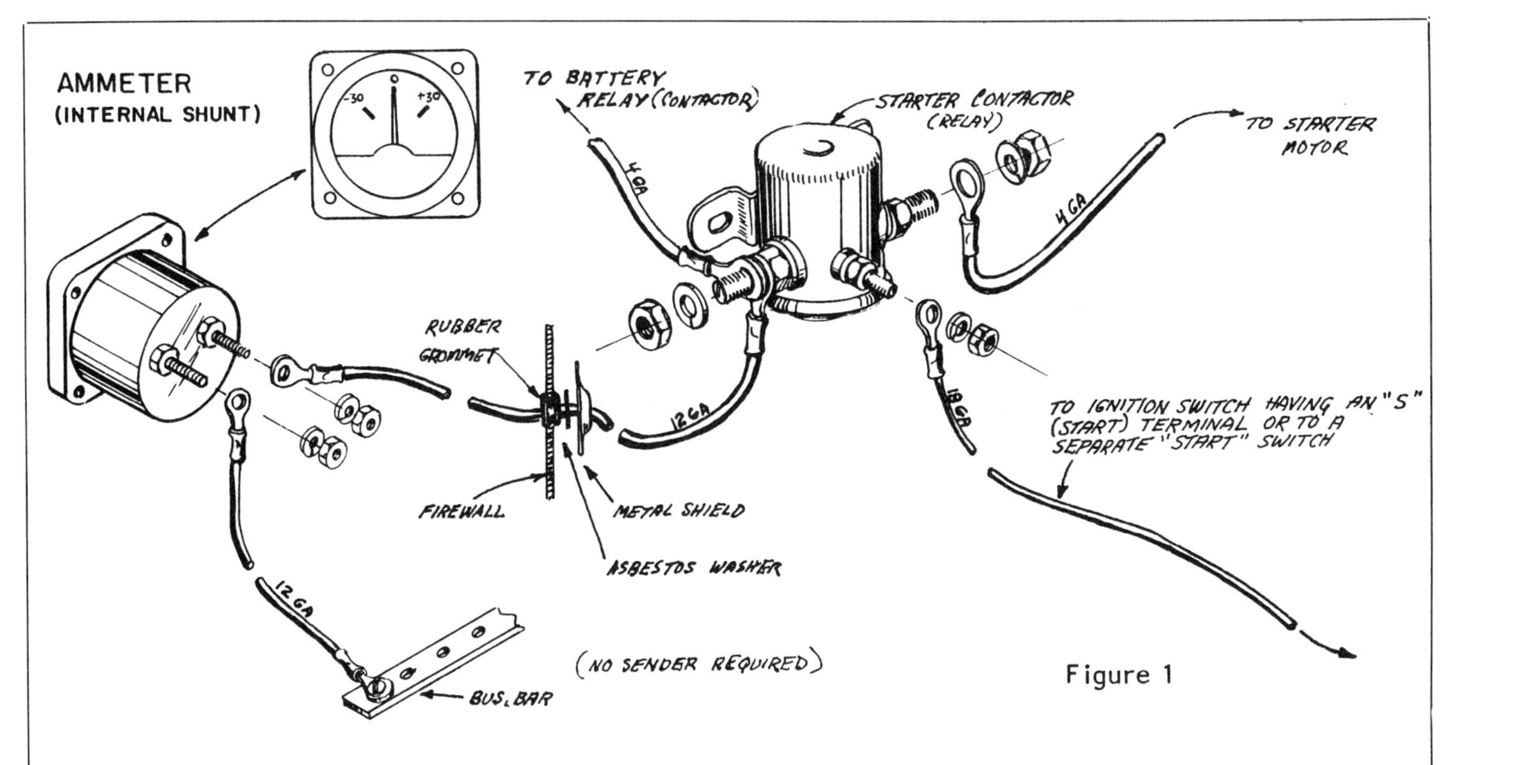

Figure 1

Ammeter

PURPOSE: To show the amount of current flowing (in amperes) from the alternator or generator to the battery, or from the battery to the aircraft's electrical system.

Some ammeters have a built-in shunt, others, require an external shunt. (A shunt is a resistor made of a special alloy that allows current by-pass.) It is not affected by temperature changes. An ammeter with an external shunt is expected to be the more accurate (and expensive) of the two types.

The current range calibrated on most ammeters is generally from a minus 30 to a plus 30. Range limits from minus 60 to plus 60 are also available. If you have a 60-amp. generator you may wish to select an ammeter with the greater range although an ammeter's indications, in service, will probably never range much past just either side of the zero indication.

When hooking the ammeter up, don't change the lead length, accuracy will suffer.

Ammeters usually require a two-inch instrument panel cut-out. Always double check the diameter of your instrument case before making the cut-out. Usually a hole 1/32nd of an inch larger than the instrument's diameter is sufficient. You will need access behind the panel to mount and remove the ammeter as most of these gauges do not use individual face mounting screws. Instead, they are usually held in place by a single "U"-shaped clamp from behind.

Connect the ammeter in series (in the line) with the load. The plus terminal is connected to the positive side of the electrical circuit in order to obtain the correct needle deflection.

NEVER, NEVER connect the ammeter across the terminals of a battery or generator!

Use a fairly large size wire, say No. 8 or No. 10 AWG, to prevent the ammeter wiring from becoming overloaded.

NOTE: Another reminder. To prevent sparking and other damage, always disconnect the negative cable from the battery BEFORE undertaking a wiring job.

With the engine off, the gauge should show a "discharge" when you load the circuit (turn on the radio, lights, etc.). If the gauge shows a "charge", reverse the wires on the ammeter posts. No harm will have been done to the circuit.

Ammeter/Voltmeter

(dual instrument - typical)

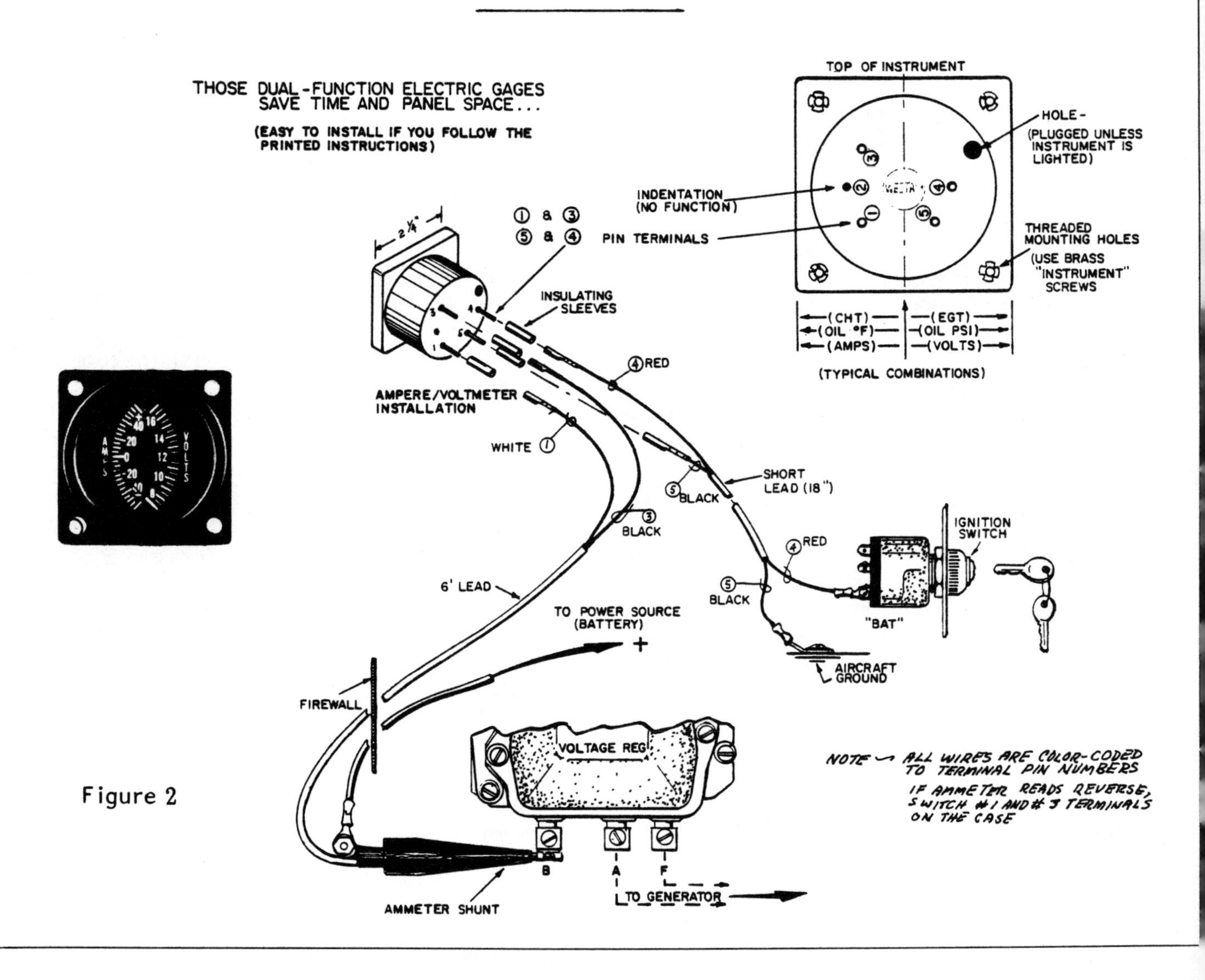

Figure 2

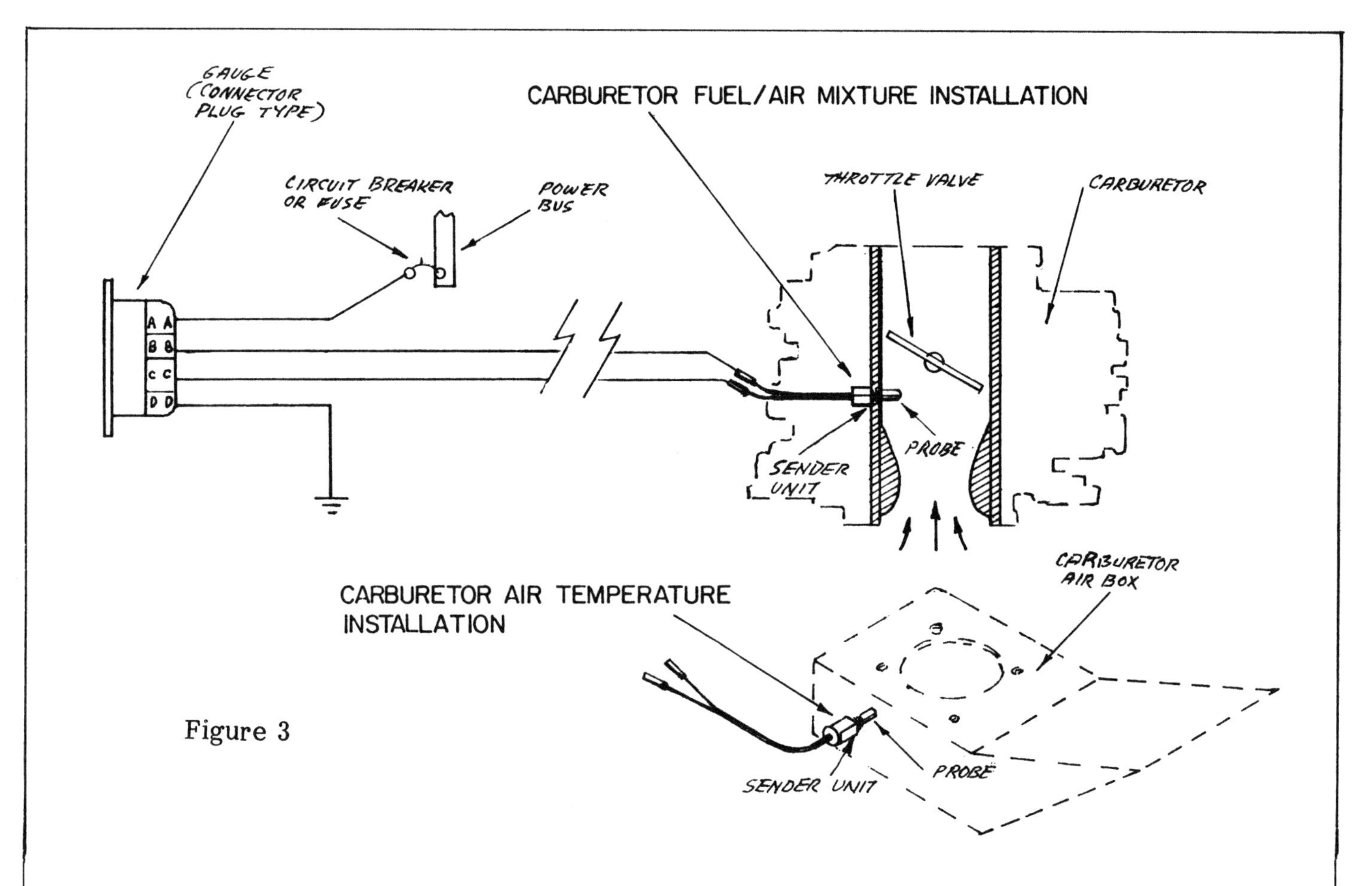

Figure 3

Carburetor Temperature

PURPOSE: To provide a means of detecting carburetor icing conditions by monitoring the inlet air temperature or the fuel/air mixture.

Strictly speaking, carburetor thermometers may be divided into two types, depending upon the location of the temperature probe (bulb):

Carburetor Air Temperature Type — It has its bulb inserted into the connection at the carburetor air inlet and measures the temperature of the air entering the carburetor.

Carburetor Fuel/Air Mixture Type —It measures the temperature at the engine side of the carburetor and gives you the temperature of the fuel/air mixture entering the engine.

Either type permits you to keep the temperature of the incoming fuel/air mixture within safe limits by the judicious use of carburetor heat.

Installation of the instrument is relatively simple, as it is usually acquired with installation hardware. Panel hole size is normally 2 1/4 inches in diameter. The installation hardware kit consists of a probe with an integral 1/4-inch x 28 thread which screws directly into the carburetor throat or the carburetor heat box. The probe has two leads (about 12 inches long each) with connectors to which two wires are joined and routed directly to the Carburetor Temperature Gauge in the cockpit. The two connecting wires supposedly eliminate possible instrument malfunction which may otherwise arise from a poor electrical aircraft bonding (common ground).

The instrument requires current from the aircraft electrical system (bus bar) but it can also be used with a magneto or distributor type installation. In such installations, the addition of a magneto adaptor and a fuse adaptor is required.

Carburetor Air Temperature gauges are expensive and, are seldom installed in aircraft used primarily for VFR flying.

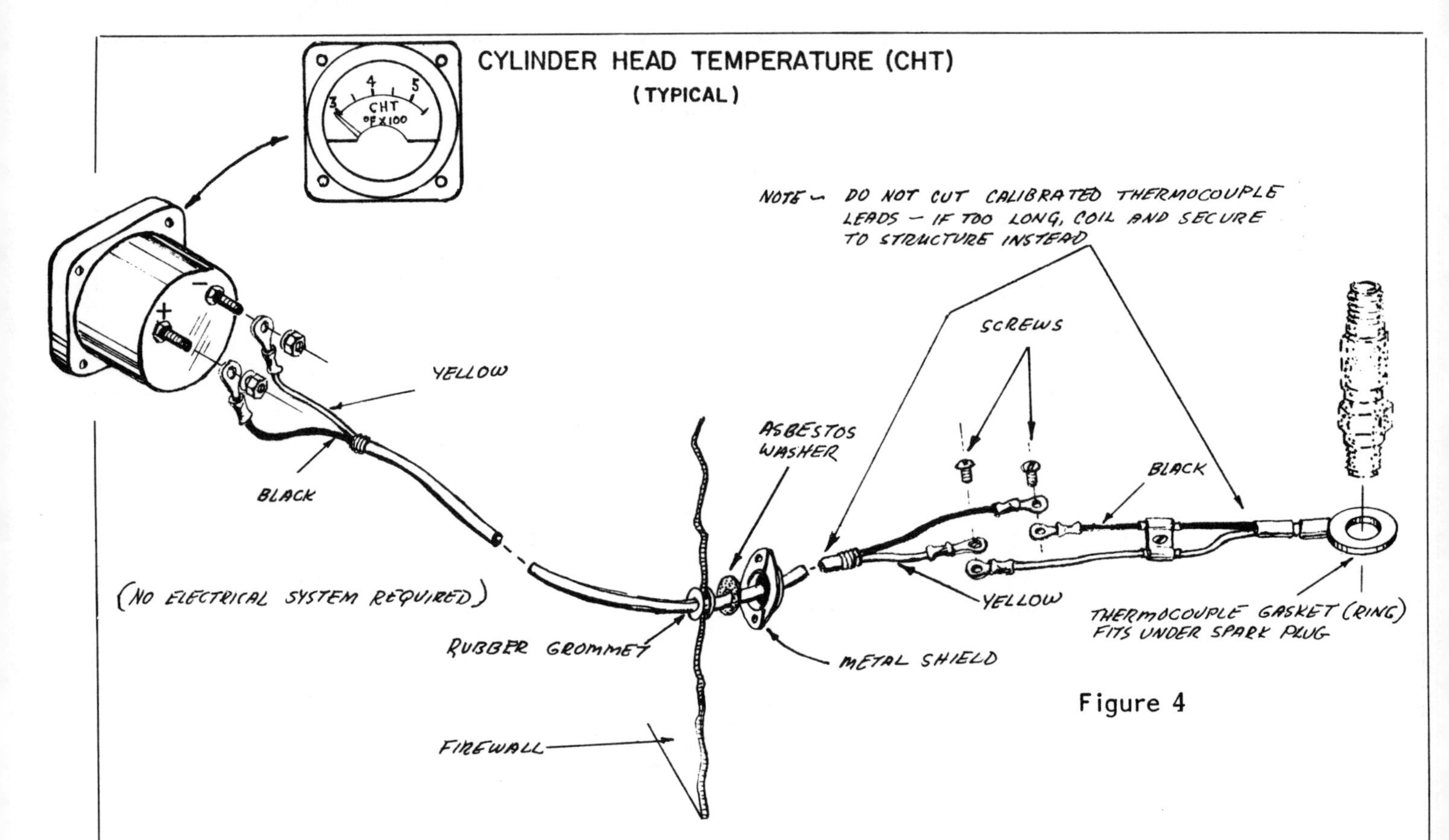

Figure 4

Cylinder Head Temperature

PURPOSE: To monitor and, thus, help safeguard the engine from excessive cylinder head temperatures.

Some cylinder head temperature gauges are resistance-change instruments and require an aircraft electrical system. They will operate only when the aircraft's electrical master switch is turned on.

Others generate their own electricity and require no electrical system in the airplane. The thermocoupling units of these devices are made of two dissimilar metals joined together at one end, which when heated, produce a small electromotive force that deflects the needle on an indicator in the cockpit.

Still another type of CHT gaining in popularity is the lightweight miniature dual-pointer series of instruments. One combination contains both a CHT indicator needle and an exhaust gas temperature (EGT) needle superimposed on the same instrument dial. The value of this type of instrument is its weight — about three ounces.

A CHT installation consists of an indicator (gauge), a thermocouple and the connecting thermocouple leads. Two types of thermocouples are in common use. One is the bayonet type and the other is the spark plug gasket ring type.

It should be pointed out that the leads used to connect the thermocouple to the gauge provide an exact amount of electrical resistance in the thermocouple circuit. Their lengths, therefore, must not be changed as the total resistance of the circuit would be altered and the instrument would be inaccurate.

Since the galvanometer in the typical gauge measures the electromotive force produced in the circuit, the greater the temperature difference between the thermocouple (hot junction) and the galvanometer (cold junction), the greater the needle deflection will be on the face of the gauge. Therefore, unless the cockpit gauge is also compensated for temperature, the accuracy of the cylinder head temperature reading will vary with cockpit temperature. A good CHT gauge is temperature compensated internally by a bimetallic spiral spring connected to the gauge mechanism.

Generally, when you order a gauge from a catalog, the proper connecting sender and hardware requirement will be listed along with the instrument you want.

When received, the installation instructions should be followed to the letter to ensure a satisfactory installation. Remember, the manufacturer knows best how his product should be installed and used.

Unless it is part of a cluster of engine instruments, the CHT gauge will be either 2 1/4 inches or 3 1/8 inches in diameter and will ordinarily have two terminals or pins on its backside. These may be embossed with a plus (+) and a minus (-) sign.

When connecting the thermocouple leads, match metal to metal and color to color. DON'T MIX THEM!

The correct size thermocouple spark plug ring for most aircraft engines is 18mm in diameter. A 14mm size is also available for use with the spark plugs installed in VW automotive engines and Franklin aircraft engines.

Installation of the spark plug ring type thermocouple is not complicated. Simply remove the spark plug from the cylinder you believe to be the hottest running, discard the original copper spark plug washer, and screw the spark plug back into the cylinder with the thermocouple gasket unit under it. Installation kits are also available with a 4-way or 6-way switch to accommodate 4-plug or 6-plug thermocouples.

A bayonet thermocouple requires a separate adapter fitting which screws into a special port tapped into the cylinder head. This adapter has a 1/8-inch pipe thread on one end and a slotted receptacle on the other to receive the spring-loaded bayonet probe.

Secure your CHT leads away from the hot engine exhaust pipes and muffler. If your CHT lead is too short for your particular installation, order a longer one. If your lead is much too long, do not cut the precalibrated lead to shorten it. Instead, coil it up and secure it to some part of the engine installation or to the structure.

Neatly grouped powerplant instruments and side mounted control panel decreases pilot's workload and helps unclutter the instrument panel in this VariEze

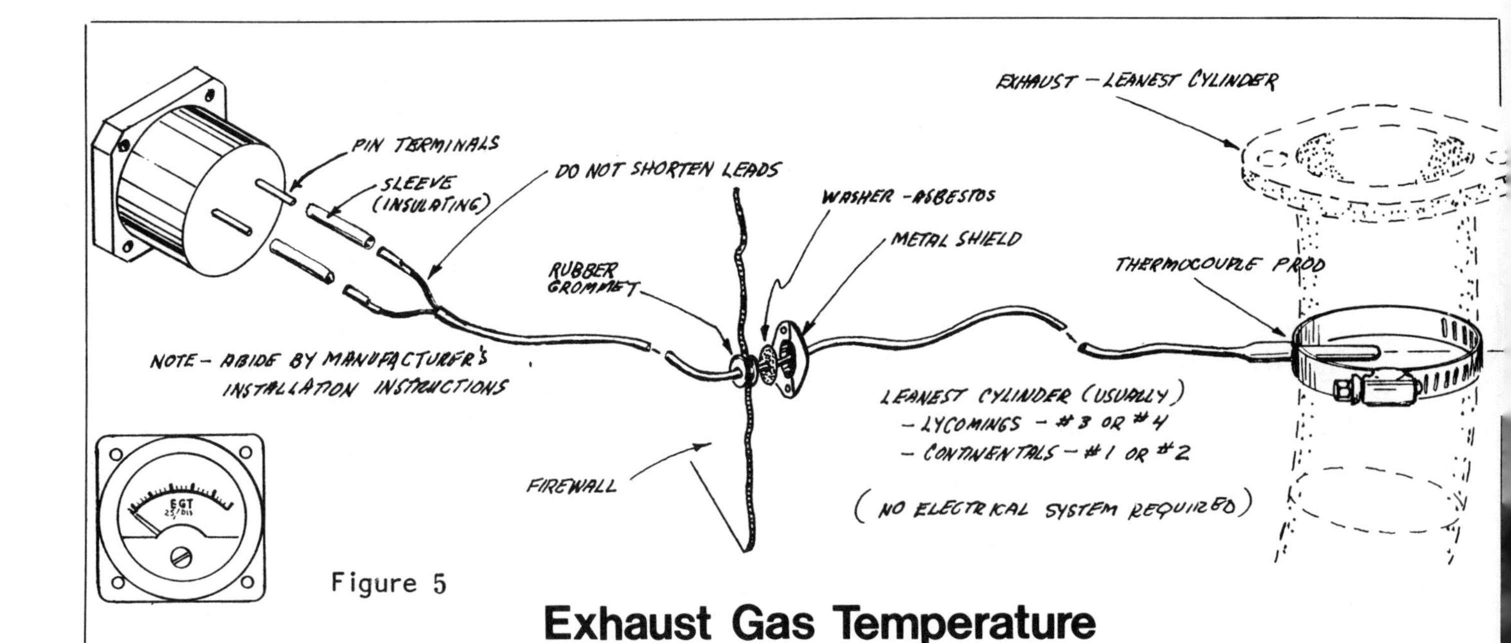

Figure 5

Exhaust Gas Temperature

PURPOSE: To provide an accurate indication of the exhaust gas temperature (EGT) of the cylinder(s) being monitored. Used as an aid in controlling fuel/air mixture.

It may be because the EGT gauge is not a mandatory instrument that many pilots feel EGT information is useless in small, carbureted engines fitted with fixed-pitch propellers. They believe that the time-tested method of adjusting the mixture by watching the rpm's (tachometer needle) and listening intently to the engine is just as efficient a process for adjusting the mixture. Perhaps, but EGT is a useful extra tool for this process.

Allowing your engine to run too rich is a needless waste of fuel. A too-rich mixture can also cause the engine to run roughly and foster a build-up of lead salt deposits on the pistons and exhaust valves. Thus, any aid in fine tuning the mixture could be a significant factor in reducing overall flying costs.

The typical EGT gauge installation consists of a panel-mounted indicator and a thermocouple probe connected by a couple of wires. Installation is very much like that of the CHT gauge. Both operate on the principle that a small amount of current will be produced when two dissimilar metals react to a heat source. Since the instrument makes its own electricity, no separate aircraft electrical system is necessary for its operation.

The better EGTs are temperature compensated for accuracy. If you install one that is not compensated, be aware of the inaccuracy when cockpit temperatures are much above or below 70 degrees F.

A multiple probe EGT unit would be a valuable aid to you in determining the engine's operating efficiency, because with this arrangement, it is possible to compare exhaust temperatures for each cylinder separately. Unfortunately, multiple probe units cost much more.

As with most engine instruments, instructions will accompany the device detailing its correct installation. These should be followed carefully in order to get the best readings from the instrument.

In general, the installation of an EGT gauge is simple. A 3/16-inch diameter hole is drilled in the exhaust pipe of the leanest running cylinder (if you can figure out which one that is). In a multiple probe installation, the hole must be drilled in each exhaust pipe approximately three inches from the cylinder's exhaust port. Follow instructions for your instrument as it may specify a location as close as 1 1/2 inches from the port. The closer the probe is to the cylinder head the greater the adverse effect on the life of the probe. On the other hand, the further the probe is from the port, the slower the gauge response.

The EGT instrument panel installation requires cutting a hole 2 1/4 inches in size. It is probably best located close to the cylinder head temperature gauge. A dual indicating instrument, showing both EGT and CHT can also be obtained. This would help reduce panel space requirements.

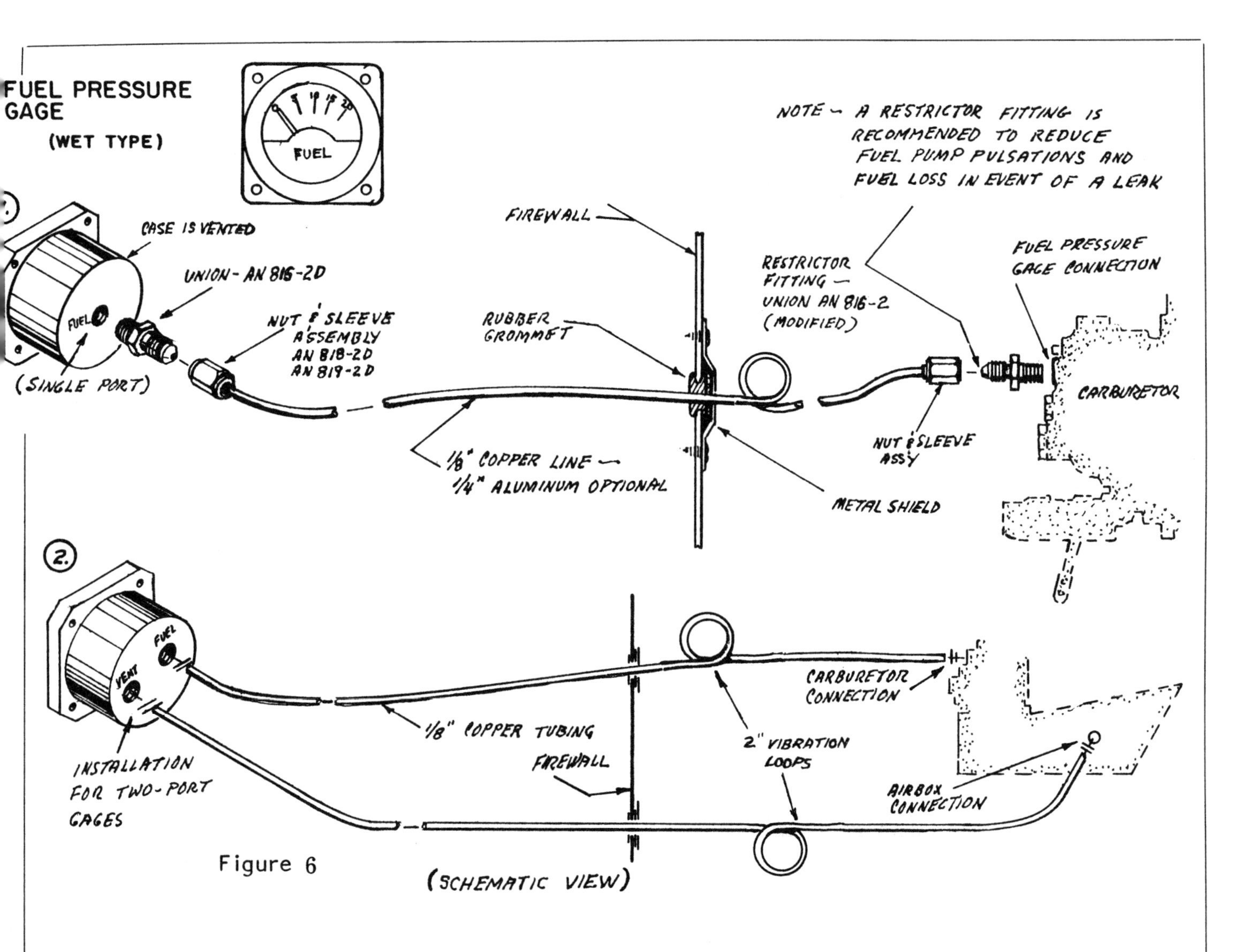

Figure 6

Fuel Pressure

PURPOSE: To monitor the amount of fuel pressure being produced by the fuel pump.

The fuel pressure gauge is a differential pressure indicator and is usually 2 1/4 inchs in diameter. It will warn you of fuel system problems such as a broken or plugged fuel line, fuel pump failure, or an interrupted fuel flow condition due to improper fuel tank selector manipulation.

No fuel pressure gauge is required or desired for gravity feed fuel installations as the pressure produced at the carburetor by the head of fuel in the tank can be very low, as low as 1/2 psi.

The gauge may have one or two ports on its backside. Those with but a single port or nipple connection have their cases vented to the atmosphere.

A line (1/8-inch copper is all right) connects one port (fuel connection) on the gauge to the fuel pressure chamber connection at the carburetor. The second port (air connection), if present, is connected to the carburetor air intake. This dual connection accurately measures the difference between the fuel pressure entering the carburetor and the air pressure at the carburetor inlet.

Some fuel pressure gauges are not encased such as those making part of an instrument cluster. They typically have but one port and are automatically vented to the cabin atmosphere. However, if this type of instrument is used, the relief valve to the engine-driven fuel pump should also be vented to the atmosphere.

A restricted type fitting is customarily installed at the carburetor for the fuel pressure line connection in order to dampen any needle fluctuations in the gauge due to fuel pump induced pressure impulses.

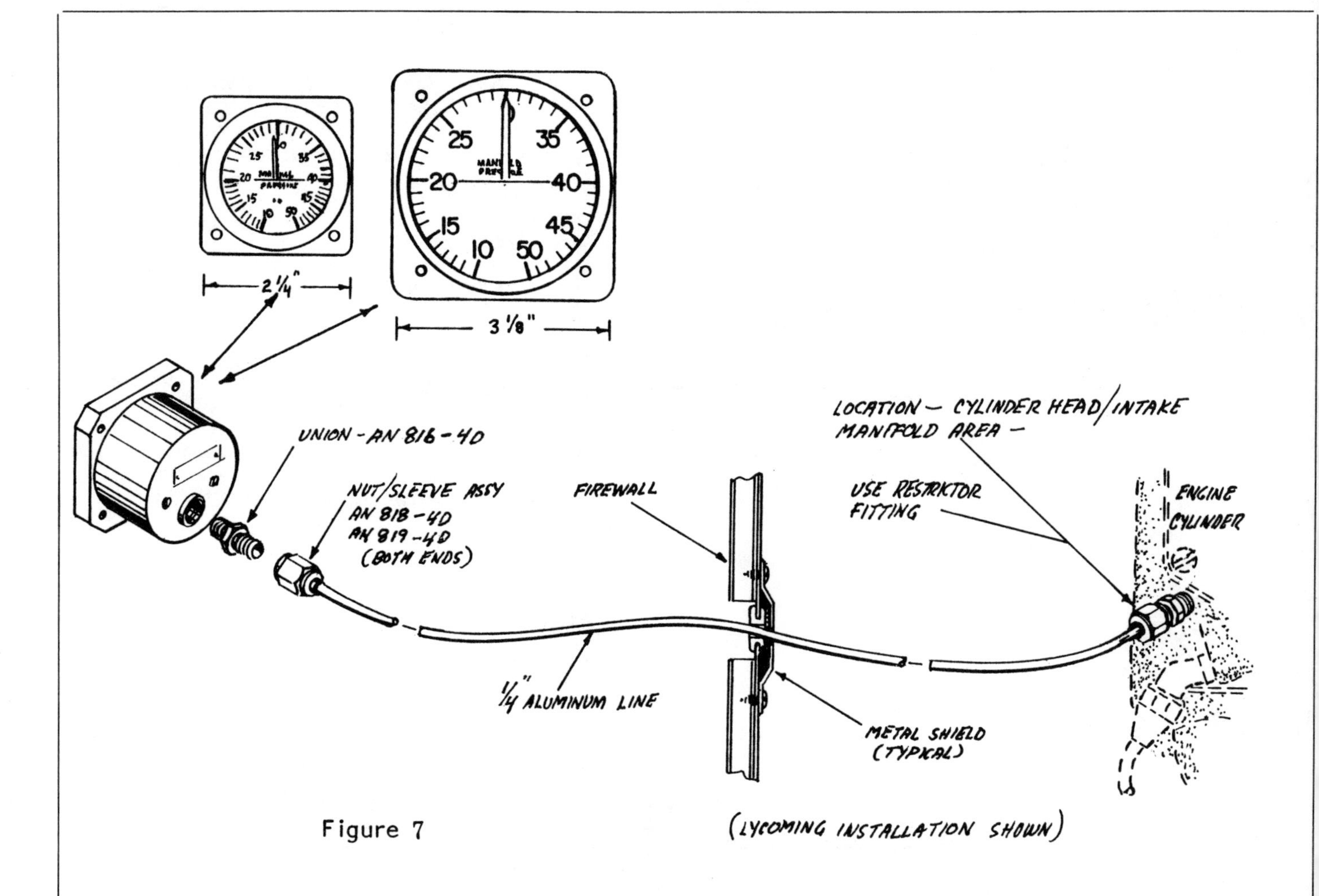

Figure 7

(LYCOMING INSTALLATION SHOWN)

Manifold Pressure

PURPOSE: To aid you in obtaining the desired and correct power settings for your engine. It can serve also as an indicator of the engine's condition during its service life.

The manifold pressure gauge will indicate the local barometric pressure (measured in inches of mercury) when the engine is inoperative or the gauge is disconnected. Actually, the gauge is really a barometer and may, therefore, be checked for accuracy against an altimeter setting or the barometric pressure obtained from a local weather station.

The gauge has a scale ranging from 10 Hg. to 40 Hg., higher on some instruments, and should be accurate within 2/10-inch Hg. it should also be altitude and temperature compensated internally.

In an idling engine, the suction created by the pistons causes the manifold pressure to drop below that of atmospheric. As the throttle is advanced, pressure builds up but it cannot exceed the pressure of the surrounding atmosphere. Of course, on a supercharged engine the fuel/air mixture is actually forced in, allowing the manifold pressure to increase above the atmospheric pressure.

The instrument panel hole required will be for either the standard 3 1/8-inch gauge or the miniature 2 1/4-inch gauge.

Use tubing to connect your gauge to the intake manifold at the point provided for your particular engine and type. This port is usually located between the intake valves of the engine and the carburetor.

The manifold pressure gauge, although useful, is of minor importance in non-turbocharged engines. Frequently, it isn't installed in homebuilts.

Lycoming engines typically have a port in each cylinder which may be used for either a primer or a manifold pressure connection as shown here by the arrows.

The arrow clearly shows the typical mechanical connection for the tachometer. The drive cable passes directly by the spin-on oil filter in this dynafocal mounted engine.

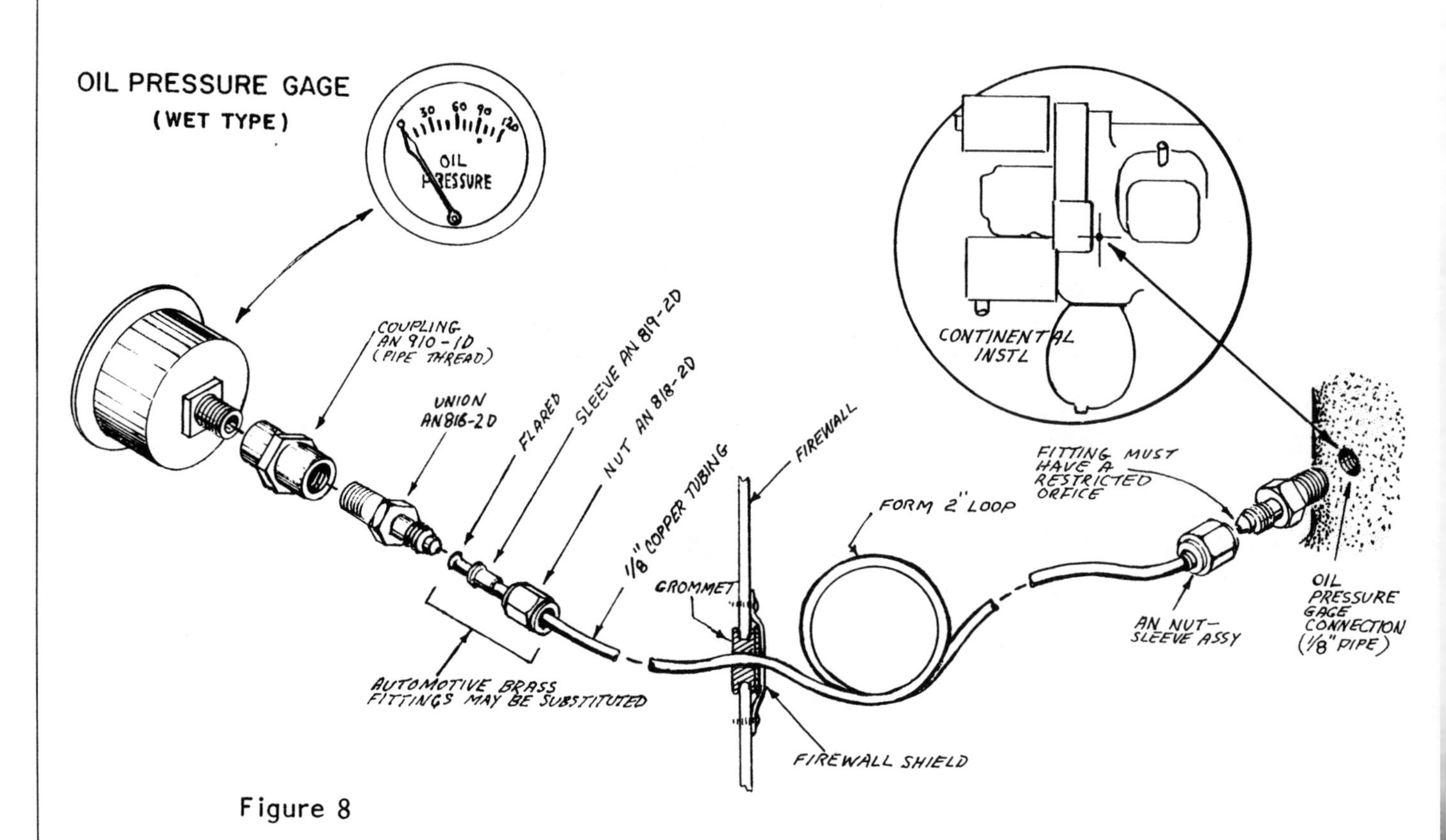

Figure 8

Oil Pressure

PURPOSE: To provide a continuous visual indication of oil pressure inside the engine's lubrication system when the engine is running.

This gauge will warn you of an oil pump failure, an impending engine failure due to a broken oil line or a depleted oil supply.

Two common types of oil pressure gauges are in general use, the direct reading (wet) type and the electrical type.

The direct reading oil pressure gauge is connected by a small tube or hose from a port in the backside of the gauge to a port in the engine's oil pressure system. The oil from the engine fills the line right up to and into the gauge when the engine is running. The pressure exerted by the oil deflects a needle on the differential pressure type gauge.

Electric type oil pressure gauges require an aircraft electrical system for power. The unit utilizes a small pressure transmitter screwed into a port opening in the engine's lubrication system. The electrical circuit is completed by wire from the engine transmitter or sender unit to the gauge and then to ground. Calibration of the gauge is in pounds per square inch.

Oil pressure gauges obtained from automotive sources are two inches in diameter. Others, identified as "aircraft type" gauges are 2 1/4 inches in diameter.

Often a length of Nylaflow tubing is packaged with an automotive gauge along with the fittings required for its installation. This is mighty convenient but some FAA inspectors won't approve it because its fire resistance qualities are poor. The preferred line to use in the engine compartment (ahead of the firewall) is 1,000 psi Aeroquip or Stratoflex tubing, or even 1/8-inch copper tubing with matching fittings.

No hose or tubing should be permitted to contact any hot engine parts, particularly exhaust pipes or mufflers. The tubing must also be prevented from chafing against any part of the structure. To prevent this problem at the firewall, route the line through a grommeted hole or a bulkhead fitting in the firewall and up to the rear of the oil pressure gauge on the instrument panel.

The oil pressure line fitting that screws into the engine crankcase must be one with a restricted orifice in it. In the event of a broken oil pressure line, the loss of oil will be considerably reduced. You can also use a standard brass fitting if you first modify it by filling the passage in the fitting with solder and then drilling a tiny hole (no larger than 1/16-inch) in it. The solder modification will work in a brass fitting but not in an aluminum one.

An aluminum AN816 fitting can be converted into a restrictor-type fitting by pressing in an AD rivet (from the flared end side of the fitting) and drilling a small N0. 40 hole through the rivet.

Hook up of the electrical oil type oil pressure gauge is made by screwing the oil sender unit into the engine port and completing the electrical connections between it and the gauge in the cockpit.

CAUTION: Do not hang oil pressures switches and senders on the end of the nipple of an aluminum tee fitting screwed into the crankcase. Such a rig will surely fail due to vibration in as short a time as 40 to 100 hours.

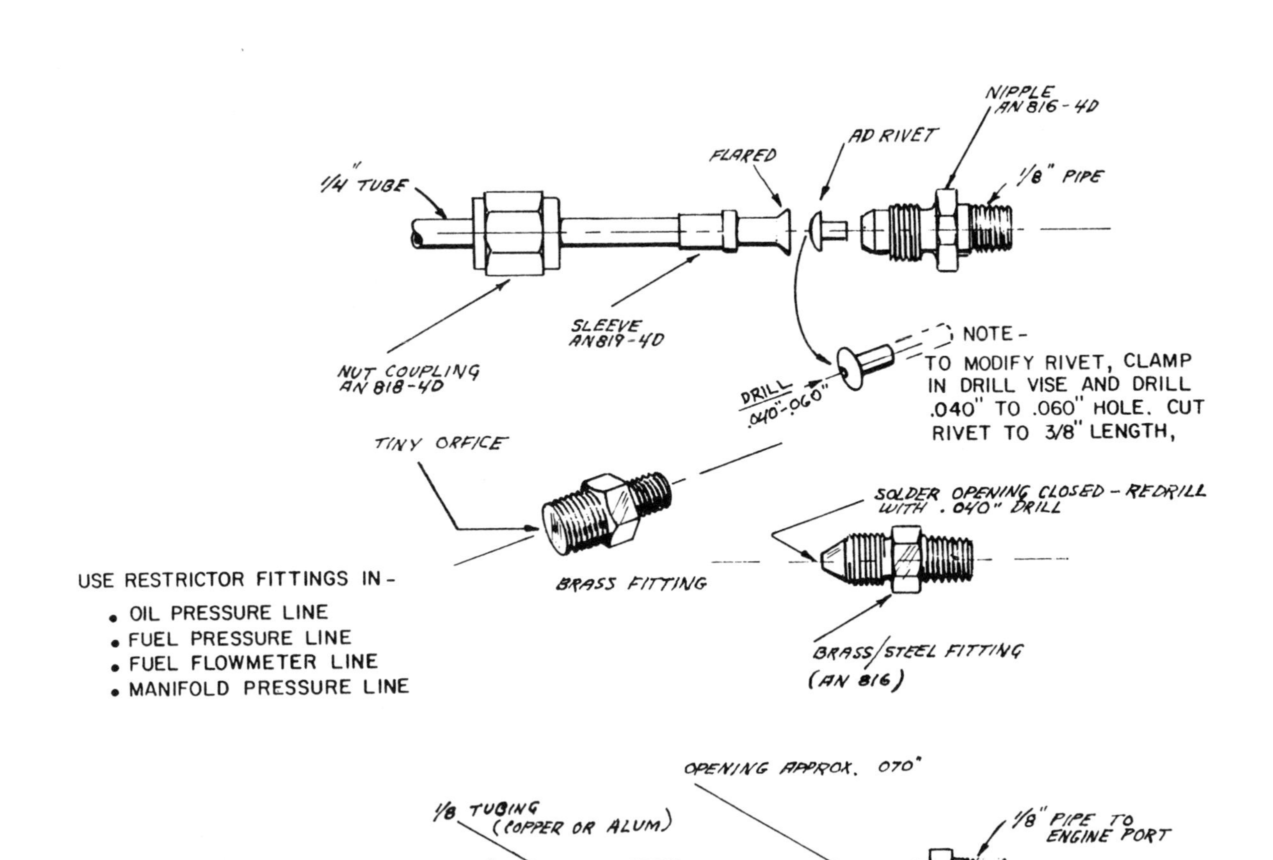

Figure 9

RESTRICTOR FITTINGS
(TYPICAL)

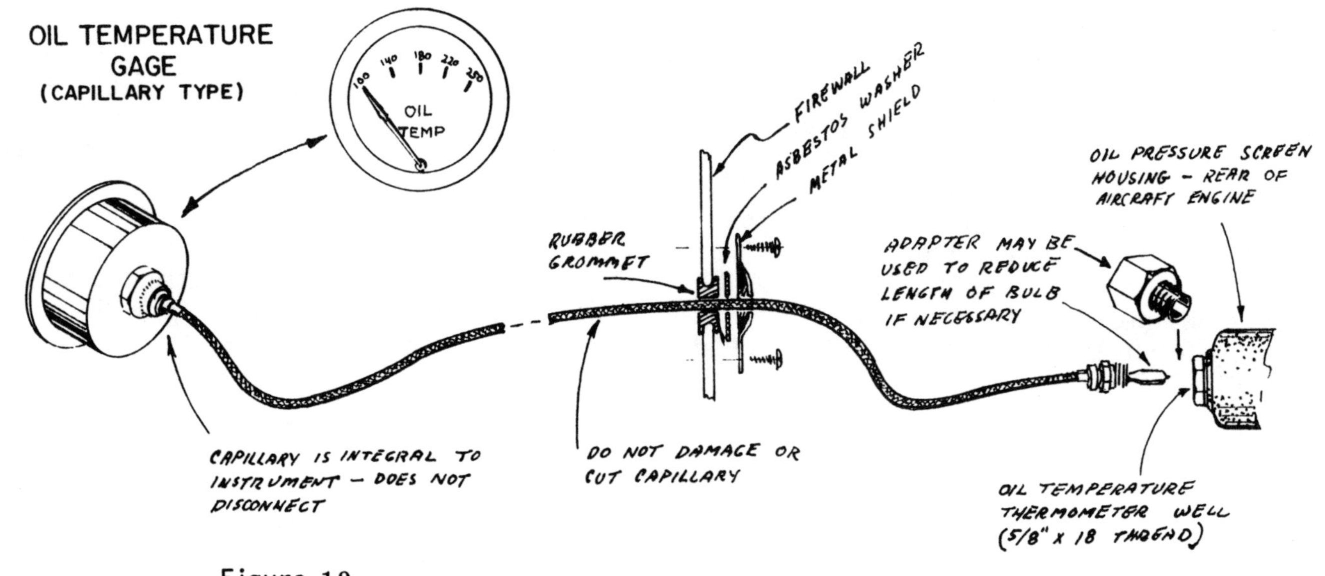

Figure 10

Oil Temperature

PURPOSE: To monitor the engine oil temperature and provide an indication of the internal cooling status of the engine.

There are two common types of oil temperature gauges, the direct reading (capillary) type and the electrical, remote reading type.

The direct reading type is a self contained unit made up of a temperature gauge, a capillary (enclosed tube) and a probe containing a highly volatile liquid. Because it is self-contained, it should always be installed or replaced as a complete unit.

The electrical type of oil temperature gauge is typically a wire-connected instrument with a probe that reacts to electrical resistance changes induced by the heated engine oil. These changes are conveyed by the wiring to the instrument panel gauge, which translates the resistance into temperature. The dial is usually calibrated in degrees Fahrenheit.

A direct reading oil temperature gauge is connected by a capillary tube to the oil pressure port in the engine crankcase. Depending on the engine and its installation, this port may be located at the low pressure inlet to the oil pump screen, or, if an oil cooler is installed, at the cooler's outlet.

The gauge is usually mounted adjacent to or near the oil pressure gauge. A dual oil temperature/oil pressure gauge would make a very compact installation.

A new oil temperature gauge will normally include installation instructions and mounting hardware. Installing the automotive type requires access behind the panel. The bulb or probe end has a standard 5/8-inch hex nut fitting. An adapter nut to mate it to the pipe threads in the engine port may or may not have to be used. Adapter nuts often are part of the oil temperature instrument package or may be purchased, if necessary, to reduce the reach of a long capillary bulb.

CAUTION: Do not kink or twist the capillary tube or make sharp bends in it during installation. A leak in the capillary will render the whole unit useless.

Route the capillary tube through the firewall after mounting the gauge in the panel. The firewall opening will have to be larger than you'd like but it must accommodate both the bulb and the hex nut. Protect the capillary at the firewall opening from chafing.

If the capillary tube is much too long for your installation, DO NOT CUT IT! Coil it and tape it to any convenient support.

Where applicable, be sure that the oil screen housing is reinstalled with a new copper gasket. Then torque and safety-wire the unit.

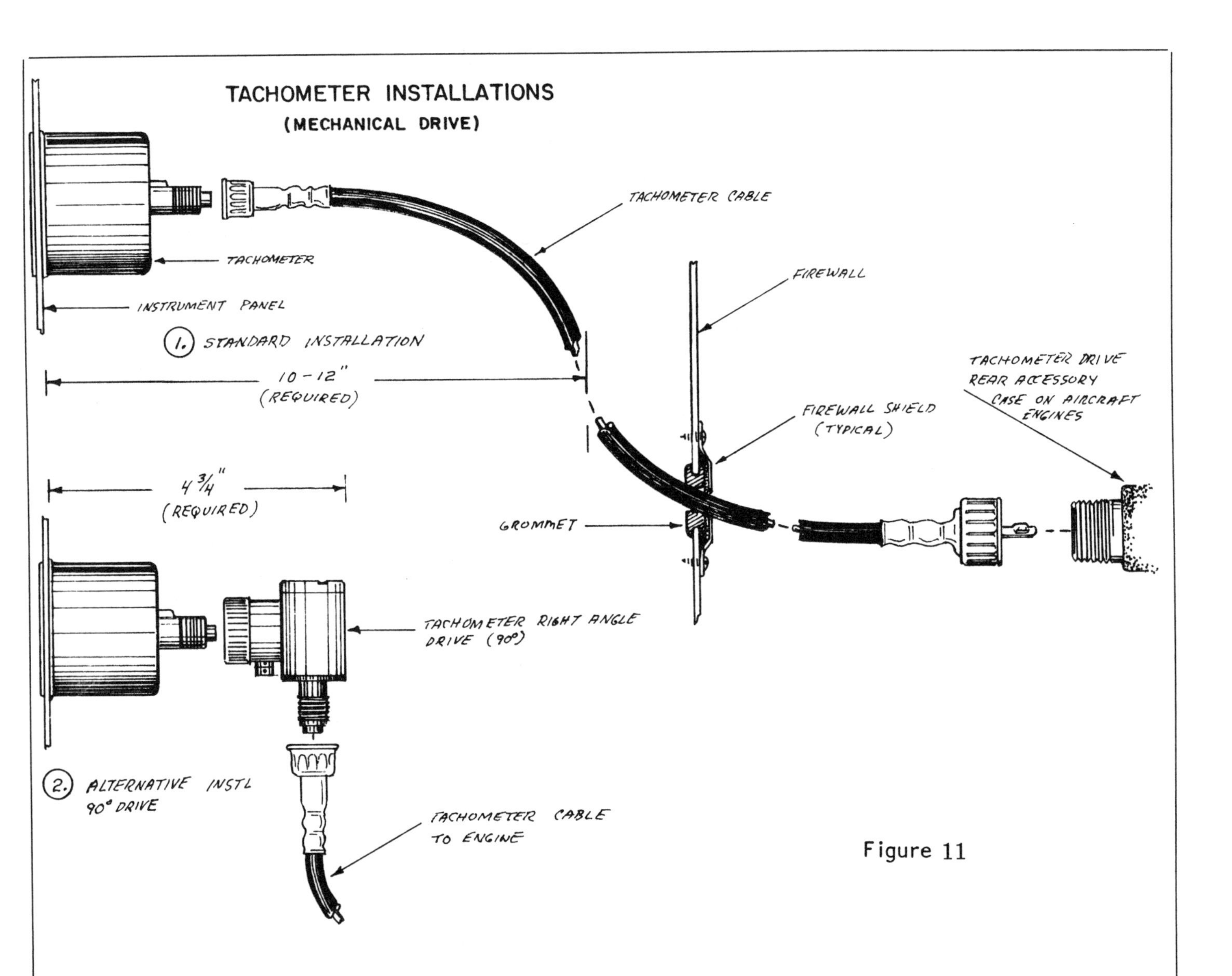

Figure 11

Tachometer

PURPOSE: To indicate engine and propeller speed in revolutions per minute.

A tachometer is simply a counter. It can be either mechanical or electrical (including magneto) in operation. Both types indicate the engine crankshaft speed, although geared engines ordinarily reflect camshaft speeds.

The mechanical tachometer is the type most used with aircraft engines. A good gauge will be one that is temperature compensated and lubricated for the life of the instrument.

Before you select a mechanical drive tachometer, you will need to know the ideal cruising rpm and the correct direction of rotation of the tachometer drive on your engine. You will also have to decide whether you want a tachometer only or a combination tachometer and hourmeter (called a tachourmeter). The only external difference between the two is that the tachourmeter has a small window in the lower portion of the dial in which an hourmeter records the accumulated hours of engine operation in hours and tenths of an hour.

The FAA accepts a tachourmeter as an official accounting for the engine time although it is really accurate only at cruise rpm.

Some engines (of non-aircraft origin) aren't fitted with tachometer drives and an electrical tachometer is the only practical means of monitoring the engine speed.

Any electrical tachometer obtained for a converted engine (non-aircraft) should have sufficient range to accomodate the engine's rpm's at maximum output. A VW engine, for example, needs a tachometer that will indicate at least 3,500 rpm, while a tach for a small ultralight engine may need one with a range of up to 6,000 rpm.

Automotive engine conversions, such as VW engines, small two-cycle engines and four-cycle industrial engines converted to ultralight or lightplane use utilize the electrical type tachometer almost exclusively.

Some builders prefer an electrical tachometer because the wired installation is much easier to make and is lighter. Any aircraft with an engine mounted a considerable distance from the instrument panel (pusher, amphibian, etc.) can benefit from an electric tach.

An electric tachometer will, ordinarily, operate from a variety of electrical power sources, such as magnetos, distributor ignitions, alternators and 12-volt or 24-volt aircraft electrical systems.

There is also an electric/mechanical tachometer. It consists of a small sender unit or generator that screws onto an aircraft engine's tachometer drive fitting and is wire-connected to the instrument panel.

An electric tachometer that is to be installed in an aircraft having no electrical system will usually have to be operated off the magneto or distributor Be sure that the one you obtain is suited for that purpose. For such an installation, a fuse adapter should also be ordered right along with the instrument.

MECHANICAL TACH HOOK-UP

The mechanical tachometer is very much like your automobile speedometer in that it is driven by a flexible cable encased in a flexible housing.

When ordering a tachometer, always provide the engine size and type and the direction of rotation needed for the tachometer, if you know it. If you are using an aircraft engine, you should have an engine manual for it. The information will be in the manual.

Adapters are also available that permit the installation of dual tachometers should you want them.

Tachometer cables do not come with the tachometer and have to be ordered separately. They can, however, be made to order for whatever length your installation requires.

To determine the cable length needed, measure from one ferrule end to the other ferrule end with a piece of rope, string or wire. You will then be assured that the tachometer cable will be long enough for the route it must take to the engine. A slightly longer cable is better than one that is slightly shorter.

Avoid forcing the tachometer cable into severe bends as that may affect its operation and will definitely shorten its life.

If your aircraft has a fuel tank mounted behind the instrument panel, you could have a bit of a problem routing the tach. The bend required for the cable to clear the tank might have to be too severe. If so, you can utilize a 90-degree adapter especially made for this problem. That device may be purchased from most any supplier catering to the homebuilt trade. Of course, with an electric tachometer you wouldn't be confronted with this problem. Some builders have been known to construct a hole, sort of a tunnel through their fuel tank simply to provide an easy route for the mechanical tach drive and perhaps the throttle and mixture control cables.

ELECTRICAL TACH HOOK-UP

Don't cut the panel mounting hole for an electric tachometer until you are sure of its size. Usually it will be either 2 1/4 inches or 3 1/8 inches in diameter, but don't count on it.

Hooking up the electric tach can be simply a matter of connecting the wiring between the gauge mounted in the cockpit and the sender unit on the other end. However, complications have been reported in some electrical tachometer installations which use a magneto as a power source. The potential exists (according to some builders) that the engine could quit if the tachometer develops an internal short while connected to the operating magneto.

Because of this danger, the installation of a fuse has become standard procedure. In fact, builders should consider it a mandatory safety measure. Install a 16-amp, 3AG fuse into the tachometer "P" lead. Then, in the event of an electrical short, the tachometer will stop but the engine won't.

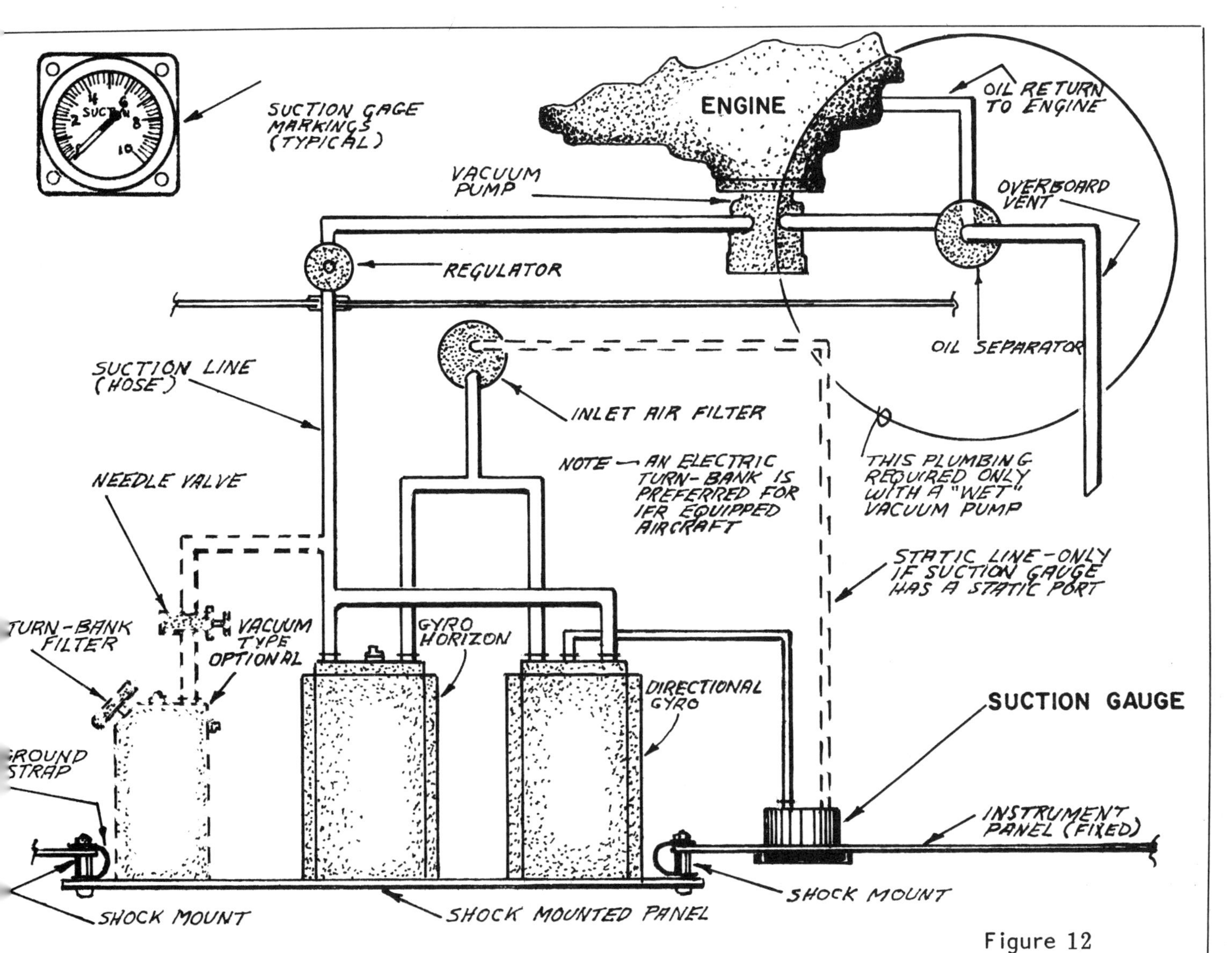

Figure 12

Vacuum Pressure

PURPOSE: To show the amount of suction produced by an engine-driven vacuum pump (or venturi) that drives the gyro instruments.

Your gyro instruments will ordinarily include an attitude indicator, directional gyro and a turn and bank indicator. If you intend to do extensive IFR flying, an electric turn and bank indicator or an electric turn coordinator is recommended. This will provide basic attitude information in the event the regular vacuum system fails.

Gyros operate on a very low pressure (suction) so vacuum gauges have scales ranging only from 0 to 10 inches of mercury. However, they will need an uninterrupted minimum pressure for reliable operation. The pressure requirement ranges from a low setting of 4.25 inches Hg. to a high setting of up to 5.8 inches Hg. An engine-driven vacuum pump is normally used to provide the constant minimum pressure required.

Vacuum pumps can be either of the wet or dry pump variety. The dry air pump has the edge on popularity as it doesn't suffer from the messy oil leak reputation of the wet pump. In any case, both types are interchangeable so it is only a matter of preference.

The back side of the instrument may have one or two inlets, depending on its internal construction. The dual inlet type will have two ports labled "Suction" and "Vent". Hook them up accordingly. The single inlet instrument will probably have a letter "P" by its port. This inlet will be connected directly to the suction line with a flexible rubber instrument hose.

Use your suction gauge for the adjustment of the vacuum system relief valve. It also warns of leaks in the vacuum system.

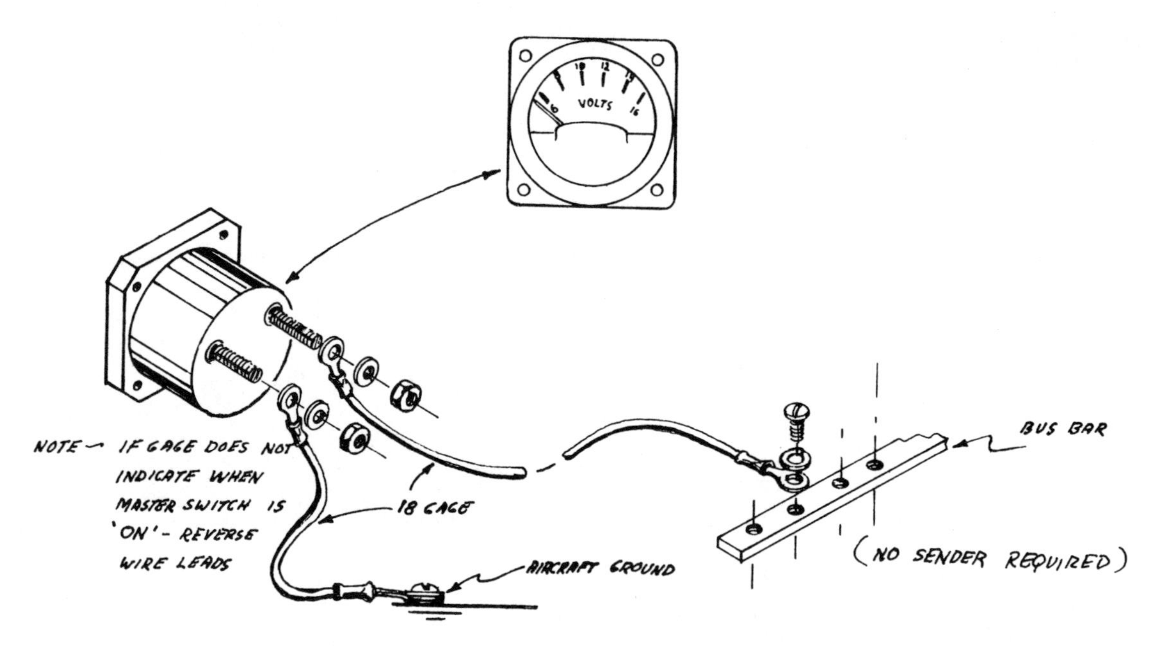

Figure 13

Voltmeter

PURPOSE: To show the level of regulated voltage being generated by the alternator or generator. When the engine is not running, it indicates the condition of the battery.

A voltmeter measures the voltage (potential difference) between two points in an electrical circuit. It would be no understatement to say that there are many types of voltmeters. The one you select must have the same voltage rating as your battery.

CAUTION: Before doing any electrical work, disconnect the ground lead from the battery as a safety precaution.

To hook-up your voltmeter, connect its plus (+) terminal to the positive side of the unit or circuit in order to obtain the correct deflection of the needle. If you don't know which post of the gauge is positive and which is negative, connect a number 18 AWG (American Wire Gauge) wire to either terminal post on the gauge and the other end to the postive terminal of the battery or unit. Connect the second terminal post of the gauge to a good ground. If the voltmeter needle does not indicate voltage, reverse the wires at the gauge connection.

Always connect the voltmeter across the electrical unit or across the points between which the voltage or difference of potential is to be obtained. This is called a parallel connection. In such a parallel connection, one wire is grounded and the other one is connected to any unit or portion of the circuit you want to monitor.

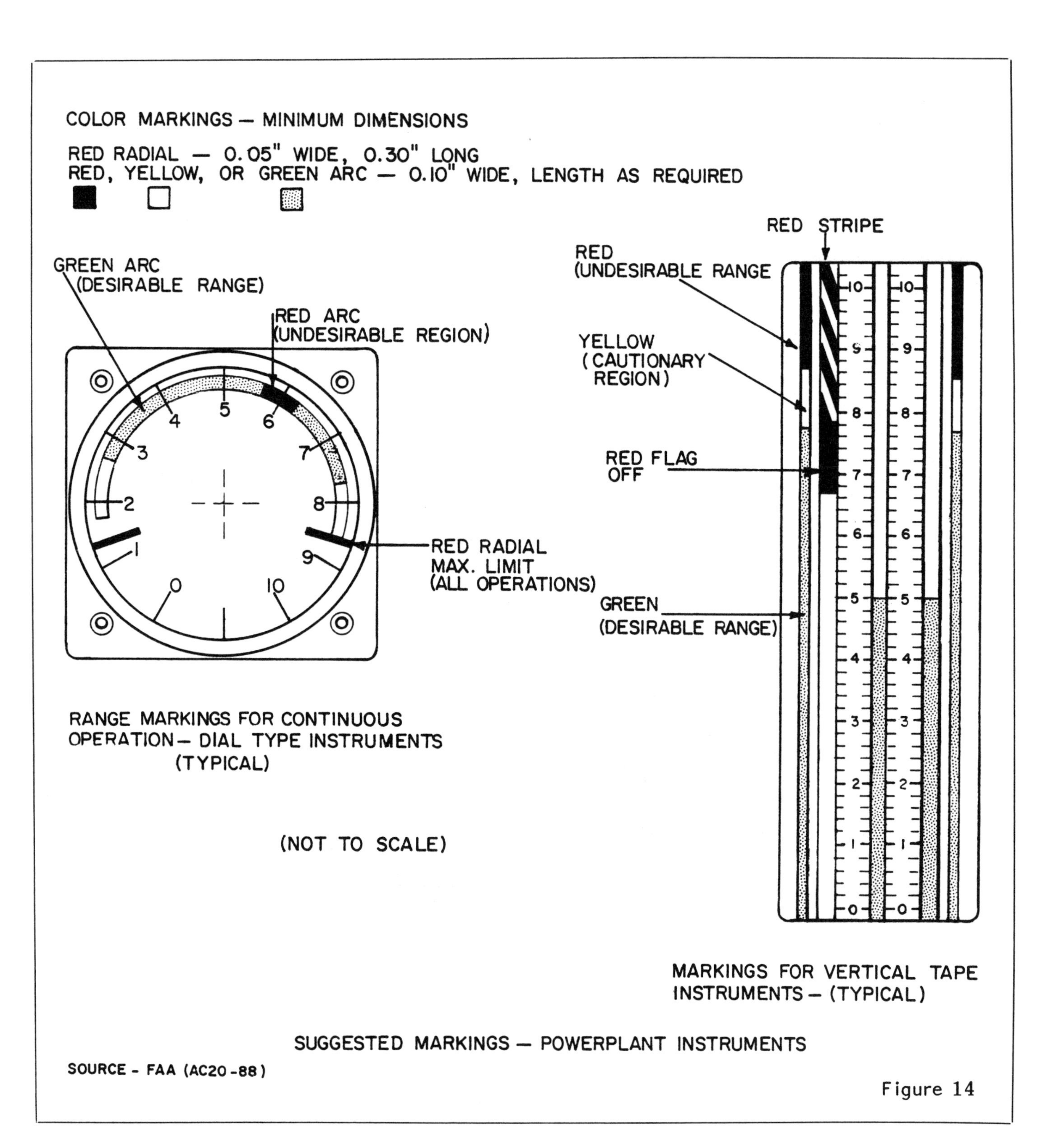

SUGGESTED MARKINGS — POWERPLANT INSTRUMENTS

Figure 14

LYCOMING
DUANE

engine installation

11

Installing the Engine

When a new builder first looks into the engine compartment of a typical aircraft, he sees what must seem to be an awesome maze of wires, tubes, hoses, controls and other mysterious gadgets. After such an encounter some builders become seriously concerned about the job of installing their own engines. They may even consider the task too complex for them to undertake. No so, amigo, even a fistfull of spaghetti is made up of individual noodles. The secret is to "divide and conquer". That is, take each wire, hose or control and fabricate and install it as a separate job. After several such individual installations have been completed you will be pleasantly surprised at the awesomeness of your own installation when viewed as a whole.

Installing an engine in a homebuilt for the first time is a bigger, more time consuming job than re-installing an engine in a store-bought aircraft. For one thing you probably don't have all of the fittings, hardware, brackets, tubing and wiring in the correct sizes or lengths already at hand. This means you must methodically undertake the fabrication, installation or connection of each unit or control as you acquire the correct parts. And (this is essential), you must complete each particular hook-up before moving on to another task.

Additionally, because space in the engine compartment is so very limited, a degree of ingenuity must be exercised if you are to provide ample clearance for all components. At the same time, don't lose sight of the need to provide an installation that will permit easy servicing and maintenance of the engine.

The stage of aircraft construction at which you undertake installation of the engine is mostly a matter of choice. You don't necessarily have to wait until all of your aircraft structure is completed before beginning the installation. Some designs lend themselves to early engine installation while others do not. A fuselage that has attached landing gear, as in many parasol and biplane designs, seems to beg for an early start on the engine installation. On the other hand, a homebuilt in which the landing gear is to be attached to a one-piece cantilever wing presents engine installation difficulties which might best be dealt with after airframe completion.

Whether you begin fairly early in the project or save your engine installation as the big finale doesn't matter. Simply follow generally established engine installation practices and you will avoid a great deal of future powerplant troubles.

It isn't necessary to remove your engine from storage if that is its present status. Actually your engine installation might drag on for longer than you anticipate and it won't hurt if your engine's innards continue to sleep a little longer. You can easily return the engine to its full service condition after installation has been completed.

HOISTING THE ENGINE

Get some help. Don't attempt to handle the engine all by yourself. You risk wiping out the largest single investment in your project if you drop it. While two men can bodily lift and install a small engine such as a VW, it is still advisable to use an engine hoist of some sort. Remember, things don't always fit exactly right the first time you try them.

Some engines have a permanently attached lifting eye situated in a central location along the upper crankcase parting line. This is good. However, you will find that the hole in the lifting eye is too small for the hoisting hook to go through. This means that you will have to devise some sort of adapter. An easy way to make one is to form a loop by connecting the ends of a short length of chain with a 3/8-inch bolt passed through its two end links and the engine's lifting eye. The hoisting hook can now be readily used to raise the engine.

If your engine has no lifting eye, you will have to resort to the use of a sling. The sling can be made of a strong rope or nylon strap passed around the cylinder barrels and connected to the hoisting hook. Be careful not to entrap any engine component behind the strap. The weight of the engine bearing against it during the hoisting could cause serious damage. On some engines it might even be prudent to remove the intake pipes before attempting the lift.

You have two hoisting options: a) roll the airplane up to the suspended engine, or b) roll the engine and hoist up to the airplane.

If your hoist is immobile, and the airplane cannot be rolled, you'll have a big job on your hands getting the two together. If you can't muster extra help, think out your problem; a solution will likely present itself.

One of the most common problems encountered in the average garage workshop is the low ceiling. Of course, the availability of the correct type of equipment will make even large problems disappear. If you can't borrow the right kind of hoist — a low profile, extended arm type — you might be able to rent one. This type of hoist is mounted on casters and eliminates the problem of immobility.

The first thing to do is to get the airplane into its flying attitude so that the engine will contact all of its mount points simultaneously. Be careful that the weight of the engine doesn't cause the airplane to nose over. This is possible with some tail draggers that are not completely fitted with wings and tail surfaces at the time the engine is installed. Just to be sure, you had better tie the tail down or weight it down with metal weights or concrete blocks. Don't forget to chock the wheels.

Next, hoist the engine up to the proper height and carefully work it up to the mount. This is where the extra set of hands and eyes of an interested assistant are invaluable. Have your engine shock mount units ready and the engine attachment bolts handy. Adjust the hoist so that the engine crankcase mounting holes are exactly lined up with the engine mount, which, of course, should have been bolted securely to the firewall beforehand.

INSTALLING THE SHOCK MOUNT BUSHINGS

It is most important, of course, that you install new rubber washers, cones or Lord mounts, depending on your engine type.

The small Continentals and the Lycoming 0-235 and 0-290 series and some Lycoming 0-320's take cone-like engine mount bushings. If your engine can utilize these in its mounts you are in luck, for they are by far the most inexpensive of the engine mount bushings.

Place one of the cone rubber bushings in the front seat and another in the rear seat of each of the engine mounting lug holes. Slip each mounting bolt in from the rear and through its engine mount hole. Slide on a steel washer. Guide the bolt through the rubber cones in the engine mounting boss and add a steel washer and castellated nut on each of the four bolts. Additional steel washers are sometimes placed between the engine mount and the aft engine bushing to alter the CG location slightly. If your engine mount bolts are slightly too short or too long in spite of your care in selecting the proper size, go ahead with the installation and replace them later. At that time it will be easy to replace them one at a time, even without an assistant. No need to delay the installation while you charge off to try to buy some new bolts. Oh yes, by all means install <u>new</u> bolts.

A center console should be made removable in case repairs to the gas tank or other unit behind it must be undertaken.

This newly installed VW engine is equipped with a Posa carburetor and a homemade air filter. Note that the gascolator is as low as it can be without protruding below the firewall.

In order for an oil cooler to be effective, air must pass through it efficiently. Mounting it up front ensures that it will be directly in the path of the cooling air flow. Flexible hoses are always used for the connections to the engine crankcase.

After all of the bolts are in place, tighten them uniformly in the same manner you would install a propeller. That is, when you snug up one bolt, move to the opposite side and snug that one. Do the same with the other two bolts. Then make a final pass in the same sequence and bring the nuts up to their proper torque.

Some builders do not know how much to torque their engine mount bolts and usually over-tighten the nuts because the rubber bushings continue to compress. This is WRONG WRONG!

> ***NOTE:*** *The engine mounts should be torqued to the values recommended for your engine. These vary considerably among engine types so check your manual. For example, the small Continentals using the rubber cone type mount nuts are torqued to 70 inch pounds (plus or minus 10) while the O-200 engine mount nuts must be torqued to 180 to 190 inch pounds. Lycoming mount nuts are secured with something on the order of up to 40 inch pounds. All this is good enough reason to be sure you know the recommended range for your engine.*

When you get up into the Continental C-90-14F and the O-200 models you must use a more sophisticated, more expensive shock mount assembly consisting of eight separate parts for each of the engine mount lugs.

These Lord mounts are installed by inserting a bushing in each of the rear counterbores (of the engine mount arm) and by placing the special cupped steel washer over its aft end and a special seat washer over its front face. The engine is then slipped onto the engine mount bolts (also inserted from behind). A steel tube with a section of rubber hose (snubber) is slipped over the end of each bolt. The front Lord bushings are similarly assembled with cupped washers.

The 'Cadillac' of engine mounts, the Lord type mounts used in Dynafocal engine suspensions are the best vibration suppressors you can buy. They are also the trickiest to install. The engine must be up against the mount and the Lord mount bushings properly seated before the bolts can be inserted from behind. Because the Lord mount rings are angled into the center of the engine, it appears at first that nothing will fit.

After the four bolts are in place, torque the nuts while the engine is still being supported by the hoist. Torque them until you detect a slight bulge in the rubber units. The rubber bonded portion of the shock mountings will take on a very pronounced sag as the weight of the engine is released from the hoist. Don't try to tighten the nuts even more in an attempt to eliminate this effect (compression set). This sag is normal. Normal, that is, provided the degree of eccentricity in the rubbers is within limits established for the type of shock mounting units you are using.

Whatever engine mount you are using, take care to select the correct type of bushings, otherwise they may be subject to excessive deflection and won't provide adequate rigidity for the weight of the engine.

PROVIDE AN ELECTRICAL GROUND

Although very few homebuilders seem to be doing it, small bonding straps (grounding jumpers) that connect the four attachment points of the engine mount to the corresponding four points on the engine should be installed. At least use a grounding strap to provide an electrical bridge across the engine mount rubbers. One end of the ground strap should be connected to some point on the engine and the other to the firewall. Be sure you get a good electrical connection .... bare metal to bare metal.

INSTALL BAFFLES

If baffles are not already installed on your engine, now is the best time to undertake that chore. Later, the engine will not be as accessible. Check over any existing baffles to assure yourself that they are in good condition. Replace cracked, worn or poorly fitted pieces of baffling.

Making a new set of baffles from scratch is a challenging undertaking. Refer to Section 6, COWLS AND COOLING, for useful suggestions.

CONNECTING ENGINE CONTROLS

There is no single correct way of installing an engine, nor is there a magic sequence

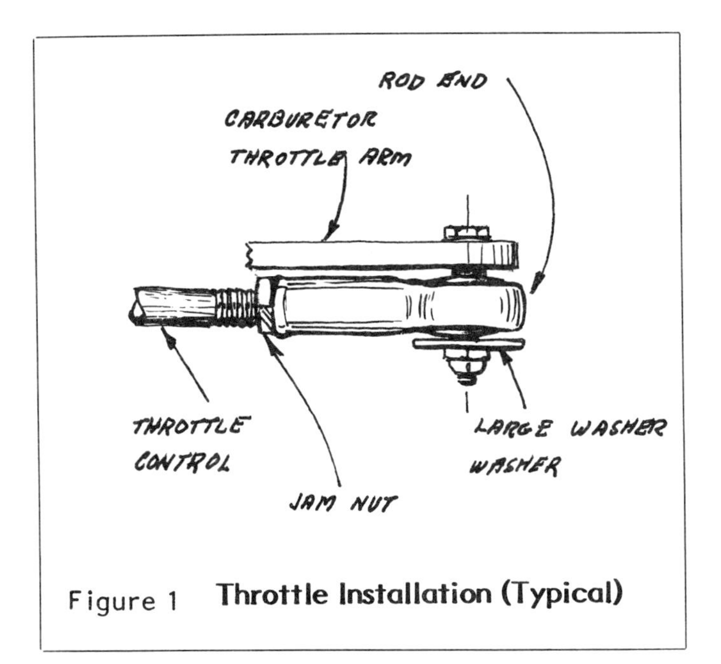

Figure 1 **Throttle Installation (Typical)**

for hooking up the engine controls. This is also true for the fabrication and connection of the numerous essential fuel and oil lines, wires, instruments and other units. One thing is certain, you must work with great care both in fabricating these items and in making their connections properly. Do this and you will avoid damaging components that are already in place. (Figures 1-3)

Here is a brief checklist showing one sequence that might be followed when hooking up the engine controls. Connect the following:

1. THROTTLE — Obtain the correct length throttle control. If possible, use a standard assembly complete with a friction lock. It is easier to install than a custom-made push-pull lever system. The end

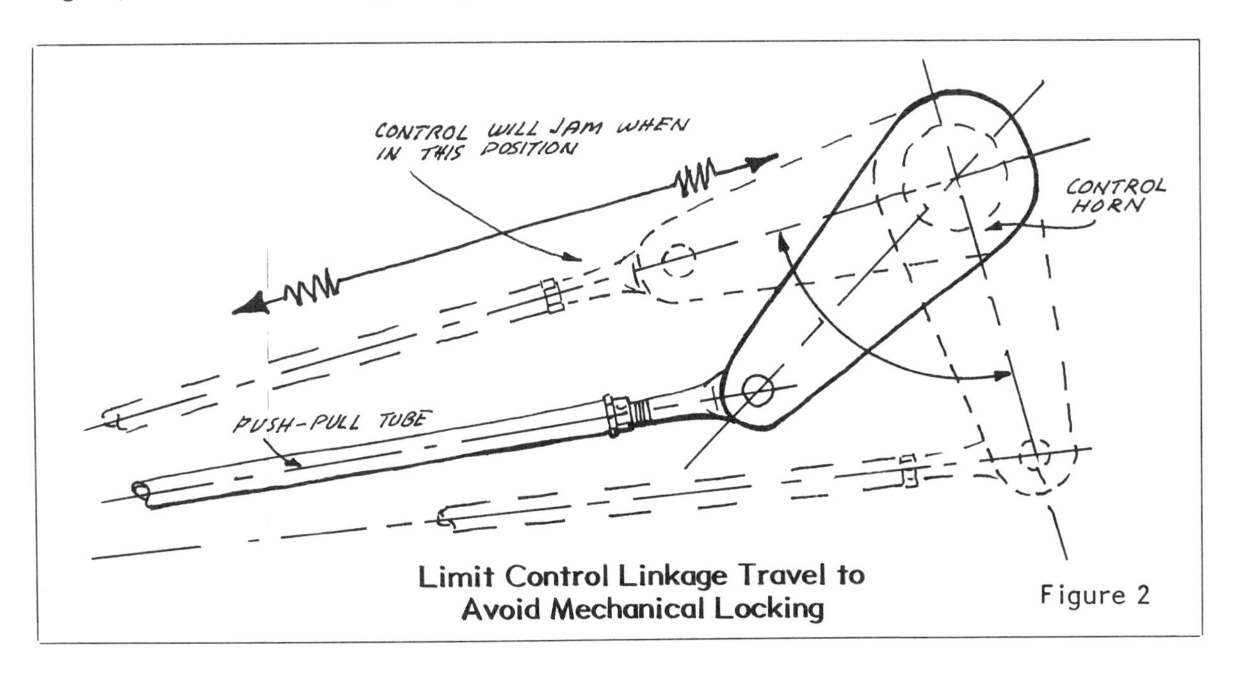

Limit Control Linkage Travel to Avoid Mechanical Locking

Figure 2

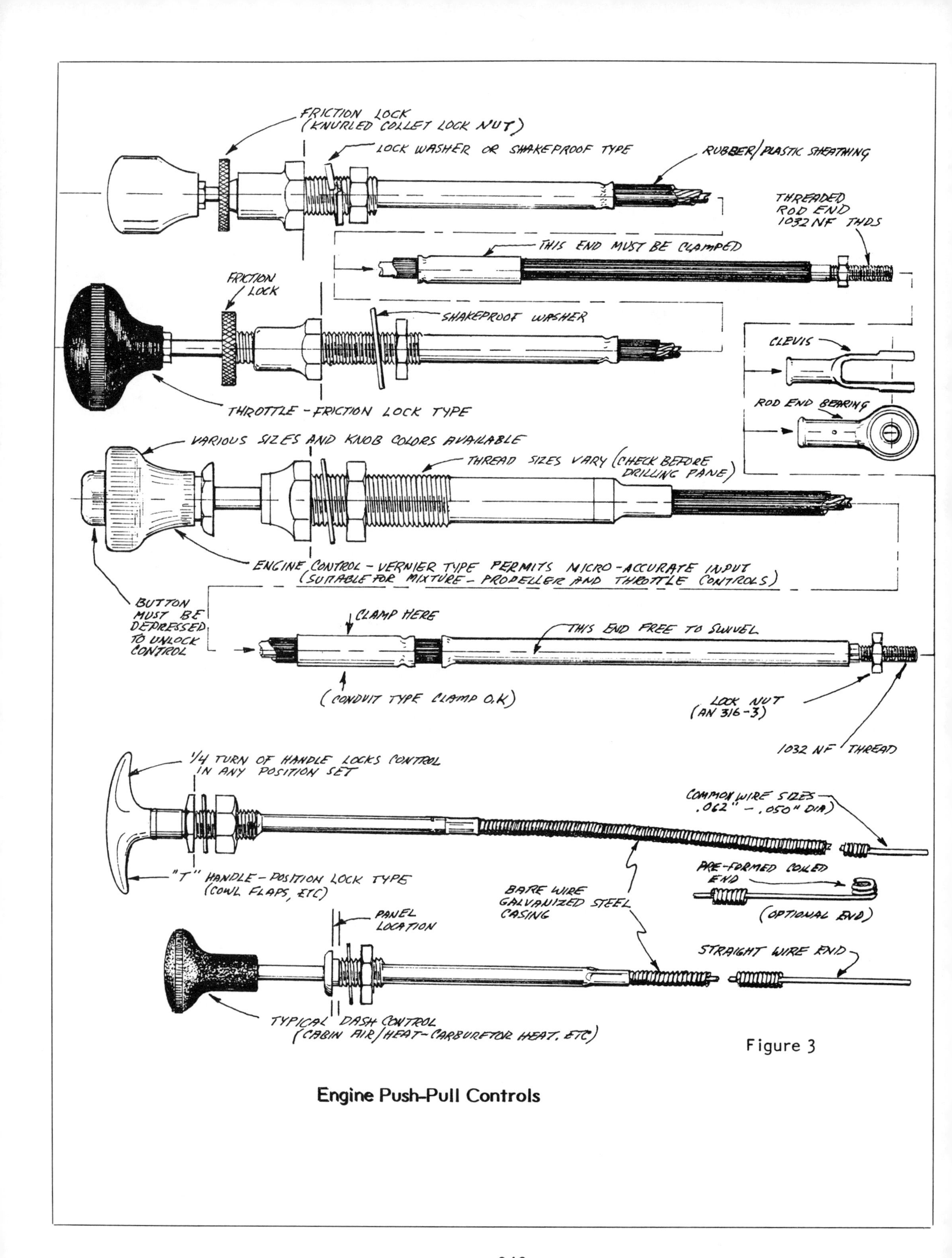

Figure 3

Engine Push-Pull Controls

that connects at the carburetor should be threaded with a 10-32 thread so that a rod end bearing can be screwed on and locked with a jam nut. The throttle is secured to the instrument panel with a lock washer and nut. Route the housing cable as directly as possible to the engine compartment and carburetor area. Do not use a unit that is too long. Most throttle assemblies are between three and four feet long. An extra foot or two will add weight and be difficult to route. The end of the flexible housing closest to the carburetor must be clamped with an Adel clamp to some convenient bracket that will hold the actuating rod in precise alignment with the carburetor throttle control arm. Be sure you obtain the full movement of the throttle at the carburetor.

2. MIXTURE CONTROL -- The carburetor mixture control cable housing must also be secured to a bracket near the carburetor so that the cable is aligned with the mixture control arm. Be sure to get the full movement of the mixture control lever.
3. PROPELLER CONTROL -- Most homebuilts will not have controllable propellers. If yours does, now is a good time to hook it up.
4. CARBURETOR HEAT CONTROL
5. STARTER ACTUATING LEVER AND CONTROL CABLE -- This is applicable only to some Continental C-85, C-90 and O-200 engines).

HOSES AND LINES

Obtain or fabricate and connect the

PLACARD THAT PANEL

Airworthiness standards require that each standard category aircraft instrument installed be labeled (placarded) as to function or operating limitations. Although standard certification regulations do not apply to homebuilts, it seems that FAA inspectors take a greater interest in this requirement than do builders.

Responsible people in the homebuilt movement are concerned over the unusually widespread practice of omitting many of the instrument markings in homebuilt aircraft. Who would argue the need to establish a safe airspeed "redline" or the "maximum rpm" limit on the tachometer" But, how about the operating range and limits for the oil temperature and the oil pressure? Which is the ignition switch? Cowl flap control? Which is the carburetor heat and which the cabin heat? See what I mean?

With the wide variety of engines in use, non-aircraft types in particular, the limits for the various gauge readings are not common knowledge. One would suspect that perhaps even some builders may not really know the operational limits of their own engines.

Take the VW engine as an example. What are its minimum, normal, and "do not exceed" oil temperatures? How about the oil pressure? Are they the same as they would be if the engine were nestled snugly in its original VW buggy? What about rpm?

Have you considered this? If you ever permit other people to fly your machine the absence of instrument and panel markings could constitute a presumption of negligence on your part should anything go wrong. Any lawyer can tell you the consequences of that determination in court.

It's time that builders individually acknowledge the merits of this need and comply with the long established practice of marking range limits on the essential instruments. This includes the important engine controls, which should be marked as to direction of movement.

The markings need not be more than 1/16-inch wide. They may be painted on the glass face of the instrument or affixed with adhesive tape.

Use red radial lines to indicate limits and short green arcs to indicate the range for normal conditions. A short vertical mark is painted on the bottom of the instrument across both the glass and the case to serve as a check that the glass has not moved in the case and thrown off the accuracy of the markings.

Labels for the engine controls and switches may be affixed using adhesive tags. Many builders prefer to use engraved plastic tabs for a more impressive appearing instrument panel. It doesn't matter how you do it, you could even use ink-marked masking tape. The important thing is to do it.

following hoses and lines:

1. Oil cooler hoses to the cooler and to the engine.
2. Oil pressure line to the engine (if a direct reading type is used).
3. Oil temperature lines and bulb to the adapter on the engine.
4. Crankcase breather hose.
5. Fuel supply hose to the carburetor.
6. Primer line discharge fitting(s) to the manifold.
7. Vacuum hose to the vacuum pump (if applicable).

ELECTRICAL WIRING

Fabricate and connect the following wiring/cables:

1. Battery power cable to the starter.
2. Battery grounding cable.
3. The wire from the ignition switch to the generator "F" terminal. Be sure to ground the shielding wires at both ends.
4. Connect the cylinder head temperature wire to the probe.
5. Connect the tachometer drive shaft to the engine adapter. Be sure it meshes properly. Torque to 100 inch pounds.
6. Connect the primary (ground wire or "P" lead) lead wires between the ignition switch and the magnetos. Be sure magneto switch is OFF.

DON'T FORGET MISCELLANEOUS STUFF

Service the engine with the proper grade of oil and install a new oil filter (if applicable). Fit and install all ducting from the baffles to the carburetor, generator, magnetos, cabin heat box, etc.

Install the propeller so that one blade is in approximate 10 o'clock position for easy hand propping. On a retractable gear aircraft you might want to install the propeller horizontally so that in the event of an emergency gear-up landing the engine could be stopped with the propeller horizontal in order to reduce damage.

Check the rigging of all engine controls and look the whole installation over carefully for security of mounting and safe routing of lines, hoses and wires. Check that everything that has to be safetied, is!

Perform initial engine run-up. Fine tune all engine controls for smooth and easy operation.

Install the cowling and hook up cowl flaps (if applicable).

This Guppy is getting its final fitting of all necessary components from the firewall forward.

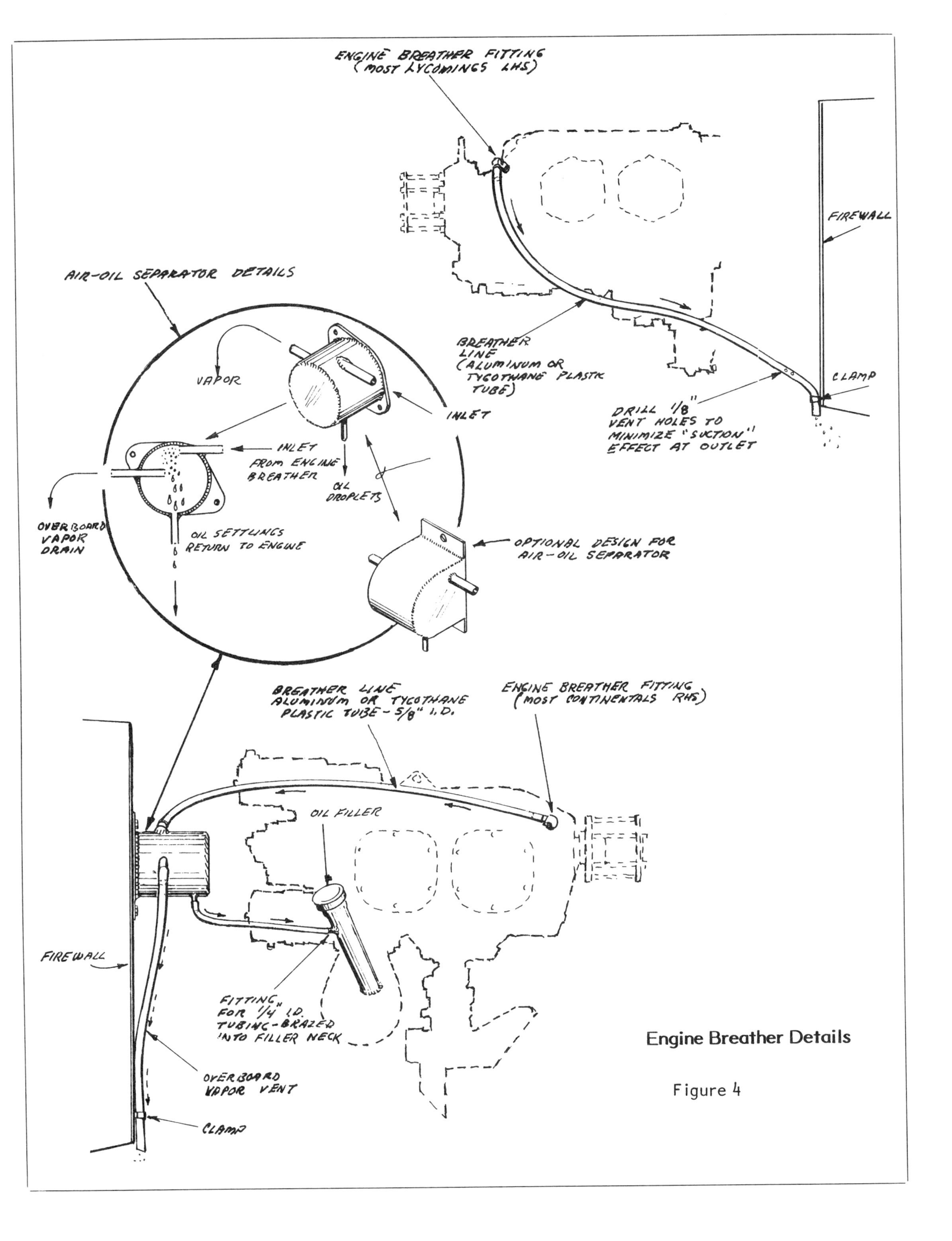

Engine Breather Details

Figure 4

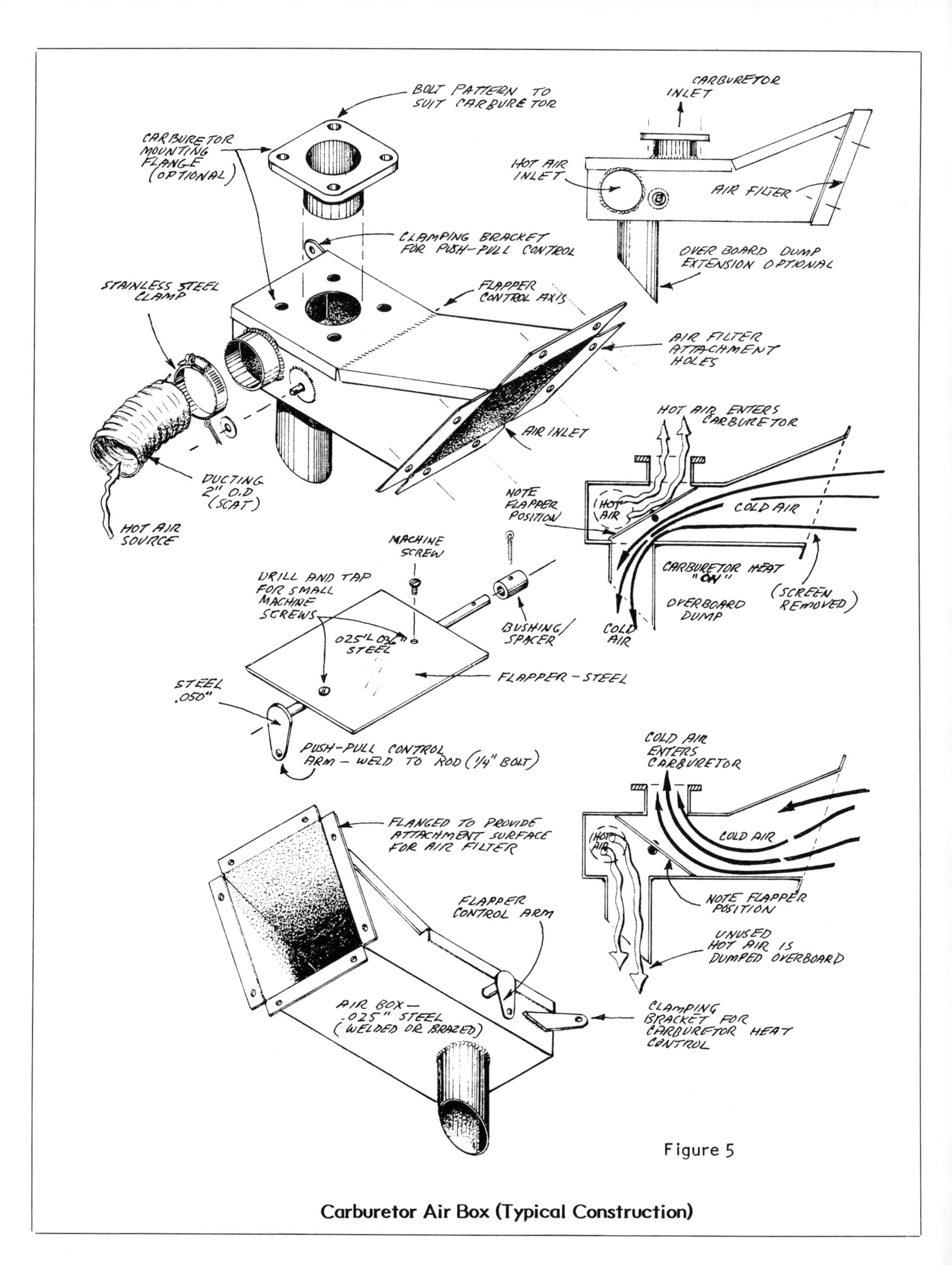

Figure 5

Carburetor Air Box (Typical Construction)

Fuel flow divider on a Lycoming with the distribution lines running to each cylinder. This is typical of a fuel injected installation.

The installation of converted auto engines invariably requires the installation of a radiator. When the aircraft is an old timer like a Pietenpol, drag and streamlining are apparently not matters of great concern.

Instruments, Controls

There is a serious lack of standardization in the arrangement and location of powerplant instruments in the cockpit panel. The lack of space remaining after positioning the essential flight instruments, often forces the designer/builder to make very unorthodox use of the remaining space. This produces some unusual instrument locations, like under the seat, behind the left elbow, etc.

The instruments should be grouped closely to minimize the cockpit workload and reduce the scanning effort. The pilot can then spend more time looking out for other aircraft and avoiding birds.

The main powerplant instruments should not be reversed relative to the positions of the powerplant controls. In other words, the location and arrangement of the powerplant instruments, such as the manifold pressure, tachometer and fuel flow should be sequenced as are the throttle, propeller and mixture controls.

In production aircraft you will sometimes see powerplant instruments that are positioned in a vertical line on the panel, although the accepted standard is a horizontal orientation. Again, instrument panel space may be cited as the reason for the deviation. Of course, homebuilders, too, have the same space problem. Ideally, though, the manifold pressure gauge should be located in the panel in line with and above the throttle, the tachometer in line with and above the propeller control, and the fuel flow indicator in line with and above the mixture control. Of course, I realize that only a few homebuilts have all of these instruments, but the omission of one or more should not in any way degrade the effort to standardize whatever instruments and controls are being used. The establishment of the natural relationship between powerplant instruments and powerplant controls can reduce pilot confusion.

POWERPLANT CONTROL DETAILS

It is highly recommended that you design your cockpit engine controls so that they operate in accordance with the following concepts of movement and actuation (based on FAR Part 23):

Each cockpit control should be located in a logical position, and (except where its function is obvious) identified to prevent confúsion and inadvertent operaton.

Locate and arrange the controls so that the pilot, when seated, has full, unrestricted movement of each control without interference from either clothing or cockpit structure.

Locate powerplant controls on a pedestal or near the centerline of the instrument panel. The location order from left to right is throttle, propeller and mixture control. (See Fig. 6) The worst and most accident prone location for the carburetor heat control is any position close to the mixture control, particularly if there is no discriminating shape and/or color coding between the two knobs. If you can't do anything to alter the shape of your mixture control knob, at least install some sort of guard to help prevent its inadvertent operation.

Aircraft with tandem seating and single-place aircraft may better utilize controls located on the left side of the cockpit compartment. The location order then, from left to right, should be throttle, propeller and mixture.

Engine control knobs, color and shape should be in accordance with established standards. (See Fig. 7)

Controls of a variable nature using a rotary motion should move clockwise from the OFF position, through an increasing range, to the FULL ON position.

The direction of movement of cockpit controls must be obvious. That is, wherever practicable, the sense of motion involved in the operation of other controls must correspond to the sense of the effect of the operation upon the airplane or upon the part operated. Here are specific examples:

Throttle/thrust — Forward to increase forward thrust, rearward to decrease thrust.

Propellers — Forward to increase rpm.

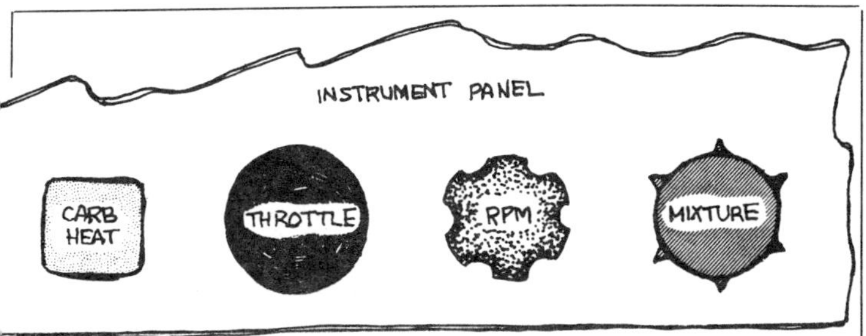

PUSH-PULL CONTROL INSTALLATION
(TYPICAL PANEL)

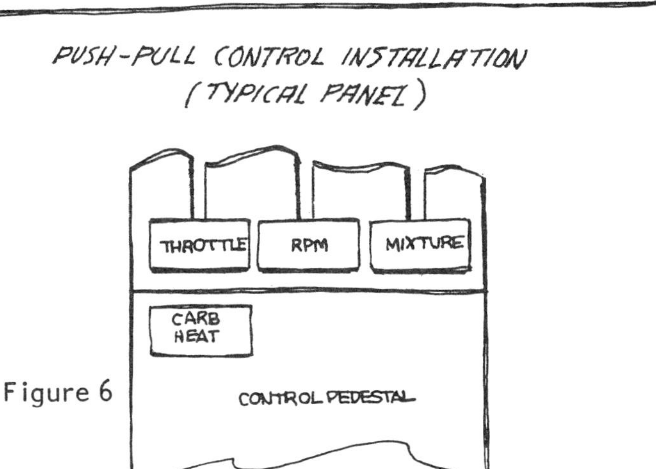

Figure 6

Standard Lever Controls for a Pedestal Installation

Fuel condition — Forward or upward for ON.

Mixture -- Forward or upward for RICH.

Carburetor air heat or alternate air — Forward or upward for cold. Forward or upward for low blower.

Supercharger — For turbo-supercharger, forward, upward or clockwise to increase pressure.

Cowl flap — Rearward or down for cowl flap open.

Landing gear — Down to extend.

The opportunities for getting involved in some mess resulting from cockpit confusion are many. It's bad enough to forget to activate the gear handle for landing, but it is possible in some aircraft to actually move the gear handle and still land gear up. How? In some planes the pilot inadvertently moves the gear handle from the UP position to the NEUTRAL/OFF position rather than all the way from the UP to the DOWN position. "Wait," you say, "you still have the warning horn/light to remind you. Right?" Well, what about the pilot who sometimes carries so much power on final approach that the gear warning horn/light does not operate?

As for checklists, unfortunately the demands for increased pilot attention and complex, involved control manipulation is a set-up for PILOT ERROR, even with checklists. It's true that most of these problems could be avoided through the proper use of a checklist, but .... pilots sometimes read or recite checklists mechanically and fail to respond to them correctly. No, the real solution is to make it difficult to impossible for the pilot to do things wrong or to receive the wrong impressions.

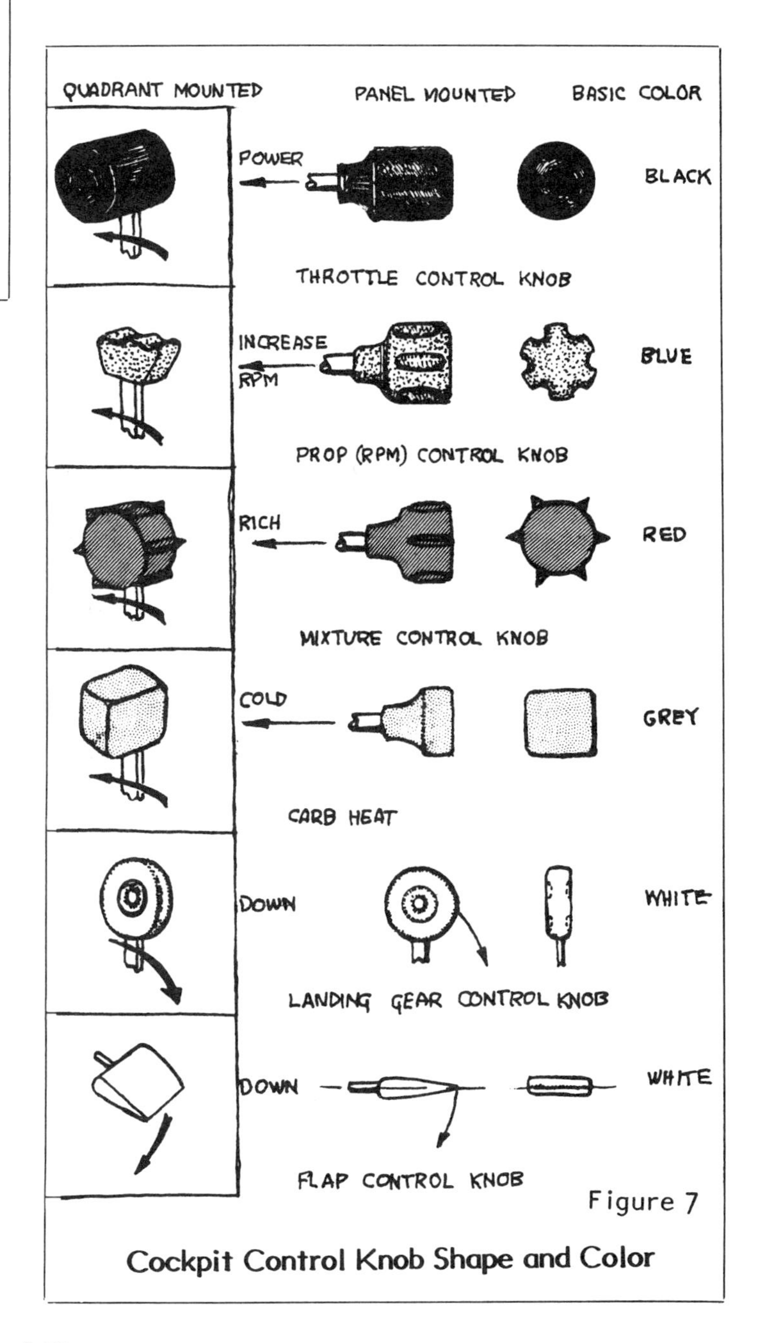

Figure 7

Cockpit Control Knob Shape and Color

ENGINE INSTALLATION NEEDS

Most of us are quite surprise to learn how much we have yet to obtain for our engine installation in the way of hardware, ducting tubing and wiring and accessories before we can complete our airplane. Would you believe it? Between 100 and 200 parts and pieces will be required for a well equipped engine installation.

It is very annoying to be busily engaged in the installation process only to discover you are lacking in some small parts that cannot be locally obtained. This means a telephone call to some homebuilt supplier and a plea to expedite the shipment of the needed item. Meanwhile, progress is delayed for a couple of weeks or longer. Don't be surprised if that sort of thing happens to you at least two or three times during engine installation.

The "PRELIMINARY SHOPPING LIST FOR ENGINE INSTALLATION ITEMS included in this book may not solve all of your work stoppage problems but it will help reduce them to a minumum. The list has an added advantage in that you can start your shopping for your installation needs early in your project. This will allow you to take advantage of the occasional bargain that frequently comes our way at the wrong time. Spreading the buying process over a longer period of time also helps reduce the heavy concentrated costs usually incurred with the purchase of an engine and the parts and accessories required for its installation.

No single list can possibly include all of the items needed for all types of projects. However, it is quite surprising that the needs are generally similar and directly related to the complexity of the aircraft design. Anyway, it is a pleasure to learn that you may not need all of the items listed.

What do we have here? Auxiliary electric fuel pump (1), engine driven fuel pump (2), fuel injector air control box (3), air filter assembly (4), oil cooler attached to rear baffle (5), and dynafocal engine mount pad (6). That vertical hose? An engine breather line (7).

PRELIMINARY SHOPPING LIST FOR ENGINE INSTALLATION ITEMS

NTITY EDED	DESCRIPTION (SELECT ONLY THOSE ITEMS APPLICABLE TO YOUR INSTALLATION	UNIT COST	ESTIMATED TOTAL COST
	CESSORIES AND PARTS		
	AUXILIARY FUEL PUMP (BOOSTER PUMP) 12 VOLT OR 24 VOLT?		
	AIR FILTER (OPTIONAL)		
	BRAKE FLUID RESEVOIR (NOT REQUIRED WITH RESEVOIR TYPE MASTER BRAKE CYLINDERS		
	CABIN HEAT VALVE UNIT		
	CARBURETOR/FUEL INJECTOR UNIT (GRAVITY FEED? PRESSURE SYSTEM? FUEL INJECTION?)		
	CARBURETOR HEAT BOX ASSEMBLY		
	EXHAUST PIPES/MUFFLER		
	FUEL PUMP - ENGINE DRIVEN (IF NEEDED)		
	FUEL SHUT-OFF VALVE/FUEL SELECTOR VALVE		
	GASCOLATOR (MAIN STRAINER)		
	GENERATOR/ALTERNATOR (TO SUIT AMPHERE REQUIREMENT)		
	GASKETS, EXHAUST (GET A NEW SET)		
	LIFTING STRAP (ENGINE)		
	MAGNETOS		
	OIL COOLER		
	OIL FILTER (OPTIONAL)		
	PRIMER PUMP		
	PROPELLER		
	SPARK PLUGS		
	SPINNER		
	STARTER		
	TACHOMETER-90 DEGREE DRIVE (IF NEEDED)		
	TACHOMETER CABLE (IF MECHANICAL DRIVE IS USED)		
	VACUUM PUMP		
	ECTRICAL ITEMS		
	BATTERY		
	BATTERY BOX (NOT NEEDED WITH A GELL CELL BATTERIES)		
	BATTERY SOLENOID (CONTACTOR/RELAY)		
	BUS BAR - COPPER OR ALUMINUM STRIP 1/16" x 3/8" x 5 to 10 INCHES		
//////	CIRCUIT BREAKERS/FUSES (AMPERAGE CAPACITIES ARE THOSE TYPICAL FOR CB INSTALLATIONS)		
	8 AUTO PILOT		
	15 BEACON LIGHT		
	2 CABIN LIGHTS		
	2 CLOCK (USUALLY A 1 AMP FUSE IS USED)		
	15 FLAPS		
	5 FUEL PUMP (12V OR 24V?)		
	30-60 GENERATOR / ALTERNATOR (DEPENDS ON UNIT'S CAPACITY)		
	8 INSTRUMENT LIGHTS		
	15 LANDING GEAR		
	20 LANDING LIGHTS		
	10 NAVIGATION LIGHTS		
	15 PITOT HEAT		
	5-10 RADIOS		
	5 STALL WARNING		
	10 STROBES		
	2-5 TURN & BANK		
//////	ELECTRIC WIRE/CABLE (MIL-W-5086A) STRANDED TIN-COATED INSULATED COPPER CONDUCTOR		
	2 ga BATTERY CABLES (ALUMINUM OPTION)		
	4 ga BATTERY CABLES (COPPER)		
	10 ga BATTERY TO GENERATOR / BATTERY TO AMMETER (NO STARTER)		
	12 ga BATTERY SOLENOID TO AMMETER / STARTER SOLENOID TO BUS		
	14 ga LANDING LIGHT / RADIO / PITOT HEAT / LANDING GEAR CIRCUITS		
	16 ga INSTRUMENT LIGHTS-MISCELLANEOUS CIRCUITS		
	18 ga NAVIGATION LIGHTS / INSTRUMENT LIGHTS / TRANSPONDER / CLOCK / MISCELLANEOUS		
//////	ELECTRIC WIRE/CABLE - SHIELDED (MIL-C-7078A USUALLY)		
	10 ga AMMETER / ALTERNATOR / BUS TO VOLTAGE REGULATOR		
	12 ga ALTERNATOR - VOLTAGE REGULATOR		
	14 ga ALTERNATOR - VOLTAGE REGULATOR / MAGNETO - IGNITION SWITCH		
	16 ga MAGNETO CIRCUITS - IGNITION SWITCH		
	IGNITION HARNESS (SHIELDED TYPE IF RADIO WILL BE INSTALLED)		

SHOPPING LIST (CONTINUED - PAGE 2)

QUANTITY NEEDED	DESCRIPTION (SELECT ONLY THOSE ITEMS APPLICABLE TO YOUR INSTALLATION)	UNIT COST	ESTIMATED TOTAL COST

ELECTRICAL ITEMS (CONTINUED)

QUANTITY NEEDED	DESCRIPTION	UNIT COST	ESTIMATED TOTAL COST
	MAGNETO FILTER (RADIO NOISES)		
	MAGNETO GROUND WIRE CONNECTORS ("P" NUTS)		
	NOISE SUPPRESSOR, ALTERNATOR (RADIO NOISES)		
	OVER-VOLTAGE REGULATOR (OPTIONAL)		
//////////	TERMINAL CONNECTORS (SOLDERLESS, CRIMP TYPE, INSULATED) ASSORTED TYPES NEEDED		
	22 ga to 18 ga (RED)		
	16 ga to 14 ga (BLUE)		
	12 ga to 10 ga (YELLOW)		
//////////	SWITCHES (ROCKER, TOGGLE, TWIST, PUSH-PULL) YOUR CHOICE		
	BEACON LIGHT		
	CABIN LIGHT		
	FLAPS		
	FUEL PUMP		
	INSTRUMENT LIGHTS		
	IGNITION SWITCH		
	LANDING GEAR		
	LANDING LIGHT		
	MASTER SWITCH		
	NAVIGATION LIGHTS		
	PITOT HEAT		
	STROBES		
	TIE WRAPS (TO BUNDLE AND SUPPORT WIRING)		
	VOLTAGE REGULATOR, DELCO-REMY (FOR ALTERNATORS, A 1961-1969 CHRYSLER AUTO REG. OFTEN USED)		

ENGINE CONTROLS (DETERMINE LENGTHS NEEDED BEFORE YOU SHOP)

QUANTITY NEEDED	DESCRIPTION	UNIT COST	ESTIMATED TOTAL COST
	THROTTLE CONTROL UNIT (FRICTION LOCK TYPE 'SAFER' THAN VERNIER TYPE) 3/16" THREADED END		
	MIXTURE CONTROL - VERNIER TYPE WITH 3/16" THREADED END		
	PROPELLER CONTROL - VERNIER TYPE WITH 3/16" THREADED END		
	CABIN AIR DASH CONTROL - FLEXIBLE METAL CASING WITH WIRE CORE		
	CABIN HEAT CONTROL - FLEXIBLE METAL HOUSING WITH STRAIGHT WIRE END		
	CARBURETOR HEAT DASH CONTROL - FLEXIBLE METAL HOUSING WITH STRAIGHT WIRE END		
	COWL FLAP CONTROL - FRICTION LOCKING DASH CONTROL PREFERRED		

FUEL LINE FITTINGS AND HOSES

QUANTITY NEEDED	DESCRIPTION	UNIT COST	ESTIMATED TOTAL COST
//////////	FUEL TANK TO FUEL SHUT OFF / SELECTOR VALVE		
	AN 816-6D - FLARED TUBE & PIPE THREAD (OR AN 822-6D 90 DEGREE ELBOW) NIPPLE		
	AN 818-6D - NUT		
	AN 819-6D - SLEEVE		
//////////	FUEL SHUT OFF TO GASCOLATOR INLET		
	AN 816-6D - NIPPLE (OR AN822-6D 90 DEGREE ELBOW		
	AN 818-6D - NUT		
	AN 819-6D - SLEEVE		
	A BULKHEAD FITTING OR SOME SPECIAL FIREWALL PASS THRU FITTING MAY BE NECESSARY		
//////////	GASCOLATOR OUTLET TO CARBURETOR INLET		
	FLEXIBLE AEROQUIP HOSE 303-6 WITH 491-6 FITTINGS OR EQUIVALENT & AN816-6 NIPPLE		
//////////	GASCOLATOR OUTLET TO AUXILIARY FUEL PUMP INLET (IF INSTALLED)		
	FLEXIBLE AEROQUIP HOSE 303-6 WITH 491-6 FITTINGS OR EQUIVALENT & AN822-6 ELBOW		
//////////	AUXILIARY FUEL PUMP OUTLET TO ENGINE DRIVEN FUEL PUMP INLET		
	FLEXIBLE AEROQUIP HOSE 303-6 WITH 491-6 FITTINGS OR EQUIVALENT & AN822-6 ELBOW		
//////////	GASCOLATOR OUTLET TO ENGINE DRIVEN FUEL PUMP INLET		
	FLEXIBLE AEROQUIP HOSE & AN816-6 NIPPLE OR AN822-6 ELBOW OR AN823-6 ELBOW (45°)		
//////////	ENGINE DRIVEN FUEL PUMP OUTLET TO CARBURETOR / FUEL INJECTOR UNIT INLET		
	FLEXIBLE AEROQUIP HOSE & AN816-6 NIPPLE OR AN822-6 ELBOW OR AN823-6 ELBOW (45°)		
//////////	PRIMER PUMP INSTALLATION (USE AN FITTINGS LISTED OR AUTOMOTIVE BRASS TYPE)		
	AN 816-2D - NIPPLE		
	AN 818-2D - NUT		
	AN 819-2D - SLEEVE		
	PRIMER FITTING FOR GASCOLATOR MAY BE NEEDED		

SHOPPING LIST (CONTINUED - PAGE 3)

QUANTITY NEEDED	DESCRIPTION (SELECT ONLY THOSE ITEMS APPLICABLE TO YOUR INSTALLATION)	UNIT COST	ESTIMATED TOTAL COST

HARDWARE ITEMS

QUANTITY NEEDED	DESCRIPTION	UNIT COST	ESTIMATED TOTAL COST
	BOLTS - ENGINE MOUNT TO FUSELAGE (CHECK YOUR PLANS)		
	BOLTS - ENGINE TO ENGINE MOUNT (CHECK YOUR ENGINE)		
	.500" DIA. (LYCOMING O-320-A1A/O-235-L2A/O-290-D2B CONICAL MOUNTS)		
	.4375" DIA. (LYCOMING o-320-E2D/O-235-L2A DYNAFOCAL MOUNTS		
	7/16" DIA. (CONTINENTAL C-125/C-145/O-300)		
	3/8" DIA. CONTINENTALS, ALL MODELS A-50 THRU O-200/LYC O-145/LYC GO-145		
	COTTER PINS - STAINLESS STEEL (ASSORTED SIZES)		
	ENGINE MOUNT BUSHINGS		
	CONTINENTAL A-65 THRU C-90 - CONICAL (8 BUSHINGS)		
	LYCOMING O-235 & O-290 (7/8" HOLE) - CONICAL (8 BUSHINGS)		
	LYCOMING O-320 (1" HOLE) - CONICAL (8 BUSHINGS)		
	LYCOMING, DYNAFOCAL - 4 ASSEMBLIES PER INSTALLATION		
	CONTINENTAL C-90-14F & O-200 - 4 ASSEMBLIES PER INSTALLATION (5 PIECES PER)		
	PIANO HINGES (COWLING ATTACHMENT, OPTIONAL METHOD)		
	SAFETY WIRE - STAINLESS STEEL PREFERRED (.030" DIA MOST USEFUL)		
	FINGER STRAINERS (TO SCREW INTO FUEL TANK OUTLET) 3/8" MALE THREAD AND 1/4" FEMALE		
	GROUNDING STRAP/CABLE- ENGINE MOUNT TO ENGINE/BATTERY GROUND TO ENGINE		
	STAINLESS STEEL HOSE CLAMPS (3 1/2" DIA. / 2" DIA. / 1 1/2" DIA.) FOR DUCTING		
	TUBING- SOFT ALUMINUM, 3/8" DIA. (FUEL TANKS TO GASCOLATOR)		
	TUBING- COPPER, 1/8" DIA. (PRIMER LINES)		

MISCELLANEOUS MATERIAL / ITEMS

QUANTITY NEEDED	DESCRIPTION	UNIT COST	ESTIMATED TOTAL COST
	ANTI-SEIZE COMPOUND (FOR SPARK PLUGS)		
	BAFFLES - ALUMINUM SHEET, 2024-T3 / 6061-T6 (.032")		
	BAFFLE SEALS - RUBBER ASBESTOS SEAL STRIPS (FELT STRIPS ARE ALSO BEING USED)		
	DUCTING - AERODUCT (SCAT) CONDUIT		
	3 1/2" DIA. - INLET AIR TO FUEL INJECTOR UNIT		
	2" DIA. - CARBURETOR HEAT DUCTING		
	2" DIA. - CABIN HEAT DUCTING		
	2" DIA. - CABIN AIR DUCTING		
	1 1/2" DIA. - MAGNETO COOLING DUCTING		
	ENGINE OIL		
	MINERAL OIL (FOR BREAK-IN OF NEWLY OVERHAULED ENGINES)		
	ASHLESS DISPERSANT		
	FIRESLEEVE, AEROQUIP (GASCOLATOR TO FUEL PUMP / ENGINE PUMP TO CARBURETOR) -12 FOR -6 HOSE		
	FIREWALL INSULATION - FIBERFRAX / ASBESTOS SHEET (1/16" to 1/8")		
	FIREWALL METAL - STAINLESS STEEL / GALVANIZED, ETC(.016" to .018")		
	HYDRAULIC FLUID - MIL-H-5606BB (ITS RED...DO NOT USE AUTOMOTIVE STUFF)		
	RUBBER GROMMETS - ASSORTED SIZES (FOR FIREWALL OPENINGS)		
	SPARK PLUG GASKETS (USE NEW ONES EVEN IF PLUGS ARE USED)		
	GASCOLATOR MOUNTING BRACKET		
	FUEL PUMP MOUNTING BRACKET		
	POP RIVETS (BAFFLE ASSEMBLY)		
	NUT PLATES, MACHINE SCREWS, AN BOLTS AND NUTS AND A VARIETY OF COMMON HARDWARE ITEMS		

SPECIAL TOOLS AND EQUIPMENT

QUANTITY NEEDED	DESCRIPTION	UNIT COST	ESTIMATED TOTAL COST
	ELECTRICAL CRIMPING PLIERS		
	FLARING TOOL FOR TUBING - AIRCRAFT TYPE (37 DEGREE FLARE)		
	JACKS (WING OR OTHER SUITABLE TYPE)		

propellers & spinners

12

About those Props....

More than one type of propeller can be used on just about any airplane. Beginning with the least expensive and lightest type, they are:

FIXED PITCH PROPELLER — As the term implies, the pitch angle of the propeller is fixed because it is made of one piece. The blade angle does not, cannot be changed. Fixed pitch props used by homebuilders are commonly made of hardwood or forged aluminum alloy. A few are handcrafted from composites (fiberglass fibers, Kevlar and/or graphite fibers).

GROUND ADJUSTABLE PROPELLER — This propeller consists of a hub and separate blades which may be adjusted to whatever angle you want .... on the ground only. Most of these propellers were originally manufactured for larger engines and see little use by homebuilders except those who restore antiques or build replica aircraft. Recently, however, a few propeller makers (individuals) have begun to make lightweight ground adjustable props for the homebuilt trade.

TWO-POSITION PROPELLER - This propeller is similar to the ground adjustable type except that the pilot may, generally through mechanical linkage, move the blades from a low pitch position (for takeoff) to a high pitch position (for cruise). This type propeller is not generally available for light planes except for a few examples manufactured many years ago. Most of these had metal hubs and wooden blades, although some examples of all-metal two-position propellers are around.

VARIABLE/CONTROLLABLE PITCH PROP — The pitch of this type of propeller can be changed manually by the pilot in flight. Changing the blade angle gives the propeller the greatest efficiency for any particular flight condition.

CONSTANT SPEED PROPELLER — This is an automatically adjusted propeller. The prop's control system (governor) adjusts the pitch to maintain a constant rpm without any assistance (or interference) from the pilot.

SOME CLARIFICATIONS ABOUT PROPS

PUSHER AND TRACTOR PROPELLERS — A propeller may be mounted ahead of the aircraft structure in a tractor arrangement or behind it in a pusher configuration .... not the same propeller, however. One propeller pulls (tractor) and the other pushes the aircraft through the air. Obviously, since engine rotation cannot be changed, the pitch of a propeller must be reversed for a pusher installation. This means, simply, that you cannot bolt a conventional propeller on a pusher. You will have to find a "pusher prop". Since pusher propellers must operate downstream of the aircraft structure, they must not only cope with a disturbed airflow, but they are often subjected to damage from rocks, dirt and miscellaneous objects.

FIXED PITCH VS VARIABLE PITCH PROPS OF ANY VARIETY — First of all, a fixed pitch propeller is safer, costs less, is much lighter and is virtually maintenance free when compare to any type of variable pitch propeller (controllable pitch, constant speed, ground adjustable, two-position, etc.). Unfortunately, a fixed pitch propeller can only be effectively designed to have good climb (takeoff performance) or a high cruise speed, but not both. A fixed pitch prop, particularly in a fast airplane is, at best, a compromise between best climb and best cruise.

In general, the optimum takeoff and climb propeller is one with a large diameter and low pitch. Unfortunately, the optimum propeller for cruise is one of a smaller diameter and a higher pitch. Obviously, there must be a compromise if you aspire to get good climb and high top speed from the same propeller. It is unfortunate, but true, that a good cruise prop will only allow the engine to turn up about 50% of the engine's rpm on takeo The double whammy of high pitch plus low forward speed simply imposes too great a load on the engine. You can imagine how "sick" the climb performance is on half the engine power. This is where the constant speed propeller really earns its keep. Although it weighs more, much more, the increased thrust available for takeoff and climb is a very important safety factor.

WOOD VS METAL PROPELLERS — Here are some generalized pros and cons which may influence your choice between a wood and a metal propeller:

No doubt about it, metal propellers are about twice as heavy as wood props. For similar size propellers, a fixed pitch metal prop (24 pounds average) can be replaced by a wooden one weighing 10 pounds. Or, if weight and cost are not factors, by a controllable pitch propeller weighing about 60 pounds and costing much much more than either.

Experts say metal propellers, particularly used, cut-down propellers, are subject to catastrophic blade failure while there is little evidence of similar blade failure among wood props. Actually wood propellers seem to be less prone to vibration problems than metal propellers and, in fact, it is most unlikely that a wood prop will ever fail from fatigue if it doesn't flutter visibly at full throttle.

PROPELLER EXTENSIONS — A prop extension is used to cope with the problem of poor airflow over the blunt, squared off nose of the typical aircraft engine. Using a propeller extension will permit you to better incorporate the sleek sweeping lines that are so pleasing in an airplane. Often, a fringe benefit of this is knowledge that the engine's cooling will also probably be improved. Sure, there is a weight penalty, but the extra weight of a prop extension is not so much as to diminish the benefit of lower drag and a nicely faired cowling line.

Some propeller extensions are unsuitable for use with wood props because they (the extensions) lack a centering boss. Wood props are, at best, difficult to center accurately without such a centering boss on the prop extension .... especially if you have to do the bolt hole drilling yourself. You really can't rely on the drive lugs and attachment bolts to guarantee your success at centering.

Ordinarily, there is no problem in using a prop extension of reasonable length, say up to three or four inches, with a flanged crankshaft. If you want to use a prop extension on an engine with a tapered crankshaft, you should remember to mention that fact when ordering your propeller extension as these engines take a different spacer installation.

Now that we have that behind us, let's look at how we go about selecting the best propeller for our airplane.

WHAT SIZE PROPELLER?

Your airplane is just about finished and you don't have a propeller for it yet. Joe, down the street, has a good metal prop but the tips are curled from a taxiing accident. He doesn't know what aircraft it came from, but he will let you have it pretty cheap. The offer is tempting because you have just found out what the prices are for new metal props. The money you have set aside may not be enough for a new propeller. Even if it is, can you afford to pay a big price for a new propeller which may not be exactly right for your bird? Of course, there's no guarantee that Joe's prop will be right for your bird either, and it still has to be repaired. What to do?

Don't think that you are alone in suffering from this particular dilemma. I would estimate that a good 20% of the homebuilts are equipped with a propeller which is a poor match for the engine and airplane. How does a guy select the right propeller? Well, let's just see if we can't reason our way through the problem. Maybe you won't get the perfect propeller for your airplane on the first try, but if you approach your problem logically, you will at least have eliminated a gross mismatch in your initial selection.

Our discussion will cover the fixed pitch propeller only since that is the one the majority of homebuilders will use.

START WITH YOUR ENGINE

The type engine you have installed will dictate the basic propeller to be used. Remember that, it is very important. The engine manufacturer has established the maximum and minimum propeller diameter and pitch permitted for each engine. If you rely on this information alone, you won't get into any serious trouble. At worst, your bird's performance may be a bit sick. But at least it should fly well enough to keep you safely airborne.

Propeller manufacturers have listings of their propellers that show the propeller models available for different engine/aircraft combinations. In addition to the diameter, the recommended standard cruise, take off and climb pitch is also given for the different aircraft using a particular propeller. This data is shown in Table 1, which can be used as a quick reference by anyone who has yet to determine what propeller his engine can handle.

The propeller pitch range shown, in some cases, is the absolute maximum authorized by the propeller manufacturer. In other cases it merely reflects the composite limits for the representative aircraft listed. In the latter instance, the authorized limits for that prop model may really be just a bit greater. However, stick to the numbers shown, or, better yet, check with the manufacturer or a propeller shop before you do anything drastic.

I guess it is generally accepted that the propeller diameter should be larger for efficient low airspeed operations, and smaller for high airspeeds. This does not mean, of course, that you can get STOL performance just because you put on a large prop any more than a small prop can make a racer out of a 1930 vintage Pietenpol, powered by a Model T Ford engine.

Have you ever heard this little 'truism': "Keep your prop as long as possible as long as possible!" That's a good rule of thumb. However, just in case you need one more argument for longer prop diameters: next time you get near a nice homebuilt with a beautifully cowled engine, just check how much of the propeller area is blanked out by the engine cowl when the prop is in the horizontal position. In spite of the utility of long props, there are situations in which the larger diameter props can't be used. Read on.

IT MUST SUIT THE AIRPLANE, TOO

You may have the correct prop for the engine but you may not be able to use it on your airplane. The diameter may simply be too large. In other words, you can't get enough ground clearance. Years ago this matter was a frequently discussed topic among homebuilders. However, with increasing popularity of the tricycle landing gear, that problem has somehow been shoved off into obscurity. Now that I think about it, I haven't heard any comments from the FAA as to a minimum propeller ground clearance for homebuilts. While some inspectors may insist on a minimum clearance, it has become more a matter of personal responsibility. A minimum of nine inches is good insurance. By nine inches, I mean with the tail up, in level attitude. This guarantees, among other things, that your prop will stay long longer!

WEIGHT AND BALANCE FACTORS

Considering the higher cost for metal propellers, I wonder why more builders are not using wooden propellers which can be obtained for about half the cost. I realize, of course, that wood props are more delicate and not as well suited to outdoor parking conditions. Another reason, believe it or not, is because metal props are much heavier. Most homebuilt designs are prone to be tail heavy and a metal prop often helps to balance things up.

Determine from your weight and balance calculations whether you need a heavier prop (metal) or a light one (wood). Wood saves about 10 pounds for the smaller engines. (See the Weight and Balance Factors section in this book for additional revelations.)

SEE WHAT OTHERS ARE USING

While you are busy searching through Trade-A-Plane and the local airports for that outstanding propeller buy, continue to gather performance data from other builders and owners of aircraft like yours.

If your airplane is a popular model, there are probably a number of them already built and flying. Ask their owners what size props they are using. Compare the results they report with the conclusions you have reached in your own personal research. However, you should realize that some of these gents may not really have accurate data to give you. Accept their information with the same reservations you may have in accepting your neighbor's claim of gas mileage for his new car.

STATIC RPM AS A GUIDE

Once you have a propeller, the static rpm that your engine can turn up with that particular prop is a good indicator of whether or not it can get you off the ground safely. By static rpm, I mean the maximum rpm the engine is capable of reaching at full throttle while the aircraft is stationary.

Static rpm is established for all certificated engines. If your engine is not one of those shown below, check with any local A&P mechanic or repair shop. What does static rpm mean to you? Well, if you give full throttle to your engine and it does not rev up to the recommended static rpm, it is quite possible that the prop's diameter is too large, or that it has too much pitch, or both. It means that perhaps you won't get enough thrust from that prop for a safe take off.

What if your static rpm is much higher than that recommended? Well, you may get off all right, but there is the possibility of exceeding the red-line limits for your engine, even during the take-off sequence.

Lycoming manuals advise that, when using a fixed pitch propeller, the static rpm should be 2300 plus or minus 50 rpm when the engine is rated at 2700 maximum rpm. This will vary slightly depending on the type of propeller. If the static rpm is too high, there is the chance of overspeeding at full throttle in level flight. (With a constant speed propeller, static rpm will be the rated rpm of the engine. This is controlled by the pitch settings of the propeller.)

Here are some well established static rpms that each type engine should be capable of achieving with any suitable fixed pitch propeller.

ENGINE	STATIC RPM
Continental A-65	not under 1960 rpm
Continental C-85	2200 rpm
Continental C-90	2125 rpm
Continental 0-200	2320 rpm
Lycoming 0-235	2200 rpm
Lycoming 0-290	2200 rpm
Lycoming 0-320	2300 rpm

MAX RPM OR RED-LINE AS A GUIDE

In straight and level flight, with full throttle, what is the maximum rpm that the engine will turn (below 3,000 feet, that is)? Is it above or below the recommended rpm red-line limits established for that engine model? If the rpm exceeds the red-line limit for your engine by more than 5%, the prop could be too short or, more likely, the propeller could absorb more pitch.

To sum this all up — if you pick a prop that is approved for your particular engine, one whose diameter still allows safe ground clearance and the proper static rpm, you probably are close to the prop/engine combination you need. You can refine the pitch requirement by checking your full throttle rpm in level flight for the fine pitch adjustments to get the most speed out of the bird. You may recall from your studies on fixed pitch prop theory (you did study that in school, didn't you?) that climb suffers with an increase in cruise capability and vice versa. Obviously, the final selection between climb and cruise performance is a compromise of your own preference. You can't have both with a fixed pitch prop.

Another thing, each time you decide to make a pitch adjustment on a metal prop, you'll also be pitching out another bunch of bucks.

BEWARE THE DAMAGED PROP

If you locate a damaged propeller and hope to have it repaired for use on your own airplane, beware. Don't attempt to straighten the bent blades yourself. It is better to hike off to the nearest authorized prop shop and have them evaluate the propeller's condition. The prop may be bent and damaged beyond repair limits. FAA Advisory Circular, AC 43.13-1 (used to be CAM 18) gives guidance for the repair of metal propellers.

Straightening, cutting, shaping and balancing a propeller is certainly beyond the equipment capabilities of the average builder. It is better to entrust these technical operations to an approved prop shop. In return, you will receive a good serviceable prop and peace of mind. In this regard, be extremely careful when buying a damaged propeller, even though it looks repairable to you. Some uninformed (or unscrupulous) individuals may have partially straightened the blades, creating the appearance (innocently or deliberately) that the thing is within repairable limits when, in fact, it is non-repairable and in a critical condition.

Reputable prop shops will not attach a "Serviceable" tag (yellow tag) on a prop that has to be cut down below manufacturer's limits. Actually, they may even emboss "EXPERIMENTAL" on the hub. This means it cannot be used on a certificated aircraft ever.

LET'S TAKE ANOTHER LOOK

The propeller people design a propeller for a particular engine and then, based on their calculations and vibration testing, establish the maximum and minimum diameters that must not be exceeded. They also determine what pitch limits must not be exceeded.

It is very important that homebuilders realize how dangerous it is to modify propellers beyond authorized limits. The continuing rash of metal propeller failures highlights this concern and has aroused the attention of the FAA, the EAA, and the propeller manufacturers.

Take another look at Table 1 and note that most of the propeller diameters listed are greater than those typically found on a large number of homebuilts. There's no doubt about it. The propellers used in production aircraft, as a rule, have larger diameters than those used in homebuilts.

At this point the representative aircraft

shown in Table 1 may or may not be particularly helpful. After all, how many homebuilts are as large as the Piper Cub or are like the Ercoupe? I guess this is exactly what the problem is. The size of homebuilts, their weight, and their (generally) lower drag characteristics, are so different from the larger store-boughts that some homebuilders may feel justified in modifying propeller and pitch dimensions in hopes of achieving better performance. It would appear that there is a direct conflict between the scientifically inclined engineering set and the sometimes not so scientifically bent homebuilders.

A few "prop sources" around the country are advertising props as "suitable for a homebuilt." Is that so? What these sellers mean is that they have a damaged prop which was probably cut down and refinished. It can no longer be used legally on a store-bought aircraft so why not sell it to the homebuilders, "they will use anything." Well, from some of the propeller accident reports I've read, it may be true. Not sensible, but true.

Some homebuilders are even playing the same game with themselves as the targets. They have a damaged prop, it will fit the hub. Why not cut it down to salvage it? The homebuilder winds up with a smaller than factory recommended diameter propeller. Furthermore, as the final step, he has a bunch of pitch cranked into it. According to his incomplete and faulty reasoning the airplane will be faster. O.K., even if that were so, what about that prop; what about that engine. How long will the prop last? How long can the engine operate above the recommended maximum red-line?

The McCauley Industrial Corporation offers the advice that all of the EAA people should be encouraged to use a propeller that is within the diameter range approved for the engine. Too many homebuilders, they say, are using short diameters on four-cylinder engines. This can get the homebuilder into big trouble.

To illustrate to you that this business of performing drastic surgery on a propeller is a matter for wide concern, still another well known propeller manufacturer, the Sensenich Corporaton, also cautions that propellers should not be modified to suit experimental aircraft design limitations. They report that the aircraft should be designed originally around the propeller/engine combination. At the same time, they also recognize that it is not economically practical or feasible to run vibration surveys on each of the many homebuilt designs. So where do we go from here?

Best advice is to go back to Table 1 and and stay within the recommended parameters. If you require propeller dimensions that differ from those available for the standard metal props, why not look into the wood prop alternative? There are many skilled craftsmen who custom-build wood props specifically for homebuilt designs.

A tapered crankshaft propeller installation. A face plate is always used with a wooden propeller.

Engine/Propeller Combinations

Engine	Propeller Model	Dimensions in Inches Diameter	Pitch	Weight (lbs.)	Wood or Metal	Representative Aircraft
	McCauley					
	CM(7154)	76-69.5	47-41	22.8	M	Piper Cub, J4-PA11-PA17
Continental A-65	Sensenich					Aeronca, Taylorcraft
(65 hp)	74CK	74-70	48-40	21	M	Champion 7AC, Ercoupe
	72CK	72	48-42		W	Interstate
	76CK	76	36-34		W	
	74F	74	40-38	10	W	
	McCauley					
	CM(7154	71-68.5	51-38		M	Piper J3-J4-J5, Stinson 10
Continental A-75	Sensenich					Ercoupe 415, Culver Cadet
(75 hp)	74CK-2	74-72	38-36	21	M	Luscombe
	76AK-2	76	48		M	
	72GK	72	48-46		W	
	70A	70	54-52		W	
	70D	70	44-40		W	
	McCauley					
	CM(7154)	71-68.5	51-42	22.5	M	Piper J3, Cessna 120, 140
Continental C-85	Sensenich					Ercoupe, Fleet 80 "Canuck"
(85 hp)	76AK-2	76	48-38	24	M	Aeronca, Taylorcraft
	74CK-2	72-70	46-40	21	M	Champion, Funk
	70AK	70	56-50	8	W	Emeraude
	72GK	72	50-44		W	
	74FK	74	49-47		W	
	74FKT	74	50-48		W	
	McCauley					
	CM(7154)	71-68.5	54-41	22.5	M	PA11, PA18, Champion 7EC
Continental C-90	Sensenich					Alon Aircoupe, Cessna 120
(90 hp)	76AK-2	74	48-40	24	M	Cessna 140, Luscombe, J3, J5,
	72GK	72	50-48	11	W	PA19
	72EK	72	54-40		W	
	McCauley					
	MCM(6950)	71-67	68-38	19.8	M	Bolkow, Cessna 150, Thorpe
Continental O-200	CF(7538)	75-71	38		M	Sky Scooter, Morane Saulnier,
(100 hp)	SCM(7146)	71-66	68-38	19.5	M	Victa "Airtourer"
	Sensenich					Emeraude
	74CK	74-70	54-40	21	M	
	McCauley					
	LM(7252)	73-70	49-52		M	J5, PA12, PA14, PA16
Lycoming O-235	Sensenich					PA20S
	76RM	76	43-44		W	
	76FM	74	50-52		W	
	76AM6-2	74	42-50	24	M	
	McCauley					
	GM(7455)	74	46-55		M	PA20, PA22, Champion 7GC
Lycoming O-290	MGM(7457)	76-70	57		M	PA22, TriPacer
	Sensenich					Champion 7GCB
	74FM	74	50-57		W	
	74DM6	74-72	48-52	29.5	M	
	McCauley					
	GM(7455)	76-74	59-47		M	Champion 7KC, PA22
Lycoming O-320	MGM(7457)	76-70	54-52	30	M	Cherokee "160", Super
	Sensenich					Rallye, Cherokee "C"
	74DM6	74-72	70-45	29.5	M	Champion 7GCA, Beagle
	McCauley					
	FA(8243)	82	41-43		M	Auster Beagle/Avion, Rallye
Lycoming O-360	MFA(7460	75-74	62-56	30	M	PA28S, Cherokee B, C, D
	DFA(8449)	82	48		M	Beech B23 "Custom III"
	Sensenich					
	76EM8	76	65-60	34.5	M	

TABLE I

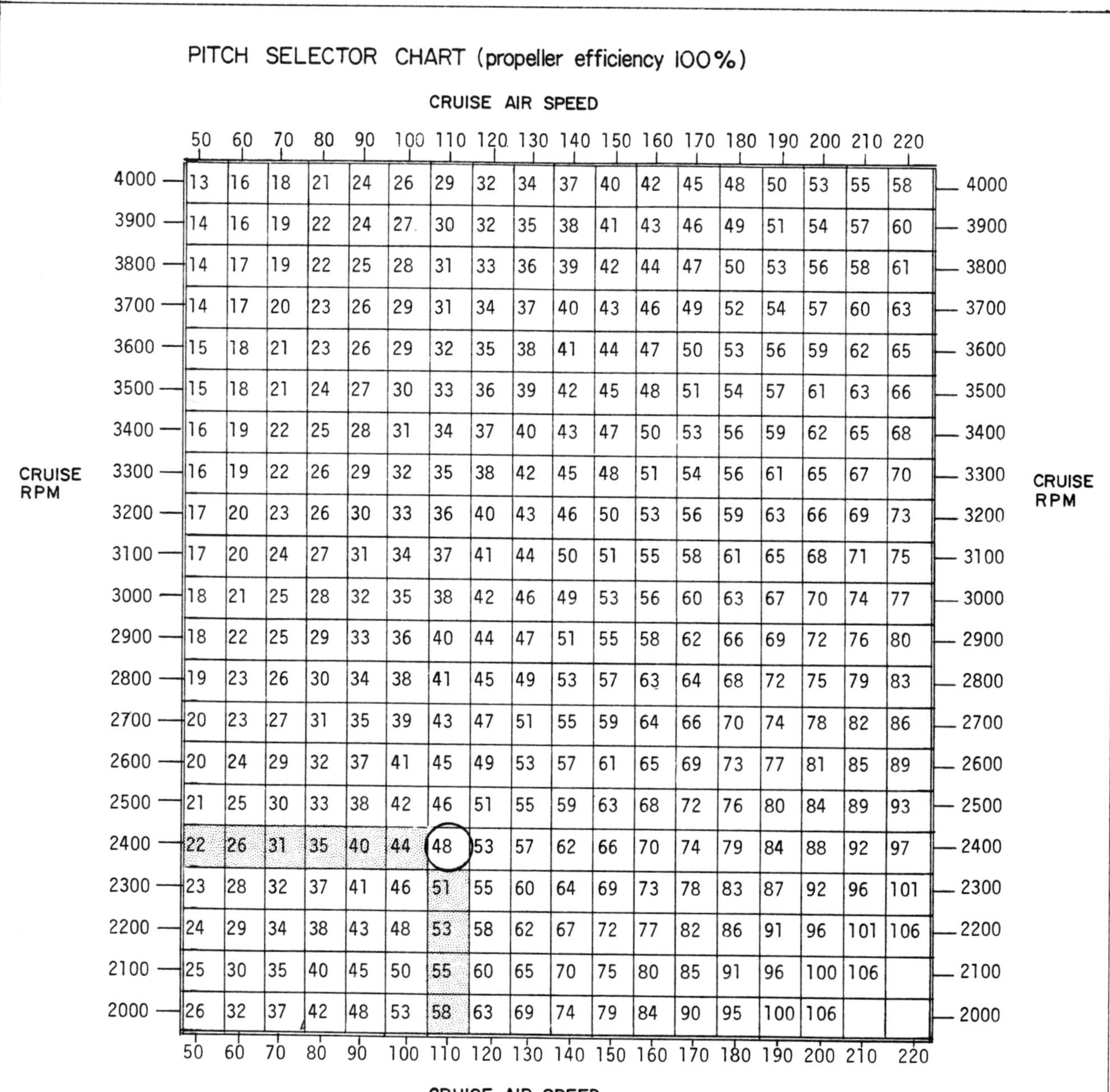

PITCH SELECTOR CHART (propeller efficiency 100%)

CRUISE AIR SPEED

CRUISE RPM	50	60	70	80	90	100	110	120	130	140	150	160	170	180	190	200	210	220	CRUISE RPM
4000	13	16	18	21	24	26	29	32	34	37	40	42	45	48	50	53	55	58	4000
3900	14	16	19	22	24	27	30	32	35	38	41	43	46	49	51	54	57	60	3900
3800	14	17	19	22	25	28	31	33	36	39	42	44	47	50	53	56	58	61	3800
3700	14	17	20	23	26	29	31	34	37	40	43	46	49	52	54	57	60	63	3700
3600	15	18	21	23	26	29	32	35	38	41	44	47	50	53	56	59	62	65	3600
3500	15	18	21	24	27	30	33	36	39	42	45	48	51	54	57	61	63	66	3500
3400	16	19	22	25	28	31	34	37	40	43	47	50	53	56	59	62	65	68	3400
3300	16	19	22	26	29	32	35	38	42	45	48	51	54	56	61	65	67	70	3300
3200	17	20	23	26	30	33	36	40	43	46	50	53	56	59	63	66	69	73	3200
3100	17	20	24	27	31	34	37	41	44	50	51	55	58	61	65	68	71	75	3100
3000	18	21	25	28	32	35	38	42	46	49	53	56	60	63	67	70	74	77	3000
2900	18	22	25	29	33	36	40	44	47	51	55	58	62	66	69	72	76	80	2900
2800	19	23	26	30	34	38	41	45	49	53	57	63	64	68	72	75	79	83	2800
2700	20	23	27	31	35	39	43	47	51	55	59	64	66	70	74	78	82	86	2700
2600	20	24	29	32	37	41	45	49	53	57	61	65	69	73	77	81	85	89	2600
2500	21	25	30	33	38	42	46	51	55	59	63	68	72	76	80	84	89	93	2500
2400	22	26	31	35	40	44	48	53	57	62	66	70	74	79	84	88	92	97	2400
2300	23	28	32	37	41	46	51	55	60	64	69	73	78	83	87	92	96	101	2300
2200	24	29	34	38	43	48	53	58	62	67	72	77	82	86	91	96	101	106	2200
2100	25	30	35	40	45	50	55	60	65	70	75	80	85	91	96	100	106		2100
2000	26	32	37	42	48	53	58	63	69	74	79	84	90	95	100	106			2000

CRUISE AIR SPEED

EXAMPLE -

A PROPELLER WITH A 48 INCH PITCH TURNING AT 2400 RPM COULD PRODUCE A CRUISE SPEED OF 110 MPH . . . . . MAXIMUM (NO SLIP).
A MORE REALISTIC FIGURE WOULD BE 99 MPH (110 x .9) ASSUMING A PROPELLER EFFICENCY OF .90%.

TABLE 2

PROPELLER SELECTION GUIDE

The speed of sound at sea level is 741 mph (approximately 1,100 feet per second).

The rotational (helical) tip speed of a propeller should not exceed the speed of sound. Above a tip speed of approximately 850 fps (950 fps for metal) propellers suffer a loss of efficiency and produce a marked increase in noise level.

If your engine's propeller flange has a run-out of more than .005-inch, don't blame the prop for an out of balance condition or improper tracking.

No propeller can be to be 100 per cent efficient. Due to propeller slippage in the less-than-solid air, its efficiency ranges from a poor 75 per cent to a typical 88 per cent efficiency. It's rarely above 90 per cent.

A blade angle change of one degree will change the engine speed approximately 50 to 100 rpm. Additionally, clipping one inch off the propeller diameter (1/2-inch on each end, that is) will cause an increase in engine speed of 50 to 100 rpm.

About static rpm (aircraft stationary on the ground and throttle wide open) -- the propeller should allow the engine to turn approximately 80 per cent of its rated rpm.

Two propellers can be designed to have the same diameter and pitch and still not produce identical performance because they were not built by the same prop maker. Each builder may have used different measurement reference points and means of fabrication.

Bent metal propeller blades may be repaired and their damaged tips cut shortened, but not beyond the limits established by the manufacturer.

Your propeller's efficiency increases as the airspeed increases. It also increases with additional diameter — up to the critical tip speed range (speed of sound).

Keep your prop as long as possible....
....as long as possible.

FORMULAS FOR A RAINY DAY -

- FIGURE YOUR CRUISE MATHEMATICALLY -

$$\frac{\text{PITCH (inches)} \times \text{RPM} \times 60}{12 \times 5280} = \text{MPH}$$

EXAMPLE - $$\frac{64 \times 2800 \times 60}{12 \times 5280} = 170 \text{ MPH}$$

ASSUME A PROPELLER EFFICIENCY OF 90 %
(many are less efficient - 75 to 85 %).

MULTIPLY 170 x .9 = 153 MPH
(corrected for propeller slip).

- FIGURE YOUR PROPELLER'S TIP SPEED -

$$\frac{\text{PROP DIA. (inches)} \times 3.1416 \times \text{RPM}}{12 \times 60} = \text{TIP SPEED}$$

$$\frac{72 \times 3.1416 \times 2700}{720} = 848 \text{ FT/SEC. TIP SPEED.}$$

TABLE 3

Wood Props

I can understand why so many of us give so little attention to that magical little stick bolted to the end of the engine. Heck, a little wood prop looks so tiny and light and so ultra simple that we are prone even to overlook the most fundamental installation requirements, much less bestow upon it the attention and care it deserves.

Some time ago I watched a fellow install a borrowed propeller on a Volksplane. He carefully backed out the bolts that were in it so that he could put in his own of a slightly different length. I saw that he meticulously fitted the propeller to the hub and gently worked the bolts through the holes in the hub. The nuts were installed and a most impressive chrome plated torque wrench was put to them. Obviously satisfied with his own work, he picked up his tools, put them away, and prepared for a short flight to try out the new prop. A couple of flips and the VW fired off smoothly .... and away he went.

What he forgot to do was track the prop to see if it was in reasonable alignment. It's considered poor form to discover that the propeller is badly out of track after you are already in the air. Besides that, it's downright scary. (Figure 1)

CORRECTING AN OUT-OF-TRACK PROP

Tracking a prop is not difficult. The process takes but a few minutes and fancy equipment is not necessary Most small aircraft can be set up for tracking by clamping a stick to the wing or fuselage so that it just barely touches the prop blade, or by placing a cardboard box or some sort of stand or fixture next to the propeller. The aircraft's wheels should be solidly blocked so that they can't move. (By the way, double check that the magneto switch is OFF.) Then turn the propeller through slowly by hand. Observe the passage of each blade with respect to the box or pointer. Each tip must track within 1/16-inch of the other. If the propeller blades are out of track by more than this it could be due to the uneven tightening of the propeller attachment bolts.

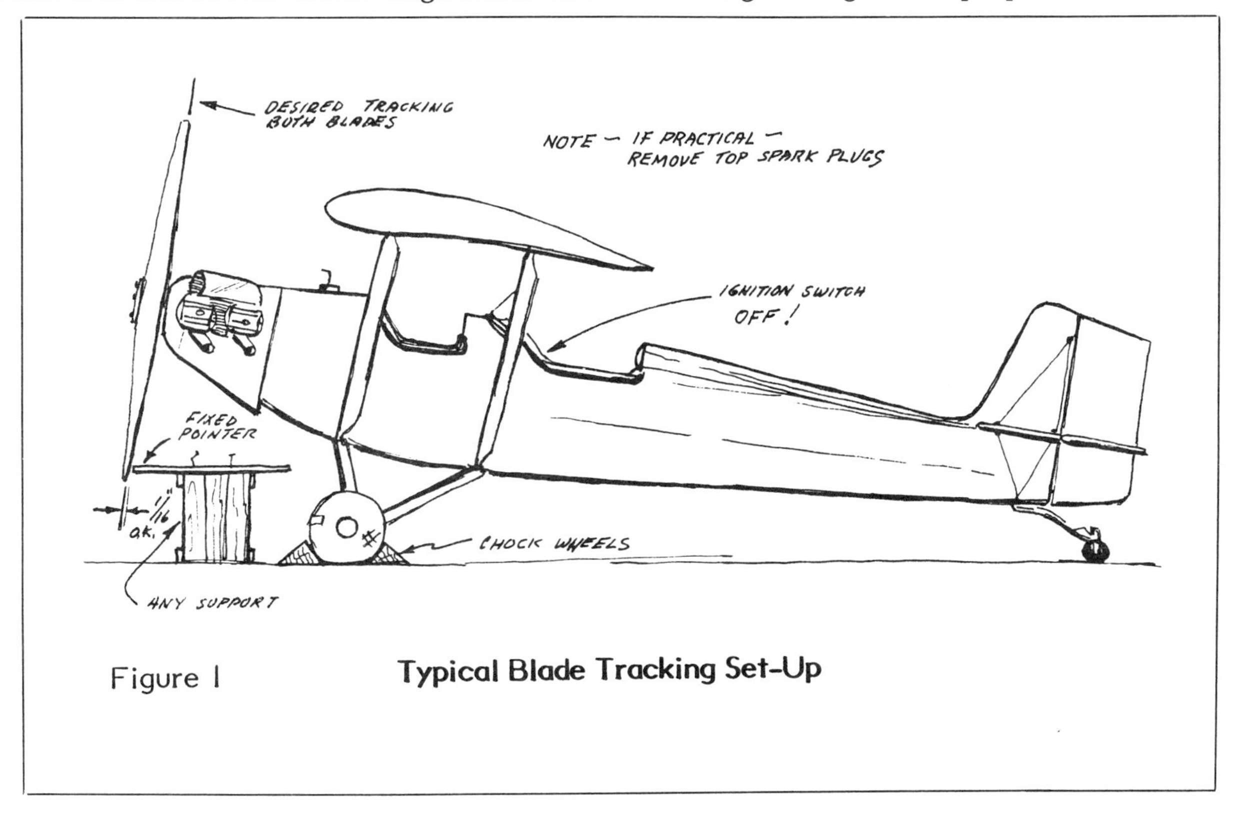

Figure 1 **Typical Blade Tracking Set-Up**

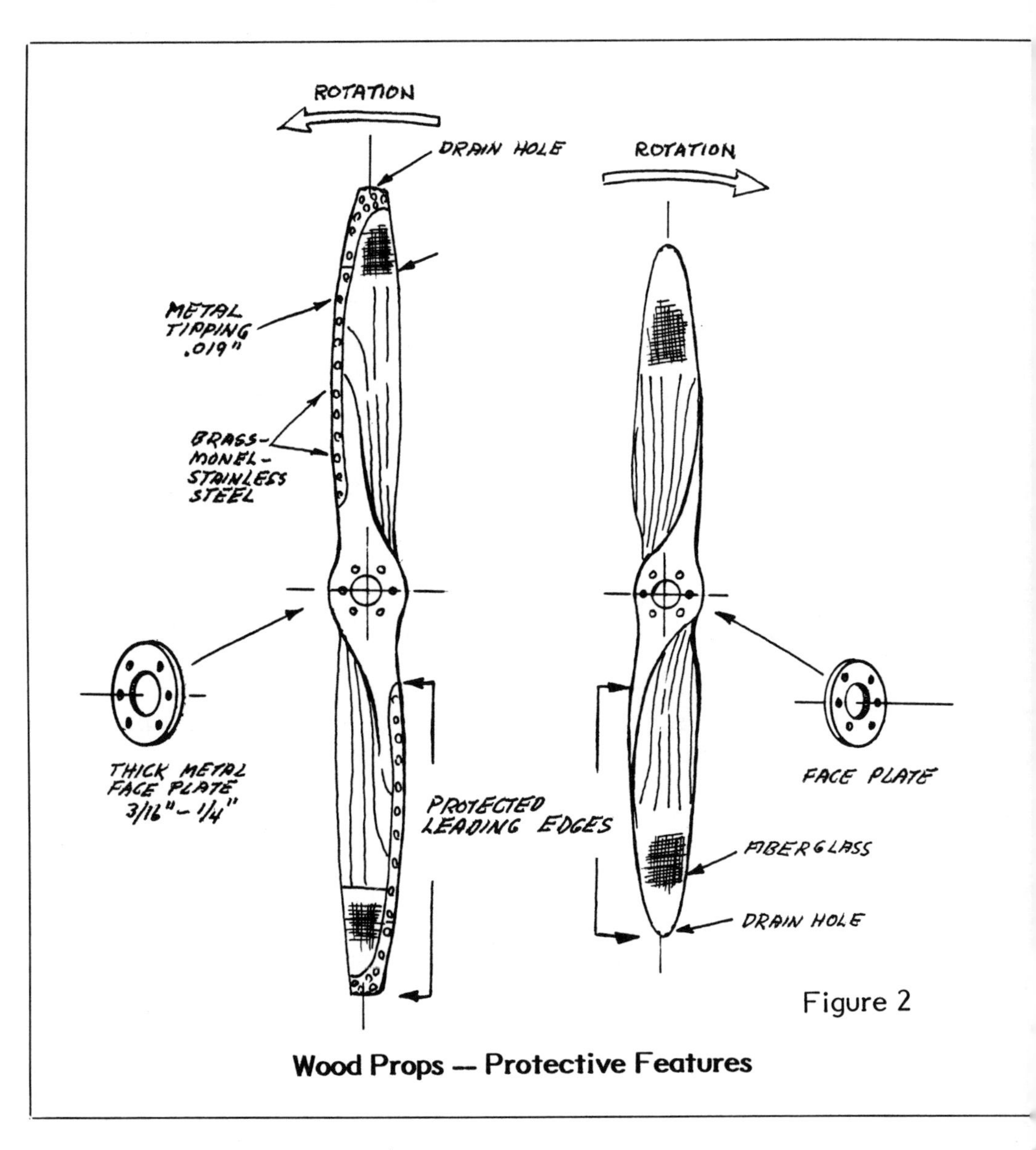

Figure 2

Wood Props -- Protective Features

If that's the case, you should re-check your tightening procedure first, since that is the most common cause. Slight out-of-track conditions can be corrected by slipping paper shims (or a similar shim material) between the propeller and the inside face of the hub's flange. This type of shimming can correct an out-of-track difference of up to 3/8-inch at the blade tips.

Sometimes, however, things go wrong during the fabrication of a propeller and the bolt holes are not drilled perpendicular to the hub. It can happen if the back spacer, glued to the aft side of the propeller intended for a VW installation is slightly untrue (tapered). After the holes are drilled, the whole thing can be misaligned and the prop will not track properly.

HOW TIGHT?

A new propeller should be torqued down until the face plate barely embosses the wood. This slight embossing must be evenly distributed around your bolt circle. Torquing the bolts "by the book" to a uniform 28 to 30 pounds, as is commonly prescribed for hardwood propellers, may be too much. (Even that old reference, CAM 18 only recommends 15-24 foot pounds.) It could cause the crushing of the wood fibers on mahogany props, for example. This is also true of the smaller propellers used on VW and ultralight engines because the face plate area is tiny compared to those big plates used to attach the comparatively huge props on antiques. So, tighten the bolts evenly, and only enough to press the plate into the wood surface ever so slightly. (Figure 2)

A FACEPLATE IS REQUIRED

For safety's sake, always use a thick metal faceplate on a wood propeller. Large washers are not an adequate substitute. For the VW engines, the faceplate may be made of

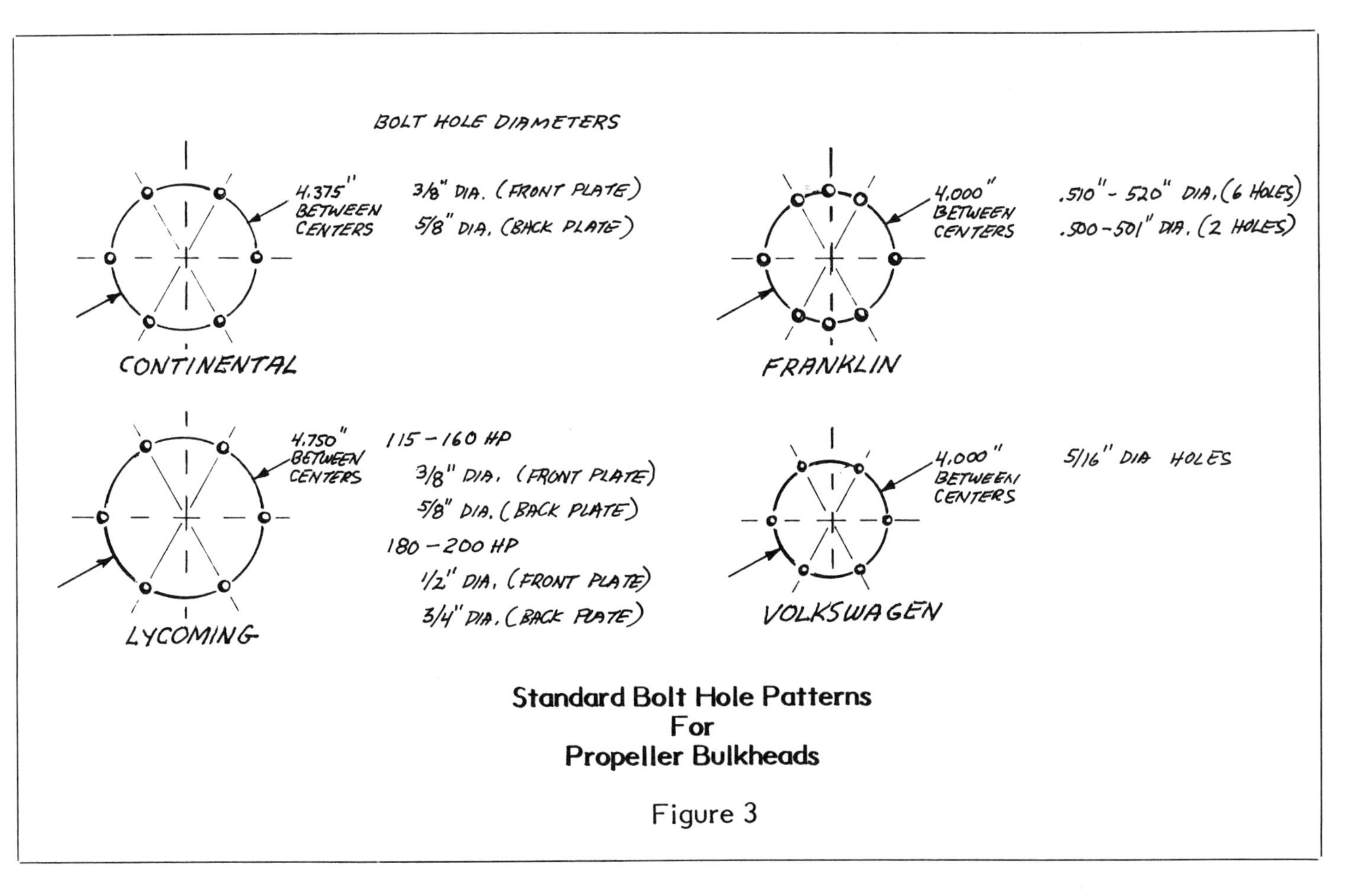

Standard Bolt Hole Patterns For Propeller Bulkheads

Figure 3

1/4-inch aluminum plate with the holes match-drilled to fit the hub plate. Larger propellers normally have a machined steel faceplate as a standard installation accessory.

BOLT INSTALLATION

Propeller bolts may well be subjected to more stresses than any other bolts used in the aircraft. Therefore, it is essential to use new bolts for your installation.

Before inserting the propeller bolts, some old time mechanics made a practice of coating each bolt lightly with oil prior to pushing it through the propeller. The wood in a propeller has a fluctuating natural moisture content which can cause some corrosion in bolts over a period of use and exposure to the elements. For my own prop assembly, I prefer to spray each bolt lightly with polyurethane varnish during installation. Of course, you can dip them as well.

Obviously, there are only two choices for the direction of bolt insertion. As a general practice, it is customery to install the bolts so that the nuts are on the front side of the prop. In some installations, however, you have no choice. For example, there is an English made hub for VW engines that is tapped to receive the threaded ends of the bolts. Naturally, these bolts would have to be inserted from the front. Bolts used in this kind of installation should be of the drilled-head variety and they should be safetied. Castellated nuts, when used with drilled bolts, are normally safetied in groups of two using .040-inch safety wire.

HOW TO INSTALL & REMOVE A PROP FROM A TAPERED SHAFT

1. To install the steel hub in the propeller:

Install the steel hub by slipping it in from the rear.

Install a face plate on the front face of a wood propeller. (Metal props don't need one.)

Insert the propeller bolts from the rear.

Add steel washers and castle nuts to each bolt. (Self locking nuts are suitable for use with a metal prop only.) Safety the nuts.

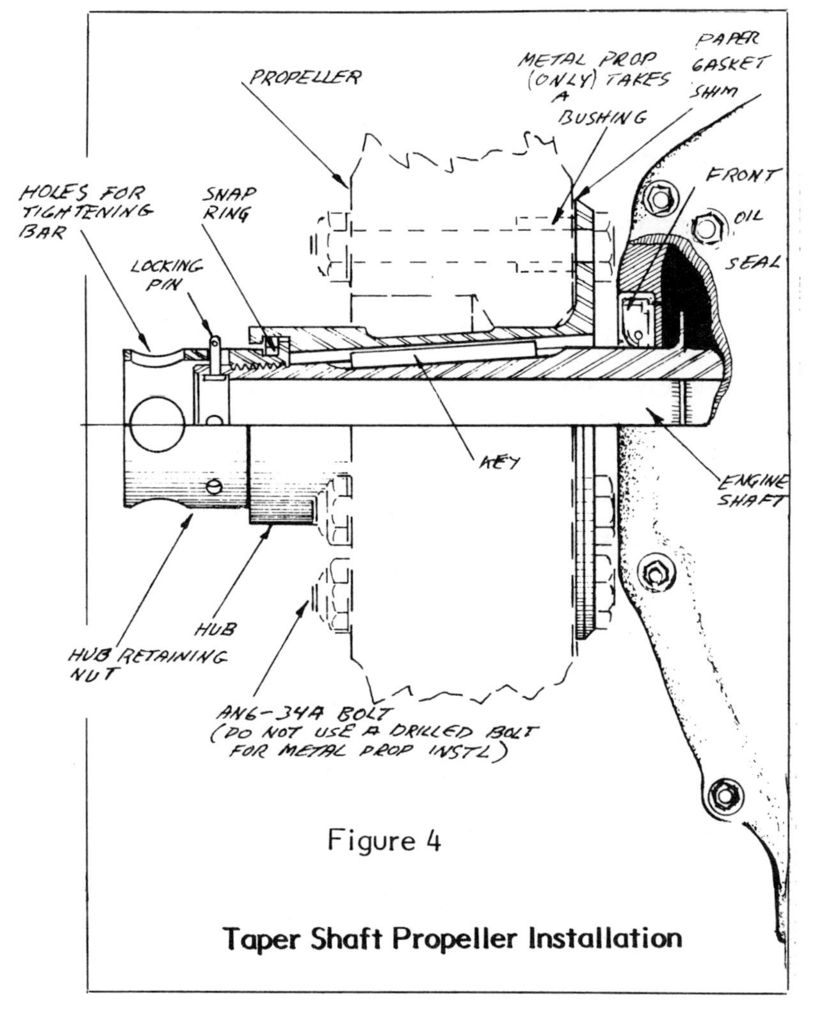

Figure 4

Taper Shaft Propeller Installation

2. To install the propeller on the engine shaft:

Check condition of crankshaft threads. Coat them with grease or oil.

As you install the hub retaining nut in the front end of the hub, compress the snap ring and push it in with the retaining nut until it snaps into the grove machined into the hub front.

IMPORTANT NOTE: The large snap ring serves as a puller because it will actually pull the propeller from the tapered crankshaft as the prop retaining nut is unscrewed.

Lift the propeller into position on the engine, aligning the keyway in the hub with the key on the crankshaft. Slide the propeller on.

Engage the hub retaining nut, turning it by hand until it is tight.

Slip a 5/8-inch x 2.5-foot steel bar through the large holes in the hub retaining nut and apply approximately 150 foot pounds of torque to one end. (This will result in about 375 foot pounds — 2.5 x 150.)

When the hub is fully tightened be sure one of the small holes in the nut aligns with one of the crankshaft holes. Insert a clevis pin from inside the shaft and safety it with a washer and cotter pin.

3. To remove a prop from a tapered crankshaft:

Remove the safetying clevis pin. (It might be a machine screw, etc.)

NOTE: Do not attempt to remove the snap ring or hub removal will be very difficult.

Back off the large round hub retaining nut by inserting a 5/8-inch steel bar and tapping on one end with an expediter (something short of a 5-pound sledge hammer). The propeller and hub assembly will back off from the crankshaft as the retaining nut is unscrewed.

Installing Spinners

Spinners have become as common as fleas on a dog. Not only does a spinner enhance the appearance of almost any aircraft, but it is generally accepted that it also assists in smoothing the airflow through the propeller hub area and improving engine cooling.

As a subject, spinners have long been ignored and, consequently, very little has been written about them or their installation. Perhaps with so many homebuilts sporting spinners these days everybody assumes that installing them must be a simple job — one that shouldn't take more than an hour or so? Oh, sure .... an hour or so per night for about a week!

Don't get me wrong, I don't want to convey the impression that fitting a spinner is difficult. It isn't, but it does tend to get a bit involved. Of course, some spinners really are easy to install. For example, the little skull cap type of spinner which is attached to a prop with a single screw. This type of spinner is most frequently seen on older light planes.

Another is the stock spinner manufactured for different standard category aircraft models. Most of these are predrilled and ready for installation when purchased. if you find one that suits your aircraft, grab it. (Figure 5)

The spinners manufactured for the homebuilt trade, on the other hand, are somewhat different. It's more accurate to describe these as "spinner kits" because, before this kind of spinner can be used, it is necessary to make the cutouts for the prop blades, align and drill both bulkheads, determine the proper spacing of the front bulkhead, drill the attachment holes so that they come out where they are supposed to, install about a dozen nutplates, and finally, install the spinner on the aircraft.

Without a doubt, the cost of your spinner will provide you with sufficient incentive to proceed slowly and with care. Therefore, expect no further admonishment from me.

DRILLING BULKHEADS FOR PROP BOLTS

If your propeller spinner bulkheads are not already predrilled for the propeller bolt pattern, your first and most crucial operation

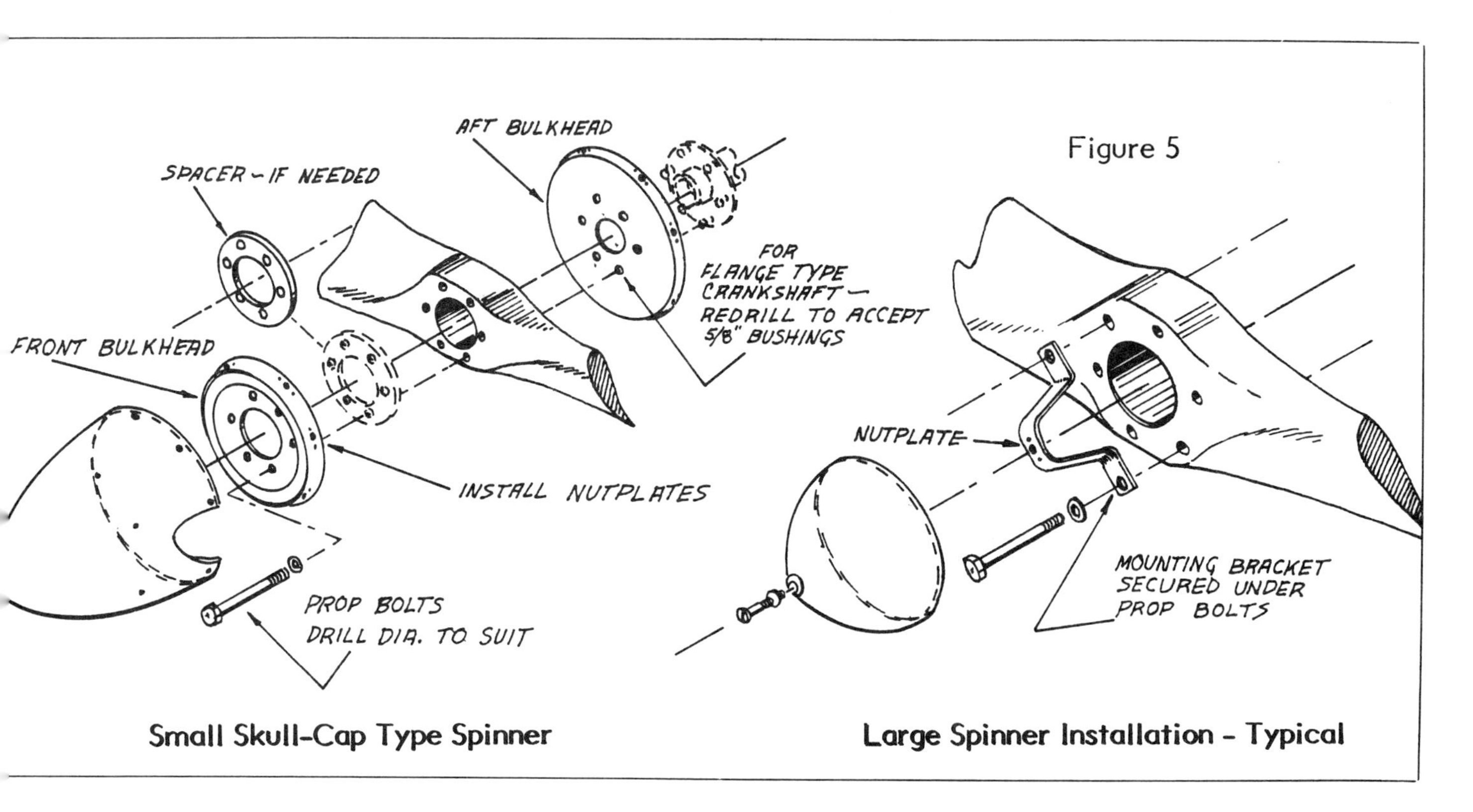

Small Skull-Cap Type Spinner

Large Spinner Installation - Typical

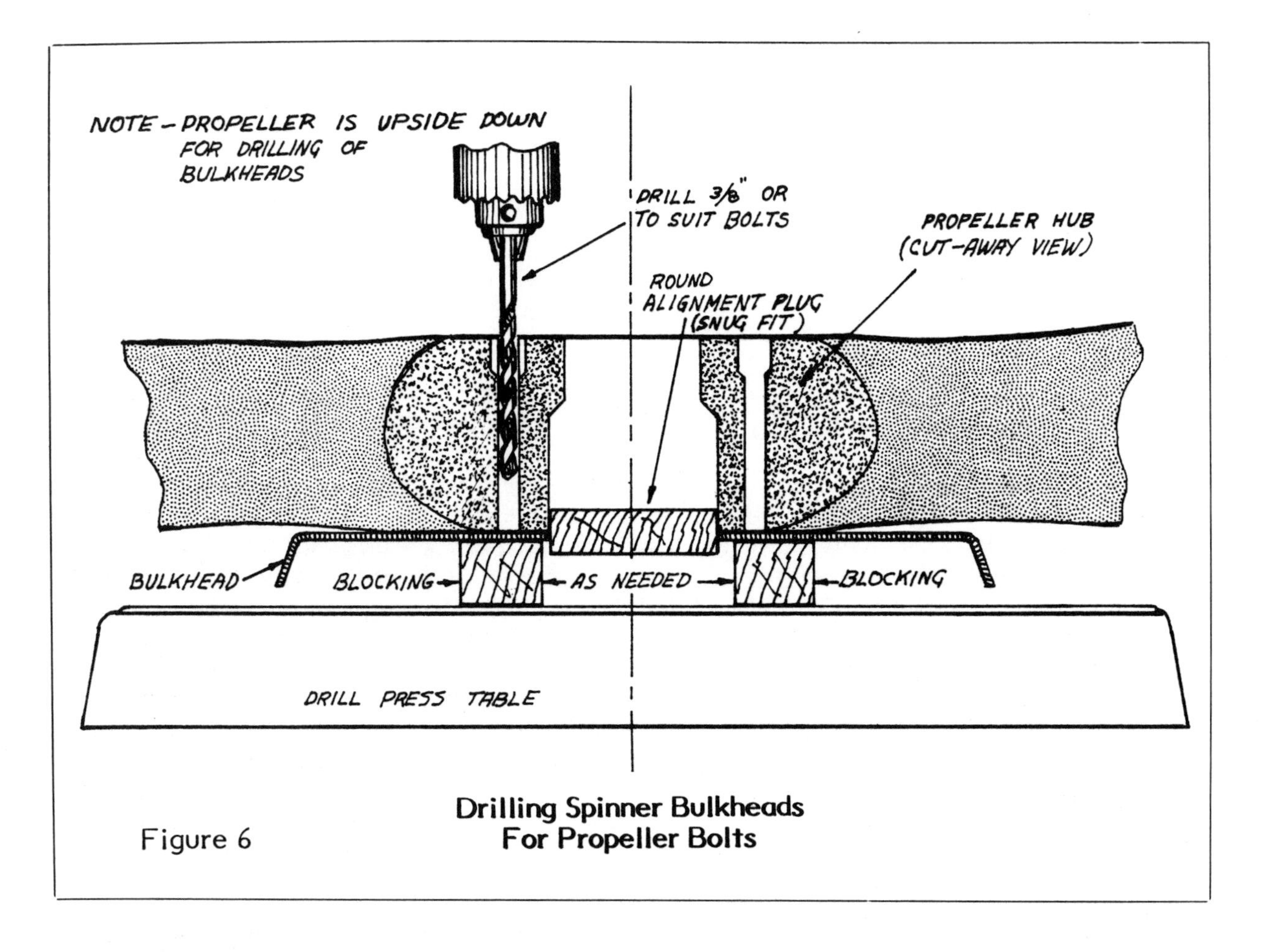

Figure 6 **Drilling Spinner Bulkheads For Propeller Bolts**

in fitting the spinner will be to drill the bulkheads yourself. Each of the bolt holes must be drilled accurately, otherwise, the spinner will rotate eccentrically.

To guarantee that the bolt holes in the bulkheads will be concentric and to minimize the possibility of misaligned or elongated holes, I would recommend that an alignment "plug" be used. (Figure 6)

This plug can be handmade of wood or turned in a lathe to make it a snug fit in the propeller hub. As far as I know, all spinner manufacturers make their bulkheads with a large hole identical in diameter to that in the propeller.

The rationale for using a plug is that since the bulkhead center hole is identical to that of the propeller, you can use this close fitting disk as an alignment jig simply by inserting it partially in the prop hub and then positioning the bulkhead over it. The alignment plug will center the bulkhead over the propeller and will eliminate the risk of slippage during drilling.

To drill the holes, position this assembly upside down in your drill press (block it up if necessary) and you will be able to guide the drill bit through the propeller bolt holes to the bulkhead surface. Always lower the drill bit into each hole and rotate the spindle by hand before turning the drill motor on. This eliminates gouging the sides of the propeller hole. As soon as the first hole is drilled, insert one of the propeller bolts before drilling the next hole. This is very important as it prevents possible misalignment during the drilling of the other holes.

Repeat this hole drilling process for the other bulkhead. This jigged drilling of the propeller bolt holes takes care of that all-important first step and gaurantees that your spinner will rotate smoothly.

TEMPLATE FOR THE BLADE CUT-OUTS

Not all aft bulkheads fit their spinner shell snugly when flush with the rim. You may find, instead, that yours slips well inside the spinner shell, perhaps as much as 1/2-inch, before wedging solidly against the spinner's walls. Check your aft bulkhead for fit to determine the distance from the work surface that the bulkhead must be blocked up during the template fitting process. Of course, if the bulkhead fits the shell snugly when its flange is flush with the rim of the spinner, no blocking at all will be required between the bulkhead and your work surface.

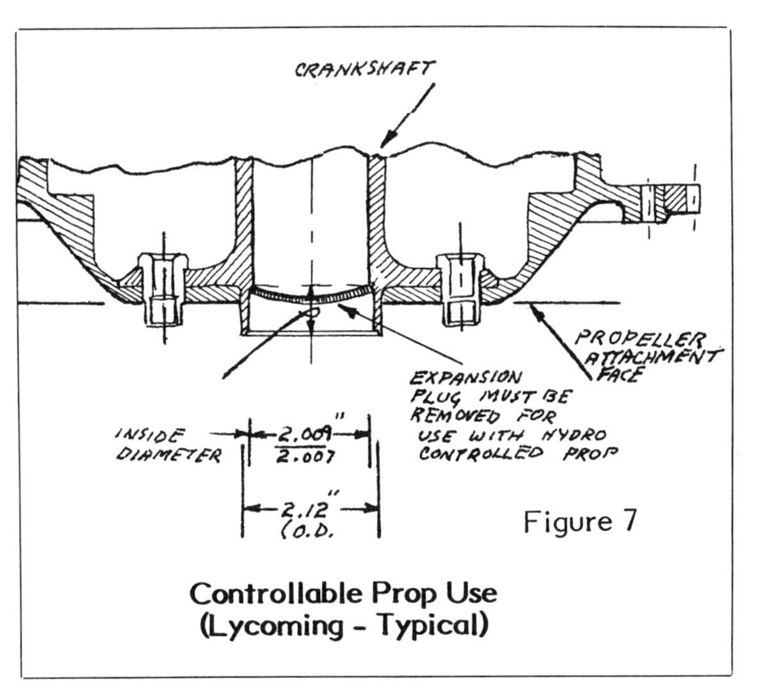

Controllable Prop Use
(Lycoming - Typical)

To get ready for the template fitting, set the aft bulkhead on a smooth work surface (blocked up as necessary) then place the propeller on top of the bulkhead and drop a couple of prop bolts into the holes to maintain alignment. You can use this assembly to make your template for the blade cutouts.

A piece of manila folder, or almost any other stiff paper, will be suitable as template material. You may be surprised to learn that you can save a lot of time by cutting out an approximate opening. You need not be too concerned with obtaining a perfect fit initially. After the hole has been roughed out in the template, use masking tape strips to bridge the gaps and voids between your template and the propeller. This remedial patching will result in an exact pattern. If you prefer a more permanent template, trace your masterpiece onto a more substantial material. (Figure 8)

Using the template, trace it onto the spinner. Measure to a point 180 degrees around the spinner and trace the same cut-out shape on the opposite side of the spinner. Note that the template is used face-side up for both cut-outs.

Use a pair of aviation tin snips to cut out the spinner openings, leaving about 1/8-inch excess metal from the line. Then, begin the "cut, file and try" process, enlarging both cut-outs so that there is an even 1/16-inch gap (no more than 1/8-inch) between the prop and the spinner. When assembled, no part of the spinner should be in contact with the prop blades at any point as it might damage the propeller.

DETERMINING FRONT BULKHEAD FIT

Because propeller hubs are of different thicknesses, you may find it necessary to insert

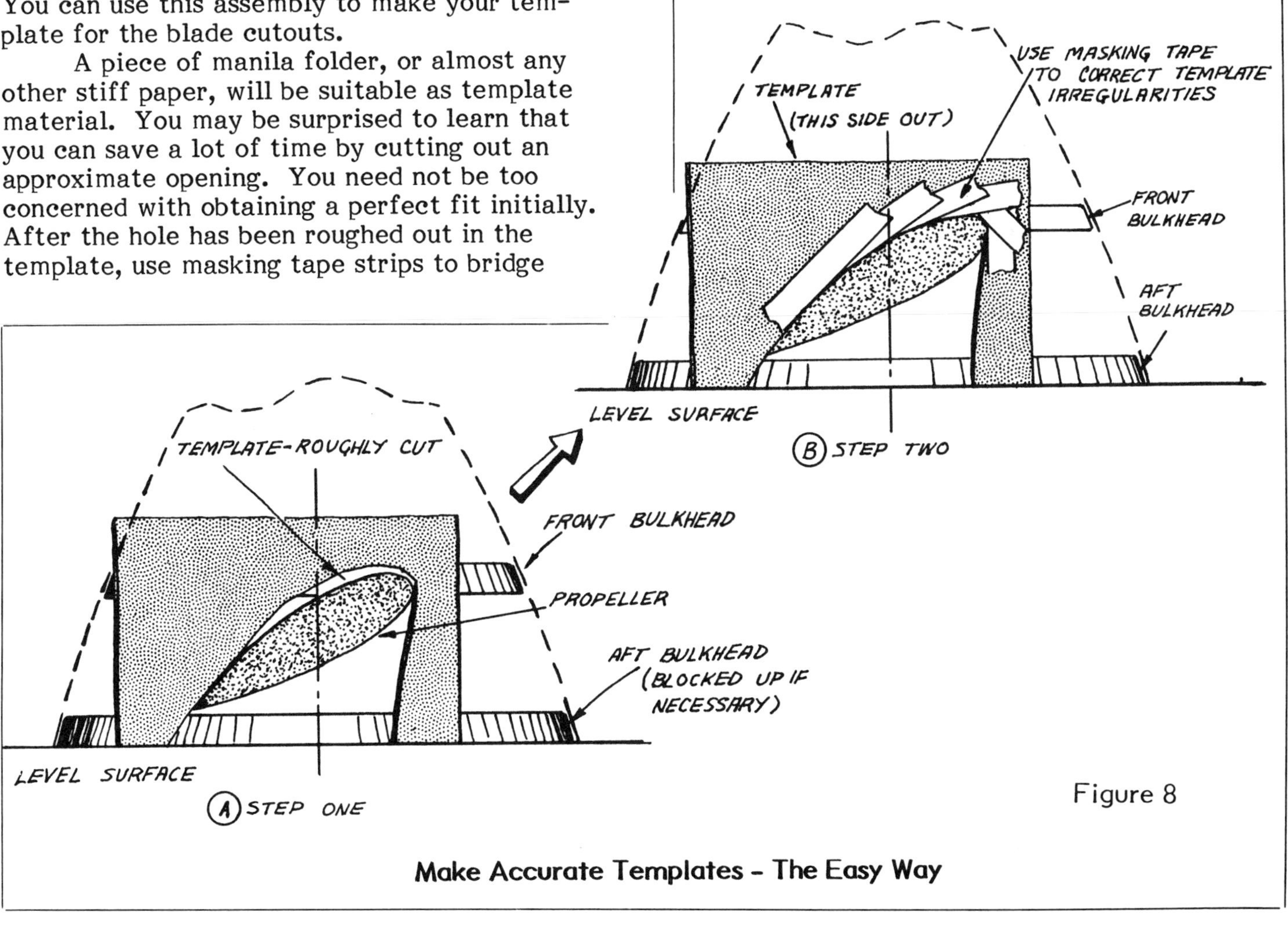

Make Accurate Templates - The Easy Way

The cut-out in the spinner behind the propeller should be fitted with a plate as shown here, particularly in high-performance aircraft.

a spacer between the front bulkhead and the prop hub if the front bulkhead flange does not bear solidly against the spinner when assembled. Of course, if it does not, how in the world can you determine what thickness of spacer you will need when you can't see inside the spinner? Here is one way: With the propeller resting on the aft bulkhead, place a 1/16-inch washer over each bolt hole. Place the front bulkhead on top of the washers and drop the prop bolts into their holes. Slip the spinner shell over this assembly and check to see if the rim of the spinner will still make contact with the work surface. If it doesn't, you know the front bulkhead fits well enough and it makes proper contact against the spinner without the need for a spacer. On the other hand, if the spinner rim does contact the work surface, add another washer all the way around. Recheck. After you have determined, from the thickness of the washers needed, how much the front bulkhead must be displaced forward to ensure a good fit, make a permanent spacer of the proper thickness to be used with your spinner installation.

DRILLING FOR ATTACHMENT HOLES

The next item is to drill the holes through the spinner that will be used to attach it to the bulkheads. Reassemble the top and bottom bulkheads (with the spacer, if needed, behind the front bulkhead) and lightly torque several prop bolts to hold the assembly together accurately. Place this assembly on your flat work surface and slip the spinner on. Press the spinner down firmly to ensure that its rim is in contact with the work surface. Mark the point where the first screw will go —approximately 1/2-inch from one of the cut-out edges. Make a light punch mark and drill the first hole with a 1/8-inch drill while pressing down on the spinner. Immediately insert a cleco fastener or a small screw. Drill one hole on each side of the two prop cut-out areas. Press down on the spinner each time you drill to assure solid contact all around its base. Space the location for the remaining holes equally. A large 10- to 12-inch spinner will require six to 10 fasteners around its base. The higher horsepower engines and the larger diameter spinners should have the 10-screw spacing. Re-drill the holes to accept 3/16-inch AN button head, or similar machine screws.

Flush head screws in countersunk holes have no place in a spinner installation. It is not a good idea at all. The screw holes will become enlarged and the security of the spinner will be jeopardized, particularly in the larger spinner/engine installations. It could even self-destruct before you notice the loosening process has started. Dimpling of the holes is also out of the question as the dimpled metal will not allow the spinner to slide over close fitting bulkheads.

After all the holes have been drilled, disassemble everything and smooth the hole edges lightly with a hand-held countersink or a larger size drill bit.

As for securing the front bulkhead, some people would rather not use any screws to snub it down. This is all right because the front bulkhead merely serves to stabilize the spinner. However, if screws are to be installed through the front bulkhead, exercise much care when drilling the holes. First, measure carefully and mark the location that you think the screws should be. Double check your measurements. Remember, you will be drilling the holes blindly and it is possible to miss the center of the flange by a large margin.

Owners of high-performance aircraft might find it aerodynamically desirable to close the rather large gaps in the spinner behind the propeller blades. Use aluminum plates of the proper thickness, cut and fitted to the openings. Install them with nutplates as illustrated (See photo). That's it.

A FEW LEFTOVER NOTES

Some spinners will have light concentric ripples in their surface. This is due to the spinning process used in their manufacture.

Expertly made, the spinner will show little or no evidence of this spinning process. You can, however, dress out these surface ripples by hand with a large 12-inch bastard cut file. Then switch to a smooth-cut file, and finally, to various grades of wet/dry sandpaper for finishing. If the ripples are rather deep and pronounced, it might be just as well to let them alone. Don't run the risk of removing so much metal that the spinner is weakened.

Incidentally, you should consider the spinner as a part of the aircraft structure in degree of importance. It can come off and do considerable damage to the aircraft in flight. And, shy away from fiberglass spinners. Their reliability record is poor.

Properly safetied propeller bolts.

center of gravity

13

That Shifty C of G

The title refers, of course, to the center of gravity — that invisible intangible that weighs heavily on the mind of anyone who builds an airplane he hopes to fly safely. Many factors affect the C of G (as our Canadian friends refer to it).

A builder who installs a heavier engine than that called for in the plans will be faced with a weight and balance change that could result in a poor, sometimes dangerous, C of G situation if the builder fails to compensate properly for major weight deviations from the original design. Here's how the situation can get out of hand.

Let's assume you would like to install a more powerful engine in order to "improve performance." This engine would undoubtedly be heavier. That means the aircraft's empty weight would be increased and the useful load that it could carry would be decreased. The bigger engine would probably burn more fuel so you would want to increase the fuel capacity by installing larger or extra tanks, adding still more weight to the aircraft —hence, C.G. problems.

Strange as it may seem, in some cases, there would not even be a corresponding increase in take-off performance and climb. This is because (in the case of a fixed pitch propeller) a propeller with a higher pitch would have to be substituted in order to absorb that increased power and to obtain that hoped-for higher higher cruise. Unfortunately, with all that extra pitch, the engine, at lower speeds (take-off and climb) cannot turn up at the necessary higher rpm to take advantage of the newly added "hosses". Even with that added horse-power, there is virtually no increase in take-off performance.

Well, wouldn't a controllable propeller increase performance? Of course, but it would also add much extra weight up front, creating another weight and balance adjustment problem.

A constant speed propeller for a 160 hp engine will weigh about 58 to 60 pounds. Compare this wth the 24-pound fixed-pitch propeller it would replace and you see the problem. Should you want to add a propeller extension to permit better streamlining of your cowling
you will be compounding the weight and balance change to a very nose-heavy condtion. Although a nose-heavy condition is considered better than a tail-heavy condition, it, too, is a problem of no mean proportion.

A tail dragger may be so nose heavy that that the pilot is afraid to use the brakes for fear that the bird will nose over when taxiing downwind under gusty wind conditions. Furthermore, three-point landings would be difficult or impossible. Of course, there would probably be just enough control left to make some nice "wheelies", but at the expense of added tire wear due to a higher touchdown speed. If the nose-heavy condition is severe enough the plane may not be able to get off the ground until it reaches a very high runway speed. Likewise, on landing it might be difficult to flare sufficiently except at high speeds.

An excessively tail-heavy condition is even worse. A tail-heavy aircraft is highly susceptible to unpredictable stall/spin behavior.

The lesson to be learned from this is obvious. During the construction of your airplane, and particularly before you start your engine installation, you should take stock of what changes you have made or plan to make that will affect your weight and balance. It is very important, for example, to consider the effect of changes in the the location of heavy objects such as batteries, strobe light power packs, the addition or removal of an electrical system, substitution of a larger or smaller engine, installation of a controllable propeller or a very light wood prop, relocation of seats and baggage provisions, and last, but not least, additional or larger fuel tanks.

It may be necessary to relocate or change the position of some of these objects in order to keep your C.G. within the allowable forward and aft limits. The C.G. range is rather limited and seldom exceeds six inches. This means a heavy engine may have to be installed on a shorter engine mount in order to keep the C.G. in limits. Sometimes a battery can be located aft rather than on the firewall to obtain the necessary correction. If you know the airplane

The long and short of it. Extending the engine mount for weight and balance reasons will require other unexpected changes --- a longer cowling and longer control mechanisms to cite just two.

will be nose heavy you should plan to locate as many weighty objects aft of the C.G. as you can.

The worst possible solution to a C.G. situation is one in which the addition of lead ballast in the in the nose or tail is mandated. This rather extreme solution is forced upon a builder who has not kept a check on the equipment he is installing. This dead weight only degrades the aircraft's performance.

Plan ahead and keep a rough weight and balance check as your construction proceeds. The most critical factor is the positioning of the engine.

Here is a simple way to get a preliminary weight and balance check for the installation or relocation of heavier equipment such as an engine, battery, etc.

RELOCATING A HEAVIER ENGINE

Let's take a typical example. A builder has a low-wing, two-seater with an installed Continental O-200 engine (188 pounds). He intends to install a Lycoming O-320 engine weighing 268 pounds. He knows he will have to shorten his engine mount (build a new one, actually) but by how much? (See Fig. 1)

But what if your preliminary estimates show that in spite of using the recommended engine, your C.G. would be too far forward or too far aft? How do you determine how much to move the engine forward or back? (See Figs. 2 & 3)

Be careful about making your calculations with unverified figures that may have been published and republished for a particular engine. Engine models are built and assembled with many variations in equipment. Lycoming, for example, reports that more than 200 variations are available in their engine models. If at all possible, use the actual weight of your engine for the C.G. computations.

Similarly, you can compute the effect of moving a battery (say) from a firewall position to some location in the aft fuselage area to correct a slightly nose-heavy condition.

Additional weight and balance computations and data are contained in my other volume, The Sportplane Builder, page 296, Weight and Balance....A Realistic Look. ($17.95 postpaid through the author)

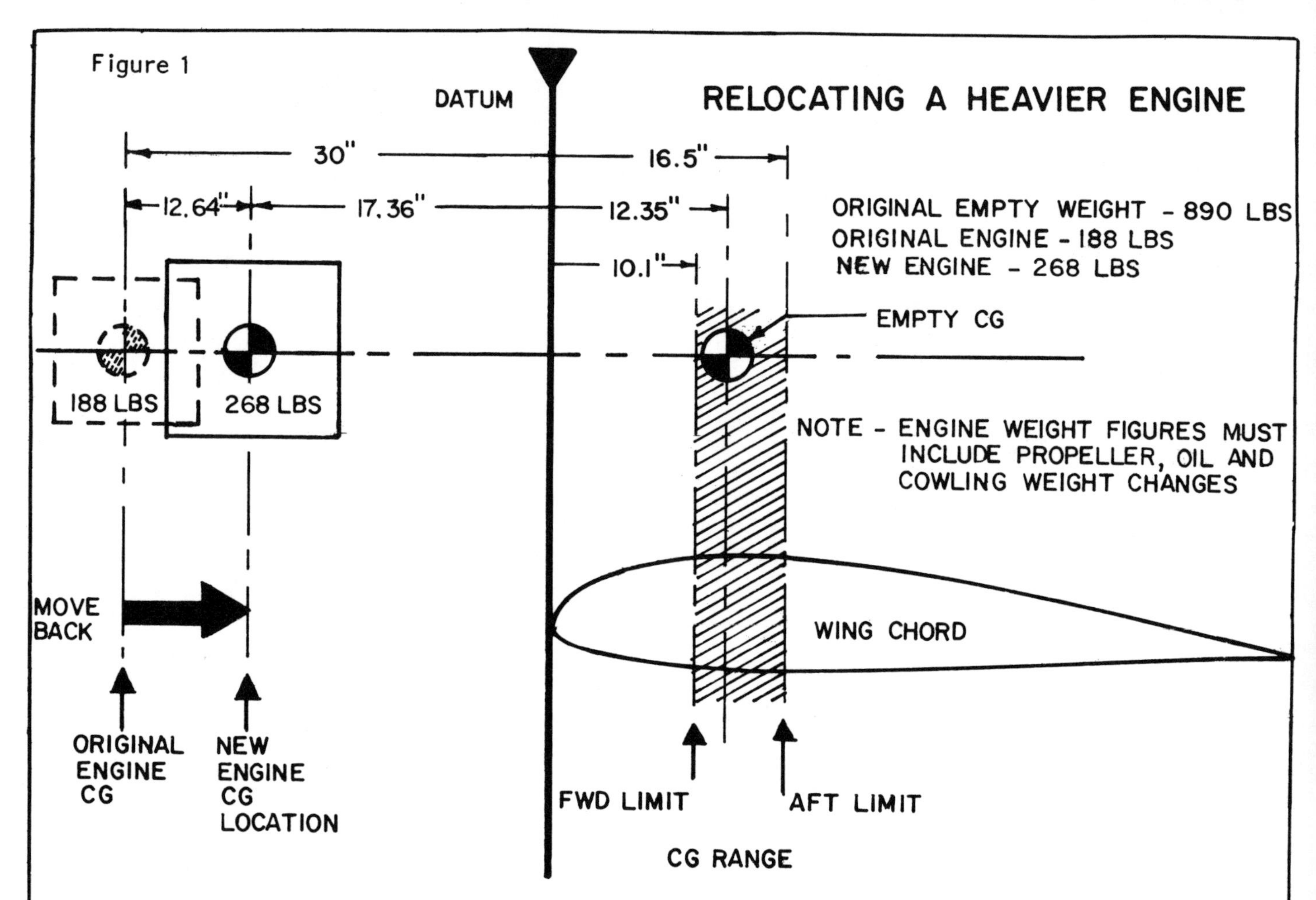

CHECK YOUR FIGURES

	WEIGHT (lbs)	DISTANCE FROM DATUM (inches)	MOMENT (pound-inches)
ORIGINAL EMPTY WEIGHT	890	12.35	10,991.50
REMOVE ORIGINAL ENGINE	−188	−30.00	5,640.00
ADD NEW ENGINE	268	−17.36	−4,652.48
			11,979.02

NEW EMPTY WGT — 890 LBS + 80 LBS = 970 LBS

11,979.02 ÷ 970 = 12.35" (CHECKS O.K.)

1. First, determine the weight difference between the two engines:
   Example: 268 lbs. - 188 lbs. = difference (80 lbs.

2. Next, compute the ADVERSE MOMENT: (New engine weight difference x old engine's distance from empty C.G.
   Example: 80 lbs. x (30 in. plus 12.35 in.) = Adverse Moment (3,388 in. lbs.).

3. Determine the distance the new heavier engine must be moved Back: Adverse Moment divided by new engine's weight = Distance to Move Back.
   Example: 3,388 in. lbs. / 268 lbs, = 12.64 in.

   (Subtract 12.64 in. from 30 in. to obtain the new moved-back location of 17.36 in.)

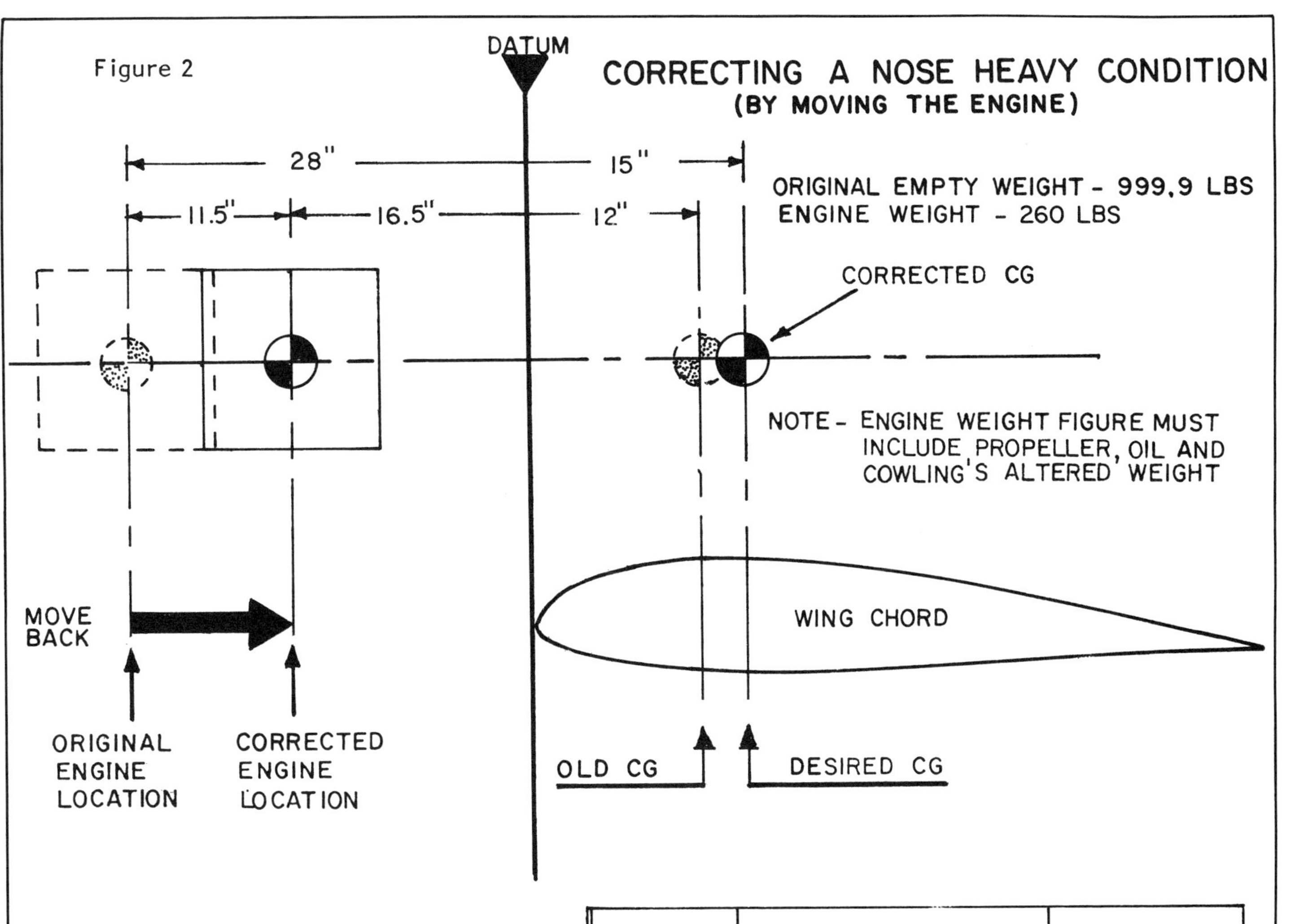

	WEIGHT (lbs)	DISTANCE FROM DATUM (inches)	MOMENT (pound-inches)
AIRCRAFT EMPTY WEIGHT	999.9	12.00	11,998.80
REMOVE ENGINE	- 260	- 28.00	7,280.00
INSTALL ENGINE	260	- 16.50	-4,290.00
			14,988.80

CHECK YOUR FIGURES -

NEW CG = 14,988.8 ÷ 999.9 = 14.99"(15") CHECKS O.K.

The calculations are similar to that for correcting a tail-heavy condition. Aircraft empty weight is 999.9 lbs. and the present C.G. is three inches forward of the desired C.G. location. You proceed as follows:

1. Find Adverse Moment
   (Aircraft empty weight x Distance between existing and desired C.G. locations)
   Example: 999.9lbs. x 3 in. = 2,999.7 in. lbs. (Adverse Moment)

2. Determine distance engine must be moved forward.
   (Adverse Moment divided by engine weight)
   Example: 2,999.7 divided by 260 = 11.5 in. (new engine mount would have to be 11.5 inches shorter)

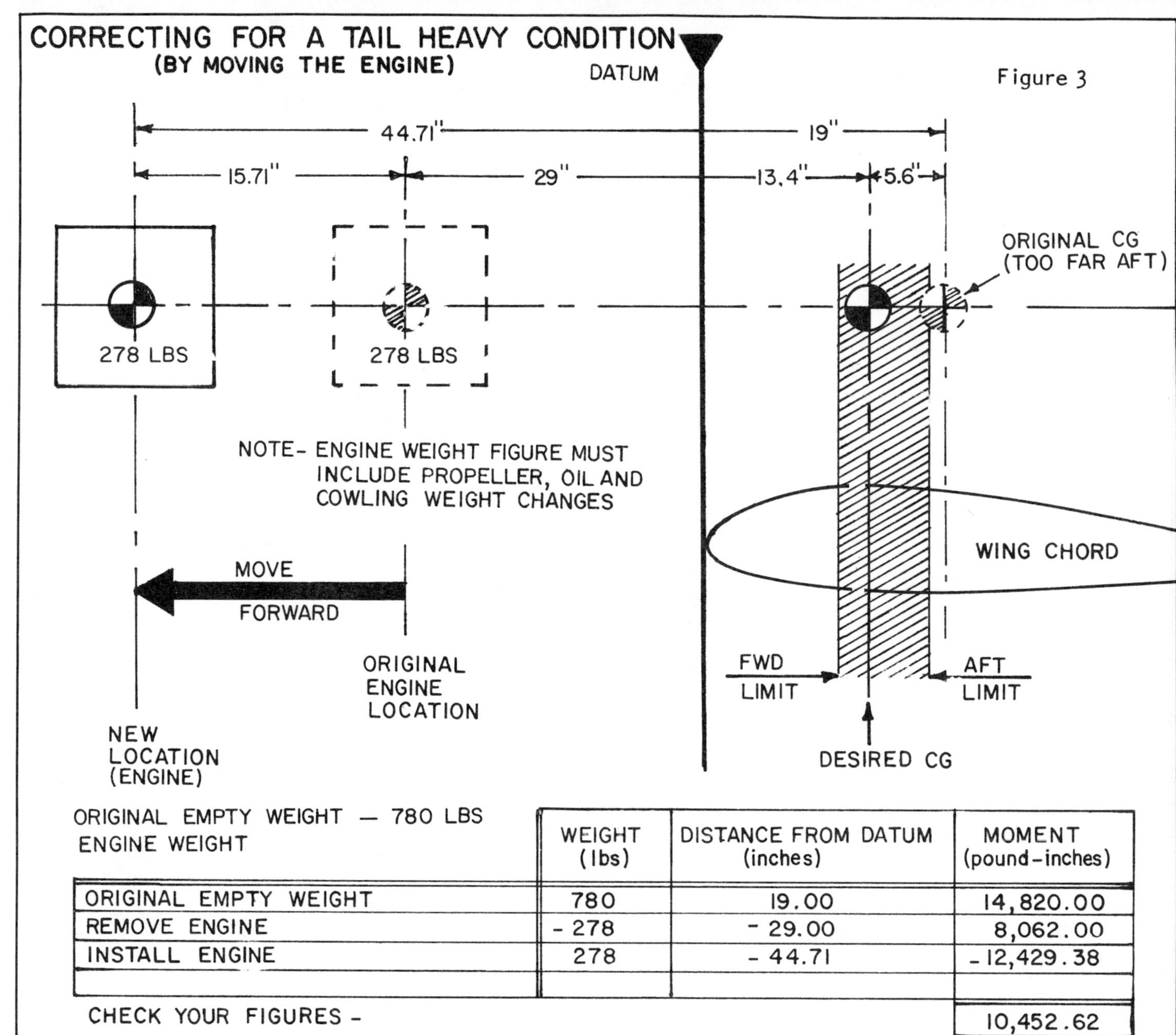

ORIGINAL EMPTY WEIGHT — 780 LBS
ENGINE WEIGHT

	WEIGHT (lbs)	DISTANCE FROM DATUM (inches)	MOMENT (pound-inches)
ORIGINAL EMPTY WEIGHT	780	19.00	14,820.00
REMOVE ENGINE	- 278	- 29.00	8,062.00
INSTALL ENGINE	278	- 44.71	- 12,429.38
			10,452.62

CHECK YOUR FIGURES -

NEW CG = 10,452.62 ÷ 780 LBS = 13.4" (CHECKS O.K.)

Aircraft empty weight is 780 lbs. it is tail heavy and its C.G. is 1.5 inches behind the allowable maximum aft C.G. You intend to correct the problem by making a longer engine mount, but how much longer? You proceed as follows:

1. Find Adverse Moment
   (Aircraft empty weight x Distance between existing and desired C.G. locations)

   Example: 780lbs. x 5.6 in. = 4,368 in. lbs. (Adverse Moment)

2. Determine distance engine must be moved forward.
   (Adverse Moment divided by engine weight)

   Example: 4,368 divided by 278 = 15.71 in. (new engine mount would have to be 15.71 inches longer)

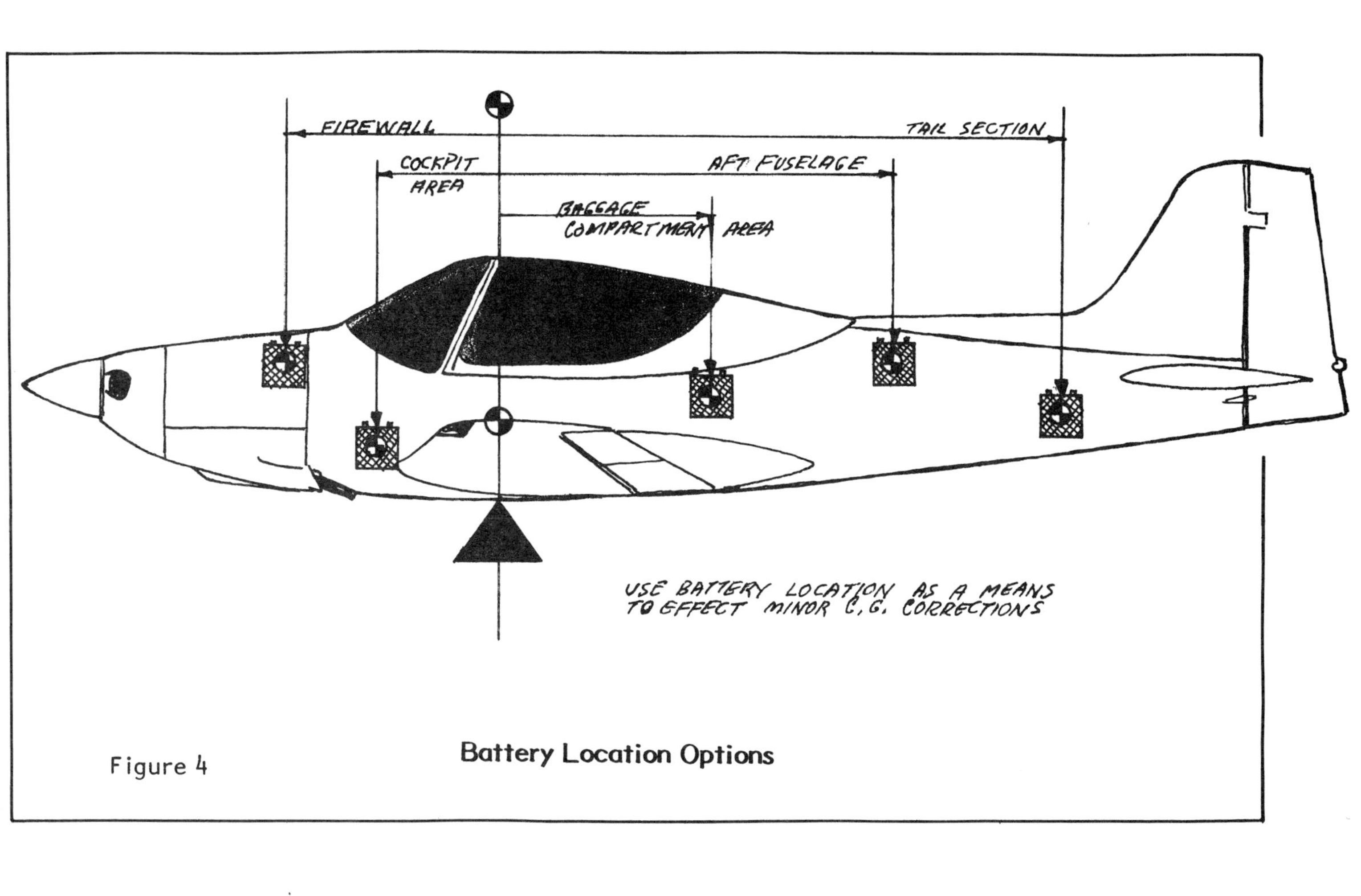

Figure 4 **Battery Location Options**

Weights: Engine Parts & Components

WEIGHT DATA

ENGINE INSTALLATION PARTS AND COMPONENTS

PART/COMPONENT	WEIGHT (LBS)
ALTERNATOR, 12V/60 AMP W/MOUNTING BKTS	10.9 to 13.0
12V SMALL, NON-AIRCRAFT TYPE	4.9 to 5.0
BAFFLES	1.7 to 3.0
BATTERY, 12V/25 AMP HR	21.0 to 23.0
12V/35 AMP HR	21.0 to 28.0
GELL CELL 12V/25 AMP HR	19.0 to 22.0
BATTERY BOX	.4 to 1.0
CABIN AIR BOX (VALVE)	.5 to 1.0
CARBURETOR (MARVEL SCHEBLER/STROMBERG)	2.5 to 3.5
CARBURETOR HEAT BOX	2.4 to 3.5
COWLING (TWO SEATER SIZE)	12.0 to 15.0
NOSE BOWL ONLY	5.0
ENGINE MOUNT (4 CYLINDER AIRCRAFT ENGINES)	5.5 to 10.5
ENGINE SHOCK MOUNTS (CONICAL)	.4 (SET)
ENGINE SHOCK MOUNTS (DYNAFOCAL)	2.8 to 3.5 (SET)
ENGINE CONTROLS (THROTTLE, MIXTURE, PROP, ETC)	.9 to 3.2
FIREWALL	3.0 to 6.0
FUEL INJECTOR UNIT	4.0 to 6.0
FUEL PUMP	1.6 to 3.5
FUEL TANK	.5 to 1.0 (PER GAL)
FUEL VALVE, BRASS ON/OFF	.5
3-WAY VALVE	.6 to .7
GASCOLATOR	.7 to 1.1
GENERATOR, 12V (DELCO REMY)	9.8 to 10.6
12V/50 AMP	16.6
24V/15 AMP	16.0
24V/50 AMP	25.0
IGNITION HARNESS (4-CYLINDER SHIELDED)	3.5 to 3.8
4-CYL/UNSHIELDED	1.82
IGNITION SWITCH (BENDIX)	.4
MAGNETO S4RN-20 (SCINTILLA)	4.8 to 5.2
S4RN-21 & S4LN-21	11.5 (PAIR)
EISEMANN LA-4	12.1 (PAIR)
EISEMANN AM-4	10.69 (PAIR)
SLICK	3.8 to 4.5
OIL COOLER, HARRISON AND SIMILAR	4.0 to 5.0
OIL FILTER (SPIN-ON)	.4
OVER VOLTAGE RELAY	.5 to .9
PROPELLER GOVERNOR (WOODARD)	3.8 to 4.5
PROPELLER	
CONSTANT SPEED	54.0 to 65.0
METAL	21.0 to 25.0
WOOD (AIRCRAFT ENGINES)	9.0 to 14.0
WOOD (SMALL...VW TYPE)	5.0 to 8.0
PROPELLER HUB (TAPER SHAFT C85)	4.4
SOLENOID, BATTERY	.8
SOLENOID, STARTER	.7
SPARK PLUGS (SHIELDED)	1.7 to 2.2 (8)
SPARK PLUGS (UNSHIELDED)	1.2 to 1.4 (8)
SPINNER, PROPELLER	2.0 to 3.3
STARTER, 12V	16.0 to 17.0
VACUUM PUMP, AIRBORNE (DRY)	3.0 to 3.4
VOLTAGE REGULATOR (GENERATOR TYPE)	1.3 to 1.7
VOLTAGE REGULATOR (ALTERNATOR TYPE)	.3 to .5
WOBBLE PUMP (WWII)	3.3
WOBBLE PUMP (NEWER TYPES)	2.6

....and yet another decision.
Should the battery slide forward
or be all the way back?

A propeller weighing 58 pounds is
definitely a weight and balance
consideration.

engine operations

14

Returning Engine, Components to Active Service

To return an engine to active service after a period of indefinite storage, it will be necessary to undo what you did to preserve it. Logical enough?

You can remove the corrosion preventative mixture from engine surfaces with a kerosene spray or a suitable commercial solvent (Gunk, etc.). A spray or brush application of kerosene may be necessary because the preservative may have become dried and caked on some parts of the engine.

1. Begin by draining the corrosion preventative mixture. Do not forget to immediately service the engine with the required amount of lubricating oil you will be using.

2. Remove all of the dehydrator plugs from the spark plug holes, all the tape, pads, plugs and silica gel bags or envelopes you used to close up the engine openings.

3. If preservative has been introduced to the carburetor, it should be removed by opening the drain and vaper vent plugs (as applicable to your engine). Move mixture control to RICH and squirt in fuel of the grade you will be using until all of the preservative is flushed out of the carburetor. Replace the plugs and connect the fuel line.

4. Since the propeller will probably be installed at this time, use it to turn the engine after removing the bottom spark plugs. This will purge the cylinders of excess preservative oil.

5. Install the regular spark plugs. Turn the propeller through the compression cycle of all the cylinders once again to reassure yourself that there is no hydraulic lock due to excessive fluid being trapped in one of the cylinders. Be sure you have removed all of the red warning streamers and tags you have previously installed. Be sure, too, that the crankcase breather is open.

6. Start the engine and, after a normal run-up, inspect it for leaks.

BATTERY

In the course of building your airplane, you sometimes purchase materials before you really need them. A case in point is the battery. Perhaps you got it on sale, or there was an offer you just couldn't refuse. Whatever the reason, if you buy the battery too soon you may have problems. There it sits, idle, for five to six months, all the while losing its charge. You may have been using it to test the navigation lights and various electrical systems time and again. Of course, that depletes its charge even more rapidly.

Naturally, any battery which has been stored or idled for an extend period of time should be load-tested and recharged, if necessary, before that final installation. Of course, it's just as important to maintain the battery properly after installation, too.

If most of your flights are short, less than 30 minutes in duration, the charging time on your battery will be insufficient. This may promote sulfating on the lead plates. The sulfate accumulation does not electrochemically convert to normal active material readily and could cause the plates to buckle. This would be followed by complete failure of the battery. Proper charging can prevent this.

While a dead battery may be a sign that the voltage regulator is not producing the required charge, what happens to a battery that is receiving too much charge? The symptoms will be excessive gassing and the development of internal heat in the battery. The water level will become depleted because the moisture and acid will escape from the cells as a fine mist. This mist is very corrosive to the surrounding structure.

Early evidence of this condition may be seen in the form of small water puddles around one or more of the battery vent caps. But don't jump to conclusions as to the cause. Investigate! It may be that the battery is "steaming" because of an unfavorably hot engine compartment location.

In many installations, even if you hand-prop the engine, the dead battery will not be charged. This is because the engine's alternator does not put out electrical power unless the battery initially has enough power to close the solenoid switch. Thus, even though the engine is running fine, electrical accessories (such as the fuel gauges, auxiliary fuel pump, radios, lights, etc.) may not work. No matter how long or how fast you fly, you simply not get a charge on the battery.

The solution? Pull the battery and recharge it.

Should your battery installation give you trouble it will probably be due to one of the following reasons:

1. A loose connection
2. The battery capacity is too small for the job
3. The battery cable is too small
4. Improper voltage regulator setting
5. Undercharging or overcharging
6. Low water level
7. A poor ground to the engine
8. A defective generator or alternator

OPERATIONAL TIPS: BATTERY

Soon after your engine is installed you will be anxious to begin checking its operation and to initiate taxi tests. All these activities will undoubtedly require your engine to be started repeatedly, imposing a severe drain on the battery. Under such conditions it won't be long before your new battery shows signs of being run down, of being drained, as it were. Don't blame your generator or alternator. They simply have not had sufficient time to function as they should to recharge the battery.

The low battery charge condition is more apt to be a problem in an aircraft that has a generator instead of an alternator. You see, the battery will begin charging only after the engine rpm's get over 1,200 or so. In such an installation the battery would not become fully charged after starting until you got airborne and flew for an hour or more.

An alternator, on the other hand, is somewhat more effective and will be trying to charge the battery even at the low engine rpm's of ground operations. However, if your battery is too low it may not have enough residual voltage to excite the alternator fields. For example, if the battery's voltage drops below six volts in a 12-volt battery (12 volts in a 24-volt battery), most likely the alternator will not charge the battery, no matter how long you fly.

Equip your aircraft with a voltmeter and the condition will be easy to detect from its reading. Without the voltmeter, you can still check the alternator's output. If your installation employs a split master switch or a separate alternator switch, turn off the alternator switch and check the rate of discharge on the ammeter (while the engine is operating). After the alternator switch is turned back on, the ammeter needle should show a "charge". If it doesn't, your alternator is not working. In some installations you can often determine if the alternator is inoperative from the "zero" reading on the ammeter (when there is no negative scale which could show a discharge reading).

Many of your early flights will probably be of short duration. This could result in a gradual decline in your battery's condition. Here are some clues:

* The voltmeter doesn't show its normal 14.2 volts

* The battery shows signs of boiling over

* Battery water level is low

* The ammeter shows an abnormally high charge rate in flight

Should any of these signs appear you had better check the battery's condition. You will probably find it needs recharging. The solution? Give the battery an occasional recharge with a battery charger, particularly if you continue to make short flights that do not give your alternator or generator a chance to do their thing. If you have no electrical outlet in your hangar, you will have to remove the battery and take it home for a trickle charge. This will ensure that your battery will be able to start your engine when you go out to fly .... but only if you remember to take the battery with you when you go to the airport.

GAUGES

The ammeter's operation is automatic and requires very little of your attention. Any time the master switch is on, it will measure the rate of charge or discharge in your electrical system. During operation, the needle would normally be near zero. Momentarily, at engine start-up, a discharge will be indicated but this should be followed by a brief interval of charge. A battery that isn't fully charged will show a higher charging rate for a somewhat longer period.

In the event your alternator (generator) is not putting out, or the electrical load exceeds the system's output, the ammeter will indicate the battery discharge rate.

A persistant needle deflection toward the 'discharge' (minus) range will soon produce a dead battery.

Older aircraft radios, those that are not transisterized, impose a heavy drain on the battery. Your ammeter will attest to this. What, with starting the engine, operating the radio and perhaps the strobe lights, coupled with prolonged taxiing to the take-off position and maybe holding for awhile, a heavy power drain is unavoidable. Moreover, a generator does not 'charge' until the engine speed is over 1,000 rpm (actually, about 1,200 rpm). In time you may have a battery that is very low and in need of a recharge. Conduct your ground operations with this knowledge in mind.

CYLINDER HEAD TEMPERATURE (CHT)

Your cylinder head temperature gauge, when properly marked for your use will have:

A RED RADIAL LINE for the maximum allowable cylinder head temperature.

A GREEN ARC showing the permissible temperature range for continuous operation. (This will be between the maximum allowable and the minimum CHT as established by the engine manufacturer.)

You should monitor the gauge frequently during the early testing stages of your aircraft. This is especially important during prolonged or steep climbs and when the outside temperatures are high. And remember, the CHT gauge does not depend on the aircraft's electrical system.

Cylinder head temperatures may be controlled to a degree by the airspeed, the mixture control and the cowl flaps, if installed. A high airspeed, rich mixture and open cowl flaps all serve to reduce the cylinder head temperatures.

Should your engine indicate an unusually high cylinder head temperature from the start, the condition should be immediately investigated because high cylinder head temperatures will shorten the life of your engine.

Don't overlook the possibility of an inaccurate gauge.

EXHAUST GAS TEMPERATURE (EGT)

Most EGT gauges have incremental markings of 100 degrees F, although the Alcor gauges are calibrated in increments of 25 degrees F. In any event, all engines operate in the general range of from 1,200 degrees F to 1,750 degrees F. Generally, these instruments display no radial markings.

When the mixture control is slowly leaned from the full rich position, the EGT will slowly increase until the temperature of the cylinder being monitored reaches its peak. Any further leaning of the mixture will cause a decline in EGT. This peak point is difficult to determine with most inexpensive EGT instruments. It is also difficult to determine in low-horsepower, float-type carbureted engines because of the uneven fuel/air distribution, characteristic of these small engines.

Your check for peak EGT should not be attempted at power settings above 75%, as detonation could occur.

Leaning an engine to peak temperature, in general, provides the minimum fuel consumption for a given power setting and altitude. Be careful, however, excessive leaning can crack a cylinder, burn a valve or cause detonation serious enough to burn a hole in a piston. Standard practice is to enrich the mixture 25 degrees F to 50 degrees F beyond peak EGT. There are exceptions, however.

Much has been written regarding the merits of utilizing an EGT indicator. I recommend that you obtain the excellent literature provided by Alcor and other manufacturers on this subject if you plan to install an EGT gauge in your airplane.

Remember, if you are operating the aircraft at or near peak EGT, you must readjus your mixture with any significant change in power or altitude. The proper leaning of your engine will help ensure less spark plug fouling and a longer life for that expensive set of spark producers.

CARBURETOR AIR TEMPERATURE GAUGE

Gauge markings should include a YELLOW ARC ranging between minus 10 degrees C and plus 15 degrees C, the temperature range most conducive to carburetor icing. A RED RADIAL LINE can be used to show maximum permissible carburetor inlet air temperature. Use a GREEN ARC for normal operating range above and below the possible icing range. Usually, the YELLOW ARC range is sufficient.

Monitor the gauge from time to time to see that the needle remains out of the icing range during possible carburetor icing conditions.

Use your carburetor heat for this purpose. The object is to maintain the carburetor temperature a few degrees, say 5 degrees C above freezing in the carburetor throat. As long as the fuel-air mixture is maintained at a temperature slightly above the freezing level, 32 degrees F (0 degrees C), the risk of icing will be minimized.

FUEL PRESSURE GAUGE

The fuel pressure gauge is marked with a RED RADIAL LINE at both the maximum and minimum allowable pressures. A GREEN ARC indicates the normal range. Although you will use the fuel pressure gauge primarily to reassure yourself that the fuel pressure is being maintained in the normal operating range, it can be useful in giving you advance warning of pending trouble.

For example: A rapid drop in fuel pressure accompanied by a loss of power may be an indication of a leak between the carburetor and the fuel pump.

A sudden and complete loss of fuel pressure that reverts to a normal indication as soon as the auxiliary (electric) pump is turned on can mean you have a cracked or broken fuel line. In such a circumstance the prudent thing to do is to immediately turn off the fuel valve to minimize the risk of a fire. Of course, you will then have a forced landing to cope with.

An oscillating needle could be a bad omen....a sign that your fuel tank is just about dry. It could also indicate a leak or partial obstruction in the line between the pump and the fuel tank.

Should the needle fluctuate rapidly upward to a higher pressure reading, the gauge could be telling you there is an obstruction in the line. In that event it is quite likely that a loss of power will quickly follow.

A sudden drop in fuel pressure, accompanied by a loss of power and engine roughness that can be remedied by leaning the mixture, could be a clue that the carburetor's float needle valve is stuck in the open position.

In spite of all these frightening possibilities, it is fortunate that the fuel pressure gauge usually just reassures you that your fuel system is operating properly.

MANIFOLD PRESSURE GAUGE

This instrument should be marked with:

A RED RADIAL LINE to establish the maximum allowable absolute manifold pressure (instrument calibrations are in inches of mercury — Hg.)

A GREEN ARC showing the range from the maximum allowable pressure for continuous engine operation to the minimum pressure approved for cruise power by the engine manufacturer.

In a typical homebuilt equipped with a fixed-pitch propeller, the throttle setting and air pressure determine the engine speed (rpm). The tachometer is the instrument used to obtain a direct indication of the power being produced. When a constant-speed propeller is installed, however, the engine rpm stays constant as the propeller automatically changes pitch to hold the preset engine speed. In this case it is necessary to have a manifold pressure gauge to indicate the engine's power.

In a constant-speed propeller installation the engine power, as indicated by the manifold pressure gauge, is controlled with the throttle while the tachometer merely shows the engine rpm as preset with the propeller control by the pilot.

When the engine is not running, the manifold pressure gauge will indicate the actual atmospheric pressure around it at the time. However, when the engine is running, the manifold pressure will show a below-atmospheric pressure reading because of the partial vacuum being created by the pistons. A supercharged engine can produce higher-than-atmospheric manifold pressure. (Some ram air installations also can increase the manifold pressure.)

A manifold pressure gauge can indicate the presence of carburetor ice, which interferes with the flow of air to the engine and causes a decrease in manifold pressure.

OIL PRESSURE GAUGE

The oil pressure gauge is automatically activated as soon as the engine starts. In cold weather your oil pressure may be a little slow to register, however, you should obtain an oil pressure reading in about 30 seconds. If you don't, shut the engine down and check it out.

The minimum and maximum oil pressures permitted will vary from engine to engine and you should get the correct range for your engine from the appropriate engine manual.

Your oil pressure gauge should be marked to show the engine's minimum idling pressure (Red Radial Line). The engine's normal operating range is marked with a GREEN ARC, and the maximum operating pressure is marked with another RED RADIAL LINE.

Monitoring the engine oil pressure is very important because a failure of the lubrication system could destroy your engine very quickly.

Here are some potentially dangerous oil pressure gauge indications and some possible causes:

Zero oil pressure — Possible causes: No oil in engine; oil line loose, not connected, broken or cracked; oil pump inoperative (rare); blocked oil line.

Low oil pressure — Possible causes: Oil pump intake screen or oil filter clogged; low oil level supply; loose bearings; warped crankcase.

High oil pressure — Possible causes: Oil lines or passages clogged; defective oil pump bypass valve; oil viscosity too high; oil pressure relief valve fails to open.

Erratic oil pressure indications — Possible causes: Low oil level in crankcase; pressure-relief valve sticking or releasing.

Before you start tearing the engine down, first check out the gauge. It could be a problem with the instrument rather than the engine.

OIL TEMPERATURE GAUGE

A direct reading oil temperature gauge provides an indication of engine oil temperature at all times....even when the engine is not running. The gauge should be marked with a GREEN ARC for normal operation, and a RED RADIAL LINE for the maximum allowable oil temperature.

Most gauges are calibrated from 100 degrees F to 250 degrees F. Some may indicate as low as 75 degrees F and a few may range above 300 degrees F.

What should you oil temperature read? Different engines have different operating temperature limits. You should always abide the limits given in your engine manual.

Troubleshooting Clues

An abnormal reading in a newly installed gauge could indicate a defective gauge. You can check it out by immersing its bulb in boiling water. The surest cure would be to replace the gauge with one of known accuracy.

An indication of rising oil temperature could be a sign that the oil cooler is clogged or that sufficient air is not passing through it in the location it was installed.

A quick rise in oil temperature attended by engine roughness could be the result of incipient detonation or preignition due to improper operation of the engine.

TACHOMETER OPERATION

The instrument markings should display the normal operating range (GREEN ARC) and a maximum (RED LINE) or never exceed rpm for your engine.

If you want something fancy for a high flying cruise machine, you can mark your instruments with a multiple stepped green stripe (different widths) to indicate a 75% engine power range at 5,000-foot and at 10,000-foot altitudes.

Should a mechanically driven tachometer become noisy or its needle begin to fluctuate excessively, it would be an indication that the drive cable needs to be lubricated. In this regard, also check that routing doesn't impose sharp bends on the cable.

Use your tachometer to verify your engine idle mixture adjustment. Do this by setting the engine for idle rpm and then rapidly move the mixture control to full lean. Watch the tachometer. If the needle rises slightly (about 50 rpm) and then drops off, the mixture is all right. Should the rpm rise more than that, say 75 rpm, your idle mixture is too rich. On the other hand, an instantaneous drop of the needle tells you the idle mixture is too lean.

The accuracy of the average mechanical tachometer leaves something to be desired. But when you get into electric tachometers,

The first run-up?
Where are the chocks?
The safety Observer?
The fire extinguisher?
Without the wings, most airplanes are very nose heavy and may nose over if the tail isn't weighted or tied down.

their batting average is often even worse. This is of little concern for most of us, flying VFR with a fixed-pitch propeller, but it could be a problem for anyone equipped with a controllable propeller or someone who does a lot of IFR flying.

VACUUM GAUGE

Use your suction gauge for the adjustment of the vacuum system relief valve and to detect a leak in the vacuum system during normal operations.

The gauge indication is the difference in pressure between the main (central) filter and the pressure relief valve. The indicator is calibrated in inches of mercury (Hg.)

A reading of between 4.5 and 5.8 inches Hg. may be necessary for the proper operation of some of the newer gyro system type instruments. The older types required only 2.75 to 4.25 inches Hg. Any indication below that established by the manufacturer for the vacuum gyro instruments you have installed will be an indication that the air flow is insufficient to spin the gyros fast enough for reliable instrument operation. Your pressure relief valve should be readjusted accordingly.

VOLTMETER

If the voltage regulator is providing the required voltage, a voltage reading of 14.2 to 14.3 for a 12-volt system and a voltage indication of 27.9 to 28.1 for a 24-volt system will be indicated on the gauge.

Excessive voltage will damage radios and other electrical equipment and can even damage your battery by making it 'boil over'.

A low voltmeter reading and a discharge or no charge indication on the ammeter signifies a malfunction in the generator or its control system.

The voltage will change during the engine's warm-up period, particularly in cold weather, showing a high reading when the regulator is cold and a lower indication when it warms up. No adjustment of the regulator should be attempted when such conditions prevail.

The Oil System (VWs)

Most VW users agree that an oil change frequency of 25 hours is a reasonable time interval for the powerplant (except under dusty conditions) and that the engine deserves a good grade of aircraft oil. Nevertheless, some builders use SAE 30 automotive oils. Mostly I suppose, because the VW engineers recommend the use of a good name brand HD oil with a viscosity of SAE 30 for use in the auto engines under temperature conditions of 85 degrees F and higher.

The majority of sportplane operators of the VW engines, however, seem to be using heavier weights of oil and lean towards the use of aircraft oils such as Aeroshell and Pennzoil 40 and 50 weights (80 and 100 under the new numbering system).

Standard engine break-in practice seems to be followed by most builders in that non-detergent oil is used, after engine build-up, for the first 20 - 30 hours or until the oil consumption tapers off somewhat (an indication that the rings are seating properly); then they switch to a good aircraft detergent or dispersant oil.

ENGINE OPERATION

Builders facing the prospect of flying a VW-equipped aircraft for the first time are usually appalled at the paucity of essential operational data for the converted engine. Maybe the following, gleaned from a recent survey of builders of VW-powered aircraft, will help:

Oil Pressure — Have you decided what the oil pressure should be for your converted 1600cc VW powerplant? The figures are similar for most models. In autos, the word is that a 5-psi to 7-psi minimum is OK but don't look for anything over 28-psi. From the survey, it was apparent that such low oil pressures were not good enough for the aircraft users of the engine. The range of oil pressures expected by these respondents range from a minimum of 10-psi to a maximum of 50-psi in flight. Most commonly reported oil pressures were 20-psi at 2,000 rpm to 35-psi in flight. Of course, you would expect to see the oil pressure gauge briefly reflect a much higher pressure when the engine is first started, particularly in cold weather.

Oil Temperatures — Survey participants report normal oil temperatures at cruise ranging from a low 122 degrees F (oil cooler installed) to 185 degrees F. The consensus is that if the oil temperature cannot be held down to at least 180 degrees F during normal cruise conditions, something should be done about it.

For some reason most of the respondents shied away from committing themselves to an absolute redline figure for the oil temperature. In general, though, the panic figure (redline limit) of 215 degrees F received consistent acceptance. As you may know, the small Continentals have 225 degrees F posted as their red line limit for the oil temperature so I guess 215 degrees F for a VW is a safe limit. Of course, you would expect the oil temperature to increase during climbs approximately 20 degrees F over your normal cruise figure.

Cylinder Head Temperatures — A cylinder head temperature gauge may not be as common a feature in a Volks-powered aircraft as in others with conventional aero engines but it is just as necessary. An engine's cylinder head temperatures should be monitored, at least until the mandatory flight testing phase has been completed. If a gauge with a single thermocouple is used, it is installed under the spark plug of what is expected to be hottest running cylinder (usually #3).

There was a great variance in the cylinder head temperature instrument readings reported, and I feel that some may have been due to faulty installations or instruments. Nevertheless, if the operational numbers obtained in your testing vary significantly from those presented here, an immediate investigation is in order. The maximum permissible cylinder head temperatures reported were limited to 400 degrees F to 450 degrees F, with "higher temperatures" in climbs.

I feel the survey info was of little help in this area as the reported cylinder head temperatures were insufficient in number to definitely establish exact guidance. Many of those reporting neglected to state whether their results were in degrees Fahrenheit or Centigrade. How are your gauges calibrated? Do you remember?

Engine timing has a significant influence on cylinder head temperatures. In general, you could say there was the common belief that when timing gets past 30-degrees B.T.C. you lose any advantage you might have had and it becomes quite detrimental to the engine's health. The range reported for timing the VW engine is from 25-degrees to 30-degrees B.T.C. You can, therefore, assume that in order to keep the cylinder head temperatures within acceptable limits the magnetos must be set to fire the plugs at or slightly below 28-degrees B.T.C.

AT WHAT RPM?

Here is where we really get confused. Some VW service manuals for the 1600cc engine show a power curve which indicates that the engine's full power is not developed until it turns 4,800 rpm. While a few builder-pilots believe that the engine can take 4,000 rpm all right, they would, at that rpm, run into the "propeller problem". A propeller's tip speed must be kept below 950 feet per second. A 60-inch propeller would exceed this tip speed by the time it reaches 3,800 rpm....far short of the required 4,800 rpm necessary to obtain the rated horsepower. A shorter 54-inch propeller can be turned at a higher rpm before reaching a dangerous tip speed. This explains why so many props for direct drive VW engines just happen to be 54 inches in diameter.

A minimum 2,900 static rpm should be obtainable to ensure sufficient power for a safe take-off. Oddly enough, a few builders reported their static rpm to be in the 3,540 to 3,700 range even though their engines apparently are cruised at between 2,900 and 3,600 rpm. In general, most pilots get a bit uneasy when the inflight rpm exceeds 3,600. Actually, for continuous operations, most VW experts like to see their conversions operated at no higher than 3,400 rpm (continuous).

Extended engine operation on the ground should be avoided. However, if necessary, the engine should be fully cowled and the aircraft headed into the prevailing wind to help maintain adequate cooling.

Props, Exhausts

WOOD PROPS

A new prop will take some time to "settle" so keep checking on it during the first few hours to see that the nuts are snugged. One reason for this is that wood is affected by changes in humidity. A propeller that is not properly secured can suddenly develop elongated holes, charred areas where slippage occurs (the friction generated by a loose propeller has been known to actually start a fire), and even bent or sheared bolts.

Oversize bolt holes and loose bolts are difficult to detect from only a preflight inspection. You may have to put a wrench to the bolts to really check them. Look closely at the edges of the washers and bolt heads to see if there has been any sign of relative movement.

Exercise common sense during all ground operations and develop the habit of performing thorough preflight and postflight inspections of your propeller. Given the proper attention, the wood propeller is pretty tough and stands up well in use.

TAKE A GOOD LOOK

During your preflight, look at the propeller carefully and be on the alert for telltale signs of:

* Cracks or breaks and deep cuts across the wood grain
* Any separation in the laminations
* A warped blade
* Any areas where there is a piece of wood missing
* Any sign that the tipping is slipping or is loose
* Looseness at the hub, loose bolts

Any of the above defects, if found, should be carefully evaluated and the proper course of action determined to correct the problem.

In the propeller tip ends are drain holes that permit any accumulated moisture to drain out. These holes should be kept open to prevent the moisture from damaging the wood. The tradition of parking aircraft equipped with wood propellers so that their blades are horizontal is indirectly related to the drain holes. This position minimizes the tendency for any excess moisture to drain into one of the blades. You may find it useful to rotate the prop one or two bolt holes on the hub so that when the engine is shut down the prop will stop horizontally and relieve you of that pesky detail.

HANDLE WITH CARE

Pulling an airplane by one blade is tough on a metal prop, but you can imagine the effect of such abuse on a wood prop. DON'T DO IT!

Whenever possible, avoid starting the engine when the airplane is parked in a gravel area. Also avoid blasting the throttle while the airplane is on loose gravel or similar loose surfaces. Particles picked up by the whirling propeller are guaranteed to cause damage.

If your aircraft must be parked out of doors, the propeller should be protected against the elements. However, rather than making a snug, tight fitting propeller cover or bootie, I recommend that the whole engine and prop be draped over with a protective covering. Putting an airtight cover over a propeller that has to stand outdoors is a good way to ruin it as moisture from rain or condensation turns to steam (and not only in Texas, amigo) inside the boot. That is why draping a tarp over the entire nose of the ship and securing the lower ends to the ground is a superior arrangement. At least it will allow air to circulate.

One more matter of importance .... a wood prop is just as lethal as a metal one. Make sure a qualified individual is always in the cockpit when your aircraft is being hand-propped. Otherwise, TIE IT DOWN!

EXHAUSTS

While the lion's share of exhaust system failures can be blamed on excessive vibration and thermal stresses, corrosion takes a heavy toll too. Prolonging the life of exhaust pipes has always been a problem.

Unfortunately, mild steel pipes (the kind that are easy for homebuilders to work with) are highly susceptible to rust. Moisture often condenses in exhaust pipes and this, combined with extreme heat, accelerates the oxidation process. Gradually the metal is eroded away and pin holes develop. You may have noticed this symptom, particularly in aircraft equipped with automotive pipes.

Some mild steel pipes were actually designed and manufactured for a few of the older models of aircraft. These pipes were given a hot aluminized or a porcelanized finish for corrosion resistance. This kind of treatment is usually beyond the capabilities of most homebuilders, however, unless they live near a large metropolitan area or can send their pipes off to be treated.

Good grades of high temperature automotive exhaust paints, on the other hand, work fairly well and will prolong the life of mild steel pipes, provided that you touch up the paint occasionally around the exhaust ports and any other place where it burns off.

Some paints are better than others, but only personal experience can tell you which. When you realize that the exhaust gases come charging out of the ports at temperatures over 1,400 degrees F, you can see that the 1,200-degree F paint is a bit shy of absolute protection.

Some hot rod buffs try to obtain a more durable, longer lasting coat of paint by preheating their exhaust pipes slightly prior to spraying them with a high temperature paint, but this is often unsuccessful unless very skillfully applied to pipes that are at just the right temperature. The basic flaw or problem in this method is the great possibility that the paint will go on dry.

Here's a better way to do it.

Use a good brand of the VHT (very high temperature) paint. It comes in regular spray cans and can be found in any auto supply house.

First, paint the header (that's the exhaust pipe, son). Then, while the paint is still wet, run the flame of a torch through the pipe as the exhaust would do if the engine were running .... not too much and not too close or the paint will burn.

The heating action really sets the paint, which then seems to become part of the metal. Or course, it is a devil of a job to weld later if a crack should develop in the metal.

A degree of skill is required for this method but it is much more effective than painting a pre-heated pipe. That, in effect, causes the paint to be applied dry and, therefore, it can't possibly last very long. So, spray it first then heat the pipe for a long lasting paint job.

DON'T FORGET MISCELLANEOUS STUFF

OK, Button 'er up and let's go!

index

INDEX

Page Numbers in Parentheses Indicate
Illustrated Reference

A.

B.

C.

D.

E.

F.

G.

P.

R.

S.

T.

U.

V.

W.

EAA
Experimenter
The "HOW TO" Magazine for the Aircraft Builder
December 1989